UNCONTROLLED SPREAD

UNCONTROLLED SPREAD

WHY COVID-19 CRUSHED US AND HOW WE CAN DEFEAT THE NEXT PANDEMIC

SCOTT GOTTLIEB, MD

HARPER
An Imprint of HarperCollins*Publishers*

HarperCollins books may be purchased for educational, business, or sales promotional use. For information, please email the Special Markets Department at SPsales@harpercollins.com.

FIRST EDITION

Library of Congress Control Number: 2021915259

ISBN 978-0-06-308001-0

21 22 23 24 25 LSC 10 9 8 7 6 5 4 3 2

To my three daughters.
You make the world a better place.

“Declare the past, diagnose the present, foretell the future.”

HIPPOCRATES

CONTENTS

UNCONTROLLED SPREAD

INTRODUCTION

In New York City, fourth-year medical students at New York University Grossman School of Medicine were given the option of graduating early if they agreed to work on the hospital's COVID wards.[1] It was an eerie throwback to events that unfolded one hundred years earlier, during the Spanish flu, when Philadelphia—critically short of healthcare workers—pressed medical students into service.

In March 2020, it wasn't only medical students who were thrust into improbable circumstances. New York City had become the initial epicenter of the COVID-19 pandemic in the United States. In its hospitals, surgical residents were running makeshift intensive care units that were fashioned out of operating rooms. Pathology residents were reassigned to the medical wards.[2] The Federal Emergency Management Agency (FEMA), which responds to national disasters, was asked to send eighty-five refrigeration trucks and personnel into the New York City medical examiner's office, where there was a "desperate need" for burial services to help manage a cascade of dead bodies.[3] To bolster its normal mortuary work, New York City added more than two hundred soldiers and airmen from the army and the national guard.[4]

New York State had issued a jarring directive urging EMS crews and other emergency service workers to forgo attempts to revive anyone without a pulse when they got to the scene of a medical emergency, amid the intolerable strain caused by COVID's surge.[5] Running short on ventilators, NewYork-Presbyterian Hospital made the extraordinary decision to retrofit its breathing machines with 3-D printed plastic tubing that allowed doctors to ventilate two patients simultaneously, using the same device.[6]

I'd completed my medical training in New York City as a resident in internal medicine twenty years earlier, and I went to medical school at Mount Sinai Hospital in Manhattan. Among my most vivid memories from training were covering the medical floors at Elmhurst Hospital in Queens. The hospital was located in one of the most ethnically diverse neighborhoods in the country, and the community's rich culture deepened the complexity, and gratification, of practicing medicine there. I knew the hospital's capable staff, and its immense capacity. Watching the scenes unfold—of Elmhurst Hospital being overrun with COVID patients, of refrigerator trucks parked outside, and of doctors and nurses describing their harrowing experiences—was hard to bear.

It was stunning, and it was shocking. But above all, it was terrifying. What my medical colleagues in the city described to me again and again was pervasive fear: Fear that they could spread the virus to their families, as each day New York hospitals were using as many masks, gloves, and gowns as they normally consumed in an entire month during usual times, quickly draining whatever stockpiles they had. Fear that they didn't know how to care properly for the sick patients overwhelming their wards, suffering from a virus that nobody yet understood. Fear that they couldn't predict how or when the arc of infection would start to ebb. And fear that a lot of lives would be lost.

It was a harrowing epidemic that brought the city's vaunted healthcare system much closer to the brink of collapse than most people, even now, recognize.

Severe Acute Respiratory Syndrome Coronavirus 2 (SARS-CoV-2 or SARS-2) is the virus that causes the disease that we've come to know as COVID-19. By the time the first cases of community spread were diagnosed in late February, SARS-CoV-2 had already rooted itself in our communities. It had been here for a while, at least since January, replicating, spreading, and doubling its numbers every two to three days.[7] Then, in March, after thousands of cases had accumulated, the virus abruptly burst into public view.

The virus didn't arrive with a group of visitors from China, where it originated, or from Italy, where it established its next major foothold. Instead, it likely rode along the breath of probably hundreds of different travelers from a variety of locations, each ferrying the infection, and evading the porous controls that the federal government had put in place at US airports. At the time nobody knew what was happening. Nobody knew how much virus was being carried by people who showed no outward symptoms of the disease. These were people who might never manifest any signs of illness but were still contagious. Without the ability to test people for the virus, we had no way of detecting its spread. We certainly had no way of stopping it.

This wasn't because the United States had never imagined it might fall victim to a deadly pandemic. We certainly had imagined the possibility. In some respects, we had been preparing for this moment through three presidential administrations, starting with George W. Bush, who warned in a 2005 speech, following the outbreak of Severe Acute Respiratory Syndrome, or SARS-1, and then avian flu, that "scientists and doctors cannot tell us where or when the next pandemic will strike, or how severe it will be, but most agree: At some point, we are likely to face another pandemic. . . . Our country has been given fair warning of this danger to our homeland and time to prepare."[8] We had a pandemic playbook on the shelf, ran exercises simulating the threat countless times, and developed the Strategic National Stockpile to store the medical countermeasures that the top experts thought the country would need. But when the pandemic we long feared finally arrived, we weren't ready. Many of the plans and preparations turned out to be a technocratic illusion. The stockpile lacked key essentials. A lot of what it contained didn't work. It was a metaphor for our fragile response.

When I worked in the federal government in public health roles, we would say that planning for medical calamities provides you with no assurance that you're prepared to deal with one. That was certainly true for COVID. The US never developed a pandemic strategy that would be broadly relevant to a range of predictable and unexpected viral threats, and the country was slow to realize the ways in which the plan we had created and tried to work from, which focused almost exclusively on the risk from flu, wouldn't apply to COVID. The federal government started off in a weak position, with plans that were ill suited to countering a coronavirus. This mismatch between the scenarios we drilled for and the reality that we faced left us unprepared. Poor execution turned it into a public health tragedy.

It was an alarming state of vulnerability for a country with the world's most technologically advanced healthcare system. Owing to mistakes in how we deployed diagnostic tests for COVID, we left ourselves blind to the virus and allowed it to spread widely and largely unchecked, so we were never able to trace its early spread and contain it. Even when the shortcomings became obvious, the Centers for Disease Control and Prevention (CDC) continued to rely on its systems for monitoring and responding to influenza, insisting even into 2021 that its flu-based interventions were the right tools in the fight against COVID. We didn't pursue an approach that closely tied our efforts to track and contain the spread of SARS-CoV-2 to the characteristics of the virus. This central shortcoming explains many of the gaps in our response to the pandemic we actually faced.

Yet COVID shouldn't have been such a surprise. There had been earlier outbreaks of new and deadly strains of coronavirus. COVID's close cousin, SARS-1, appeared in 2002 and spread threateningly in 2003, and another dangerous coronavirus, the Middle East Respiratory Syndrome, or MERS, emerged ten years later. Moreover, the scientific literature over the last decade is riddled with reports of SARS-like coronaviruses that were found in bats and appeared to have the potential to sicken humans.

In 2012, six people developed an illness with symptoms matching COVID after clearing bat feces from an abandoned copper mine in Yunnan, a province in southwestern China that's one of the country's most biologically diverse regions.[9] Three of the patients died. Chinese government scientists sampled from the caves coronaviruses that may have caused the outbreak, but those specimens were never shared, and officials maintained that the culprit was an unspecified fungus.[10] And while the results of official investigations into the outbreak were never fully revealed, a group of Chinese researchers, working independently, would later conclude that the probable culprit was a SARS-like coronavirus that originated from Chinese horseshoe bats.[11] A coronavirus that later became known as RaTG13, judged to be the closest known relative to SARS-CoV-2, was sampled from the same mineshaft.[12]

In 2013, scientists reported on two novel coronaviruses isolated from bats in China that closely resembled SARS-1 and bound tightly to the same angiotensin-converting enzyme 2 (ACE2) receptor that lines the human respiratory tract, the same route through which the SARS-CoV-2 virus would gain entry into our cells.[13]

In 2016, scientists reported on another novel coronavirus, also closely related to SARS-1, that also showed the ability to infect human cells.[14]

There have been other accounts of outbreaks of unusual respiratory illnesses among people who frequented the caves in southern China that were home to bats known to carry coronaviruses highly similar to SARS-1.[15] In 2018, researchers sampled the blood of Chinese citizens who lived near these caves and discovered that about 3 percent of the local population had antibodies to coronaviruses that had never been previously identified.[16] Any one of these spillover events could have been a predecessor to SARS-CoV-2, testing humans for the first time.

Scientists issued repeated warnings that one of these novel coronaviruses could start to circulate widely. They cautioned that a disease could emerge that had the same deadly features of SARS-1 but was more easily spread, threatening nations. But

SARS-1 had disappeared, and MERS never developed the capacity to transfer easily between people. So the warnings prompted only passing actions that sputtered once the immediate threats seemed to subside.

And even if a disease like COVID could have been foreseen, we still wouldn't have been ready for it. We needed an approach that prepared us for unforeseen risks and focused on establishing core capabilities and not just trying to guess which virus will threaten us. We shouldn't accept that we'll be able to anticipate the next threat or that even a predictable risk (like a pandemic strain of flu) won't adapt in some sinister way that allows it to slip past our countermeasures.

So, instead of assuming that actions designed to combat flu would be effective in countering any pandemic, we should have drawn from our experience with SARS-1, MERS, Ebola, Zika, and other infections, to remember that strategies must be closely tied to the biology of diseases that we're trying to mitigate. These include the biological features related to the way pathogens spread and the social factors that contribute to transmission.[17]

These insights should have shaped a preparedness agenda where our policies corresponded to some of the common characteristics shared by different viral threats. We needed to create plans and countermeasures that aligned with aspects of risk that were threaded across a broad range of the potential dangers, rather than wrongly assuming that our tactics could be safely adapted from the preparations we had made for a pandemic involving influenza. We now must learn from our mistakes and approach future pandemics with an altogether different mind-set. We need to tie our future strategies to the epidemiology and biology of diverse categories of potential threats and the social construction of disease. This will arm us with the capability to implement a more flexible response that can counter a fuller set of conceivable threats, including new viruses as well as new strains of known viruses that may have evolved in dangerous ways. Then we need to fashion interventions that target the social and geographic communities where the advance of a novel disease is most likely to occur.

In contrast to diseases spread largely via droplets or contaminated surfaces, where transmission from each infected person is to a smaller number of individuals, for diseases like COVID—with some degree of aerosol transmission, where a lot of the spread is from a small number of superspreaders and where the risk is typically from indoor environments with poor airflow and filtration—these dangers will require a different set of interventions. Pathogens with different incubation periods will need to be planned for differently. So will diseases that can spread through asymptomatic transmission. This is just some of the foundation on which a flexible approach to pandemic preparedness will have to be constructed.

The night of September 11, 2001, after the attack on the World Trade Center, I was providing medical care at a triage center at Chelsea Piers on the West Side of Manhattan, close to the site of the toppled buildings. I was a third-year medical resident at Mount Sinai Hospital, and had been assigned to help staff the facility.[18]

When I got to the site, I was stunned. In that single day, an enormous field hospital had been erected, with hundreds of treatment rooms and dozens of fully equipped surgical suites. After the 1993 attack on the World Trade Center, and countless other terrorist incidents, someone had decided that New York City needed to be able to handle a mass-casualty event that could exhaust its existing hospitals and medical resources. Federal and state agencies had stored away enough medical equipment in the city, or nearby, along with the logistical support needed to create an enormous field hospital in a single day. Tragically, it turned out we didn't need the facility that day—there were very few survivors pulled from the rubble of the fallen towers. But we would have been ready if more casualties had come our way.

By contrast, when COVID struck and we needed certain capacities in New York and across the nation—whether it was stockpiles of ventilators and masks, the ability to dramatically expand our diagnostic testing or create intensive care beds, or the capability to manufacture drugs and vaccines in huge amounts—nothing comparable would be in place. We weren't ready. We didn't prepare for this kind of viral contingency.

Hardening ourselves against future threats means elevating our public health preparedness as a matter of national security. That means that we'll need new domestic capabilities, including faster ways to develop and deploy countermeasures. We'll also need to strengthen key supply chains. As COVID struck, it became every nation for itself. Countries hoarded masks and turned off exports of critical medicines to grow their own stockpiles.[19] In the future, we cannot rely on global manufacturing for key components. We'll have to build domestic capacity for making commodity products like gloves and masks and greater resiliency for making high-technology products like biological drugs and vaccines.

We need government institutions that are consistently focused on preparing for these risks just as we prepare for other low-probability but high-impact events that can threaten our survival. The National Aeronautics and Space Administration (NASA) scans for and maps large meteors that could strike the earth. The Federal Emergency Management Agency (FEMA) prepares recovery plans for a radiological attack. There's no comparable contingency planning for pandemics. When COVID struck, all eyes turned to the CDC, which, it was assumed, would quarterback our response. However, the agency doesn't have an operational capability to manage a crisis of this scale.

It also means building global capabilities to uncover new viral threats when they first emerge and being able to gather the intelligence we need to know how to counter them. China ignored commitments it had made for sharing information in a public health crisis. It grudgingly revealed incomplete material only after the risks were largely apparent. China never shared samples of the early strains of SARS-CoV-2.[20] These behaviors proved we cannot rely on global cooperation alone to alert us to emerging threats. Going forward, we need to cultivate more effective ways to monitor potential hot zones and gather information on these risks.

And most especially, it means addressing protracted social challenges that leave many of our most vulnerable communities excessively exposed to the dangers posed by infectious disease.

As for the political response, President Donald Trump and his advisers grew increasingly weary of the economic impact of the measures recommended by public health authorities, and they were wrongly convinced that uncontrolled spread was inevitable regardless of what steps were taken. This partly underpinned an ambivalence by the president on how forcefully to embrace mitigation. That doubt was reflected in public statements, and actions, that were contradictory and harmful. At the same time that the president was berating Georgia Governor Brian Kemp for opening tattoo parlors and bowling alleys too soon, he was tweeting dangerously to his followers to "Liberate Michigan."[21] At the same time that many of the president's top health officials were urging people to wear masks at the start of the fall surge of infection, the president stood on the balcony of the South Portico of the White House and, in a theatrical gesture, took off his mask while he was still contagious with the virus.[22]

These high-profile political shortcomings—the inconsistent messaging and shifting fortitude from the White House—get much of the public attention when we think about the problems with the federal government's response. But a lot of corrosive failures were at the agency level, inside an ill-prepared bureaucracy. Years of inattention to these risks and underfunding of the key institutions, turf wars between federal agency heads, a lack of strong leadership that should have brought health agencies together in a coordinated way to solve problems, and bureaucratic hubris, all created a system failure. The pandemic was not inherently uncontrollable. I spoke to federal health officials and White House staff often about the nation's efforts as these events were unfolding, which gave me a close-up view of the response, and I'm convinced that if we had adopted a more coordinated national response and stronger surveillance at the outset, we would have had a better outcome.

Operation Warp Speed was a bright spot: one of the greatest public health achievements in modern times. The unprecedented research effort helped to deliver safe and effective vaccines and to secure new efficiencies in manufacturing.[23] The Trump administration deserves credit for helping to facilitate that accomplishment, which will eventually end the pandemic. The success of Operation Warp Speed proved what government can accomplish when it functions well, to improve our preparedness and protect the nation. Other parts of the system were unable to deliver a more synchronized and effective response.[24] We wouldn't have been able to prevent a US epidemic, but with better preparation, better alignment, better leadership, and better engagement of our tools of public health, we could have delayed its onset and reduced its scope and severity.

To see how the outcome might have been different, we have to put aside the often-shallow politics of our present moment. This book will strive to do that. One of the reasons why I believe I was successful as FDA commissioner is that I avoided looking at things through a political lens, as is so often the temptation in Washington, and tried instead to view our challenges from a public health perspective. This is not a book about President Trump, or about the governors you cheered on or those you disliked. I'll let others delve deeply into the personal stories and the politics. I'm convinced that America's failures and successes in this pandemic were not merely the function of any particular person's or party's ideology or behavior. Even with different leadership, many of the same system failures would have persisted. To help safeguard our public health, we need to identify those more fundamental weaknesses—the root causes of our vulnerability—to make sure this doesn't happen again. My intention is to reveal and examine the systemic breakdowns that can and must be addressed. Those who argue that all of the shortcomings flowed from a failure of political leadership by the president or a particular governor overlook the deep-rooted challenges that must be corrected to assure that we have a more resilient and effective preparedness for the future.

But this is not to say that the sickness of our political culture didn't contribute to our weakness in the face of this virus. Suppressing the spread and limiting the scope of the pandemic required constant action using the full complement of public health tools that we could assemble. Instead, we argued over the use of masks, over the merits of testing those who were infected and tracing their close contacts, and over taking steps to limit activities like indoor social gatherings that we knew were sources of spread. The mitigation caused severe economic adversity that hit certain segments of society especially hard. These measures were onerous and controversial, and we were right to debate them. Honest debate is one thing, but our response was demoralized by a sizable enterprise devoted to manufacturing skepticism about any steps that could potentially reduce the scope of the spread, even obviously effective measures like masks or vaccines.

In this book, I examine the shortcomings that exposed America to the worst effects of this virus, and I try to draw some lessons on how we can make ourselves more resilient to similar threats. We must be better prepared when another dangerous pathogen inevitably emerges and threatens a global pandemic, but our purpose shouldn't be simply to make sure we're better equipped to respond, but to secure a greater assurance that a global contagion on the scale of COVID can never happen again. Reflecting on where our response fell short can provide a roadmap for how to be more effective when the next virus emerges. And there will almost certainly be a next one, maybe sooner than we care to believe.

The COVID pandemic changed the course of human history. We saw very clearly that pandemics are sources of outsize instability and suffering, worsening vulnerabilities left by economic and social strain and exacerbating these hardships. COVID was not an unexpected event, yet it wasn't properly anticipated. The weaknesses and failures it revealed cost America dearly.

In one of the largest studies done to estimate the number of New Yorkers stricken with coronavirus at the height of the city's epidemic, researchers found that between February and July 2020, about 1.7 million people were infected. About 1 percent of them died.[25] During that dreadful spring wave of infection, *The New York Times* described "an apocalyptic coronavirus surge" of victims streaming into the city's hospitals.[26] The winter was even worse for the country, when the epidemic widened. Yet at the same time, on many days, the nations of China, Japan, South Korea, Taiwan, Singapore, Hong Kong, Thailand, Malaysia, Vietnam, New Zealand, and Australia were collectively registering fewer daily COVID cases than the city of Los Angeles.[27] Some of these nations, especially China, had employed draconian tactics that would have been firmly rejected in the US. But others had used mostly wide-scale testing and tracing to contain their spread. The Pacific Rim had figured out how to control the virus. We had not.

The pandemic devastated New York and left America with a palpable sense of vulnerability. To those who suffered and died, and to the rising generation in our country, we owe a clear-eyed review of the facts. What did we get right and wrong? What did we miss and why? What can we build on, and what must we build anew? What should we have known at the outset, and what must we never forget? These are the questions that guide this book. And along with them, undiminished by my effort to provide a comprehensive accounting of what went wrong, is a sense of awe at America's capacity to rise to a challenge, our innovation and ability to quickly develop highly effective drugs and vaccines, and the willingness of men and women in all walks of life in our society to take on terrible burdens for the sake of helping others.

Those NYU medical students pressed into emergency service, and in whose shoes I walked two decades ago, were certainly scared in the early days of pandemic. But they showed up. Like the doctors and nurses around them, like their fellow Americans who kept lines of shipping and logistics moving, or who took awful risks to go to work, or who set up online schooling on the fly, guarding their children from fear through an unprecedented crisis, or who made other painful sacrifices so that others might be safer, answering a call most never imagined or expected. It's in recognition of their spirit and their sacrifice that the nation must learn from its failures and must be prepared to do better by its people the next time a new contagion tries to crush us.

CHAPTER 1

AMERICA THE VULNERABLE

"I'm not saying we're at the edge of a pandemic, but the problem is it's hard to tell."

"I'm not saying this will spiral worse but it could and it represents an existential risk."

Those were two messages I sent on Saturday morning, January 18, 2020, to Joe Grogan, the head of the Domestic Policy Council in the White House. Grogan and I had worked together at the Food and Drug Administration (FDA) in 2005, when I served as the agency's deputy commissioner. Now he was running the domestic policy portfolio for the president of the United States.

I had also worked at the FDA in 2003, when the nation was still grappling with the SARS-1 outbreak. I returned as the agency's commissioner in 2017, at the start of the Trump administration and served in that role until April 2019. Over my time serving in different federal public health roles, I had seen these threats come and mostly go. But this new virus looked ominous. A mysterious pneumonia was spreading in the Chinese city of Wuhan, and from the sketchy accounts, it was making people very ill. I was worried. I knew that reporting from China was unreliable, and the details we had could be just a small shard of what was really under way. That had certainly been our experience with SARS-1, where the initial reporting on the outbreak was dangerously incomplete. Much later, we would learn that the Chinese government had withheld material facts about the scope of its SARS-1 outbreak from global health authorities, and from its own citizens.[1]

The events unfolding in Wuhan worried me for other reasons. One was the velocity of the spread of this new infection. Chinese officials initially reported just a dozen known cases. Now, a few days later, on the morning I reached out to Grogan, the case count had quadrupled overnight, to 198 patients.[2] Seeing the news, early that morning I had conferred with scientific colleagues in the US and Australia on new reports that most of the patients in China were severely ill. I texted to Grogan that he ought to make sure the administration's health team was "coordinating a response to Wuhan." I suspected that whatever the case numbers were, the actual scope of the outbreak could be much larger. Chinese health officials were saying that the source of the infection was an unspecified animal that had been sold in a local market, where many of the cases had allegedly originated. But some of the new cases had no discernible links to that market.[3] The notion that one outdoor food market was the site of the spread, and that the origin was a single zoonotic source, seemed to be unraveling, and yet Chinese officials were still clinging to this theory. The US Centers for Disease Control and Prevention already had evidence, in the form of clusters of cases that occurred in the same families, to suggest that the virus was spreading between people. Yet several days earlier, on January 14, the World Health Organization (WHO), relaying China's assurances, tweeted that there was "no clear evidence" of human-to-human transmission.[4]

"I could be wrong, but I would be surprised if China has shared actual virus. It will make developing screening tests harder," I texted to Grogan. "Ask HHS for a briefing," I suggested, adding, "my worry is it looks humanized, it looks human-adapted, but it's very hard to tell."

Joe Grogan was among a handful of White House officials who

were focused on the virus early and he was appropriately apprehensive of the potential danger it posed. Acting on our exchange, that same day he contacted the Department of Health and Human Services (HHS) and asked that briefings be set up for Monday, January 20. He also reached out to the CDC and discovered that lower-level conversations had already begun between that agency and the National Security Council.

Direct engagement from the head of the Domestic Policy Council, this early in a public health crisis, was not typical. But it fit the playbook during the Trump administration; one that emerged partly as a consequence of a perception that voids had been left by some of the administration's top health officials at the Department of Health and Human Services in responding to earlier outbreaks of dangerous viruses. In 2019, the Domestic Policy Council had helped organize the response to an outbreak of measles, and the previous year it had been involved with the response to a large cluster of cases of a mysterious polio-like illness that was causing paralysis in children. Flaccid paralysis (as it came to be called) was later attributed to an epidemic caused by an enterovirus (EV-D68), one of a family of viruses typically spread by touching surfaces that have been contaminated by an infected person and then touching one's mouth, nose, or eyes.[5] Illnesses caused by enteroviruses are usually mild, but if the virus infects the central nervous system, which happens in rare cases, it can cause serious illness. A spike in cases of acute flaccid myelitis, where the enterovirus directly infects the motor neurons of children or causes these nerve cells to become inflamed, had occurred in 2014, and a bigger spike recurred in 2018, with 238 confirmed cases in forty-two states.[6] The children, around five years old on average, had previously been healthy. In most cases, they would report having a mild respiratory illness a week before the weakness started. Months later, many were left with some disability. Nobody knew why the infections were suddenly causing the syndrome or how to prevent them.

There had been frustration in the White House and later on

Capitol Hill with the sluggish response of HHS to that 2018 outbreak. While I was FDA commissioner, I had met privately with one senior Republican senator who excoriated the HHS secretary's office for not taking the cases more seriously and not mounting a coordinated effort to investigate and contain the outbreak and isolate its cause. It was a leitmotif of the frustrations that would reemerge with COVID.

The government's slow response to the measles outbreak and later to the epidemic of flaccid paralysis had left Grogan convinced that he couldn't wait for information about Wuhan to be reported through the usual chain of bureaucratic communication. The Domestic Policy Council, Grogan believed, had to get more directly engaged in managing the White House response.

Alex Azar, the secretary of Health and Human Services, called President Trump the same day, after Grogan had asked for the briefing from HHS, to update the president on the outbreak for the first time. The CDC director, Dr. Robert Redfield, had first briefed Azar on the novel virus two weeks earlier, on January 3, and Azar in turn alerted the National Security Council. Now it was the Martin Luther King Jr. holiday weekend, and Azar reached the president on the golf course in Florida that afternoon. The health secretary received a tepid response when he assured Trump that federal health agencies were ready to handle any cases that could emerge in the US.[7]

By the time China revealed those 198 cases, the coronavirus was already spreading out of control. Researchers at Imperial College London had models estimating that there were probably thousands of cases already circulating in China by the middle of January.[8] Later events would prove that there were actually many more. The next day, Sunday, January 19, I sent Grogan another note. "Where are the non-pneumonia cases?" I asked. It was a reference to the reports coming out of China, which disclosed only those patients with severe pneumonia. I told Grogan it was unusual that a virus would manifest itself just as severe pneumonia and not with milder symptoms in some of its victims. I worried that there must have been less serious cases that were not getting identified. If true, then the severe cases could represent the tip of a dangerous iceberg. The actual number of infections, and the scope of the outbreak, could be much larger than what was being observed. "Are milder cases being sought in China with their new diagnostic kits? They need to be able to get a better idea. We need to have screening capacity ready here too," I wrote to Grogan.

By this point, the National Security Council had begun discussions about setting up temperature-screening checkpoints in airports, for passengers inbound from China.[9] Around this same time, global awareness of the virus had been amplified by a video posted to Twitter on January 17 that showed medics in Level D hazmat suits boarding a flight outbound from Wuhan and scanning the seated passengers for fever using infrared thermometers.[10] The video was a jarring visual, suggesting that there was more danger than was being reported.[11] It no longer looked like just a single outbreak of pneumonia from an infected animal in a market. It suggested that Chinese officials were already worried that the virus might be spreading person-to-person.

On Monday, January 20, Dr. Zhong Nanshan, a pulmonologist and leading Chinese authority on respiratory disease, who in 2002 had gained wide distinction inside China following his role in helping identify SARS-1 and galvanizing the country's response to that earlier epidemic, confirmed in a television interview what was now obvious.[12] The disease was spreading between people. China said that the virus had already infected fifteen healthcare workers in Wuhan. It was a tragically belated disclosure and a critical turning point in the global conflagration.[13]

By late January, the contagion was getting more attention inside the White House. Staff members told me that they were receiving reports that other countries in East Asia, including North Korea and Vietnam, were becoming anxious over China's growing outbreak. The North Koreans had shut their border with China, voluntarily closing a major lifeline for the cash-strapped regime. The panicky state of these regional players added to growing concerns inside the White House: perhaps these nations, with closer ties to China, knew more about the risks than we did?

White House staff told me that, up until that point, they had little sensitive information to inform themselves on the scope of the risks. They were relying mostly on open-source material that they were pulling off the Internet and reports from friends on the ground. More ominously, they didn't have a reliable plan to work from. For all the talk of a pandemic playbook that was purportedly on the shelf, initiated in the George W. Bush administration and refined under the administration of President Barack Obama, it was written for a pandemic involving a novel strain of influenza. The Trump health team would use this blueprint nonetheless, and exacerbate its shortcomings by making some fateful mistakes in how it was implemented.

By January 20, a Chinese medical group tracking daily developments had put the country's case count at 321. Four days later, in the first published study of the outbreak, Chinese researchers reported on the first forty-one patients admitted to a Wuhan hospital between December 16 and January 2.[14] The doctors described an aggressive infection that caused high rates of respiratory distress and critical illness. Fifteen percent of those diagnosed with the disease died. These were some of the first definitive medical reports to make clear the severity of the outbreak in Wuhan.

On January 21, the White House started to actively plan the legal, policy, and logistical requirements for implementing an airport screening program to detain travelers from China and subject them to temperature checks and extra surveillance for symptoms.[15] The next day, President Trump got his first question on the virus during an interview on CNBC from Davos, where he was attending the annual gathering of the World Economic Forum. Asked by *Squawk Box* anchor Joe Kernen if he was worried about the outbreak, Trump replied, "No. Not at all. And we have it totally under control. It's one person coming in from China. . . . It's going to be just fine."[16]

Whatever reassurance policymakers held that the virus could be contained, it would be shaken a day later, on Thursday, January 23, when China imposed a cordon around Wuhan, a city of 11 million people—a full lockdown restricting movement into and out of the city that would last two months. In a short period, China had gone from political calm to a draconian step that put the grave scope of the crisis on full display. The hospitals in Wuhan were becoming overwhelmed, and the city was making arrangements to quickly construct field hospitals. When you saturate your healthcare system in the setting of an outbreak, you can be certain that community transmission is already under way. You are only measuring the people who are sick enough to present for medical care, and it's likely that many more people are infected, but not sick enough to seek out healthcare providers. China's epidemic was now widespread.

Inside the White House, the lockdown in Wuhan gave more ammunition to those who were pushing for travel restrictions on people coming to the US from China. These tighter restrictions were already under active discussion, but the specter of China barring the movement of its own citizens made the administration's effort to do something accordant seem more obvious. The US travel ban that was ultimately announced, which went into effect on February 2, would deny entry to foreign nationals who had visited China in the fourteen days prior to their arrival to the US.[17] The World Health Organization would largely endorse China's lockdown of Hubei Province (where Wuhan is located), one of the most sweeping travel restrictions in modern times by any major economy, and yet the WHO would be critical of the restrictions imposed by the US, and other countries, on foreign nationals traveling from China.[18] In other words, Chinese nationals couldn't move around inside China, and the WHO approved of that policy. But if these same individuals somehow managed to get on a plane and travel to the US, and were subsequently not permitted entry into America, the WHO found that policy objectionable.

This contrast between the WHO's embrace of China's harsh

travel bans on its own citizens, and the organization's criticism of the softer restrictions the US imposed on travelers coming into America, would become political fodder inside the White House for critics of the WHO. Trump officials saw the WHO's perceived duplicity as proof of the organization's bias, and its softness to China. It would color future White House decisions and add to a feeling within the administration that the organization was biased.

That same day that China was locking down Hubei Province, on January 24 the US Senate held a closed-door session where Dr. Anthony Fauci, the longtime director of the National Institute of Allergy and Infectious Diseases at the National Institutes of Health; Dr. Robert Redfield, the director of the CDC; and Dr. Robert Kadlec, the assistant secretary for preparedness and response at HHS, jointly briefed about twenty senators on what they knew about the virus.[19] Fauci had helped manage the nation's response to Ebola and swine flu. He was also a key figure in the efforts waged in 2005, when public health experts worried that a novel strain of bird flu would trigger a deadly pandemic, and President George W. Bush had directed health officials to craft a playbook that the nation could use to respond. Fauci was highly regarded on Capitol Hill and had seen the nation through other crises with infectious diseases. He and Redfield were, for now, reassuring. Redfield said that the CDC was prepared. However, anyone in the US who still doubted the threat that the novel coronavirus presented, and there were many, now needed only to look at China, which risked its entire economy in a bid to control the contagion. China had crushed one of its major cities. The severe lockdown was the most visible evidence that the situation was much worse than many people initially realized. China's draconian action startled some of the White House staff and sent a wave of unease through their ranks.

It was around this time that negotiations were under way to repatriate the roughly one thousand American diplomats and businesspeople working in Wuhan.[20] The Chinese were initially belligerent, worried that the appearance of Americans fleeing the city would be an indirect criticism of China's effort to contain the outbreak.[21] White House staff were also exchanging emails with the CDC to press the idea of more screening checkpoints in major airports. This was different from a ban on all foreign travelers, a measure that would be taken later. For now, the idea was to stop travelers who had symptoms so they could not enter the country. It was more show than tell. Studies showed that airport screening was notoriously leaky. It had been tried in past pandemics, with poor results, and prior modeling had found that such screening would miss most of the cases that passed through an airport.[22] However, if the goal was to provide visible evidence that the government was taking action, the checkpoints would serve a role.

What we needed next were readily available diagnostic tests to identify cases, along with a whole-of-government mobilization. We needed to produce supplies like masks to protect healthcare workers and ventilators to care for the sick. We needed to invest immediately in the development of countermeasures like vaccines and drugs. We needed to shore up our healthcare delivery system to expand its capacity. We needed to protect vulnerable communities like nursing homes. And we needed to expand the capacity of local health departments to track, trace, and isolate those with the infection. At this point, events unfolding around the world provided a strong indication that the contagion could soon reach America. By the end of January, all of these capacities should have been adequately planned for but were not. Absent these measures, we would never catch up with the spread. For the next year, we never did. In that critical first month, officials at HHS were focused mostly on implementing the travel bans and repatriating Americans from abroad. Many of our political leaders, and some health officials, still harbored doubts as to whether the virus would even become epidemic here.

It was especially important to get diagnostic tests in place quickly, so we could implement widespread testing for the virus and detect cases early, before they led to large outbreaks that we wouldn't be able to control. I feared that the only way to avoid a US epidemic would be through a massive testing capacity that we didn't yet have and would be hard to field without concerted action. We needed to act immediately to develop this capability.

On January 23, five days after my first texts to Grogan, I wrote an opinion piece for *The Washington Post* in which I spelled out my concerns for a wider public:

> The illness may be more widespread than we realize in the fog of viral war. Because its incubation period is likely to be as much as a week, travelers can be harboring the infection and show no symptoms. A key to containing the virus's wider spread will be developing effective point-of-care diagnostics to implement more widespread screening of patients. We learned that lesson with the mosquito borne Zika virus when it was first reported in the United States in 2015. Initially, most blood samples had to be shipped to a central lab operated by the Centers for Disease Control and Prevention. That slowed diagnosis and limited the ability of doctors to intervene more quickly when patients were infected. Health-care providers urgently need easy access to effective diagnostic tests.[23]

Zika had shown how unprepared we were for these risks. As I'll describe later, the CDC had badly managed the development and distribution of a screening test for Zika, and so we didn't have enough testing in place to be able to use it as a tool to limit Zika's spread. Similar mistakes would be made with COVID, with even more tragic consequences.

I continued to make the case for stronger and swifter action. Four days later, on January 27, in a post for CNBC's website, I wrote, "Given past experience, we know that the public health labs performing the more sophisticated RNA tests that CDC is aiming to deploy will quickly become overwhelmed in the setting of multiple outbreaks here in the US. In that case, to prevent wider spread, what may be most needed are simple, reliable tests that let us screen more widely for the spread of this novel virus."[24] The next day, in *The Wall Street Journal* (in an op-ed titled "Act Now to Prevent an American Epidemic"), I pointed out that "the novel coronavirus now epidemic in China has features that may make it very difficult to control. If public-health authorities don't interrupt the spread soon, the virus could infect many thousands more around the globe, disrupt air travel, overwhelm health-care systems, and, worst of all, claim more lives." Among the needed measures, I said that "if the number of cases increases, experience from the 2009 swine flu pandemic and the 2015 Zika epidemic suggests that the CDC will struggle to keep up with the volume of screening. Government should focus on working with private industry to develop easy-to-use, rapid diagnostic tests that can be made available to providers."[25] I had coauthored the article with my former FDA colleague Dr. Luciana Borio, who had also previously served as director for medical and biodefense preparedness at the National Security Council.

Testing would be a major gap in our response and the most visible symptom of the capabilities that we lacked in confronting a public health crisis of this magnitude. We just didn't have the resiliency in our system to quickly step up the production of test kits and testing equipment at the scale that was required. We couldn't organize a collective effort to get key industries all working in a coordinated way to plug these gaps in time. Our health officials were overwhelmed early on with repatriating Americans from China and focused too little attention on preparing the homeland for the inevitable spread of the virus. Many publicly clung to a tragic view that SARS-CoV-2 wouldn't become epidemic here. By March, when the evidence that it would spark a global pandemic could no longer be discounted, it was too late.

That relative complacency was being fed by reliance on faulty information that had US officials believing that the virus wasn't spreading here, even long after it had actually arrived in our communities. It was driven largely by a failure to field a test that would let us screen for it. COVID hurt and humbled us. In the richest healthcare system in the world, with all of our advanced technology, our political and public health systems were overwhelmed. We would soon find that we were far more vulnerable than we'd expected.

CHAPTER 2

CONFUSION AND SUBTERFUGE

Wei Guixian would get sick around the same time every winter. Working as a merchant at Wuhan's Huanan Seafood Market, where she sold shrimp, she caught a lot of colds.[1] Her symptoms started the usual way, but this time they quickly worsened.[2] With increasing congestion, she went to a local clinic, where she was given an antibiotic and sent home. But overnight her condition deteriorated. So, the next day, she visited Wuhan's Eleventh Hospital in search of more care. Again she was sent home. "The doctor at the Eleventh Hospital could not figure out what was wrong with me and gave me pills," she would later tell a Chinese newspaper.[3] But the pills didn't work either. Five days later, she went to Wuhan Central Hospital, one of the region's busiest medical centers, with more than three thousand patient beds and employing about four thousand providers and staff.[4] Wei was now gravely ill with severe pneumonia in both her lungs and was admitted to intensive care.[5] It was December 15, 2019.

Wei was told by the doctors that her illness was "ruthless." They were seeing patients with similar symptoms, who had also frequented the nearby Huanan market. Many of the patients, like Wei, were in severe distress.[6] The market was located in the Jianghan District of Wuhan, a city of 11 million and the capital of

China's Hubei Province. With its 653 stalls, it was a bustling outdoor food bazaar popular for its live animals and wholesale seafood. Like the other patients at the Wuhan Central Hospital, Wei was at first screened for common respiratory viruses, but the tests all came back negative.[7] Something novel was making her sick, and by every measure, it was spreading among the city's residents. By December 21, local doctors counted at least three dozen hospitalized patients. Many were seriously ill and had matching symptoms.[8]

Unable to identify the cause of the menacing illness, on December 24, the doctors at Wuhan Central Hospital, where Wei was now critically ill, took fluid from the lungs of another patient with similar symptoms and sent it for deeper analysis. He was a sixty-five-year-old deliveryman who also worked at the seafood market and was first admitted to the hospital on December 16. A CAT scan showed extensive infection in both of his lungs, even though, as with Wei, a standard battery of tests for known viruses and bacterial infections had also come back negative.[9] Doctors tried prescribing a milieu of antibiotics and antiviral drugs, but the medicines didn't help. With his condition declining, doctors sent his sample to Vision Medicals, a genomics company in Guangzhou. They needed answers. They needed to find the cause of the contagion.[10]

Vision Medicals offers genetic sequencing directly to doctors. The company can read the genetic code of the individual pathogens that it finds in a patient's body fluids to try to isolate the virus or bacterium that was causing the illness.[11] By combing through all of the viruses found in the man's sputum, Vision Medicals sought to determine if he was harboring a novel pathogen.

The genetic instruction set that directs the activities of a virus are contained in its ribonucleic acid, or RNA. Much as human beings carry our entire genetic code in our DNA, for some viruses it's the RNA that conveys this information. Sequencing RNA is the process of determining the precise order and contents of these genes. This application of genetic sequencing to the routine diagnosis of infections is relatively new and, until recently, was largely confined to specialized research. In this case, doctors suspected that a novel respiratory virus could be causing the cluster of infections. Using "metagenomic sequencing" to read all of the genomic material present in the patient's sample, doctors aimed to zero in on the pathogen that was making him sick.[12] Many of the machines used to perform this sequencing are made by the American company Illumina, on whose board of directors I serve.

Viruses are not technically alive. But they're not dead, either. They operate in a pseudo corporeal state, where they hijack our cellular machinery to carry out their cycle of replication and spread.[13] In doing this, they engage in most of the activities that give an organism the features of life. But viruses cannot do it alone. They need a host to model the gestures of life. They need a specific species that they can invade, pirating its cells and using them to invade and propagate.

Once inside a host, often a virus will have a specific "tissue tropism," meaning it has a fondness for infecting and replicating inside a certain part of our body. Some viruses can infect many types of cells. Other viruses live primarily inside a certain organ.[14] The rabies virus travels along our neurons. Hepatitis hides in our livers. SARS-CoV-2 likes to live in the lining of our respiratory tract.

Often the harm from a virus results from the injury it causes directly to these tissues. In other cases, the peril it brings owes to how our immune system responds to it. The same inflammatory response that our body mounts as a way to attack a virus can also damage our organs if that process gets out of whack. That's true in hepatitis. It's the inflammation that hepatitis triggers in our liver that does much of the damage.[15] It was also true with COVID. For many patients, the most severe consequences of COVID weren't the direct effect of the coronavirus on their lungs, but rather the exaggerated immune response their bodies unleashed to kill the virus.

Because the SARS-CoV-2 virus was novel, meaning that human beings had never encountered it before and thus had no established means to combat it, the human immune system tended to overreact. In the most severe cases, it could trigger a "cytokine storm," a cascading activation of immune cells, which in turn excrete small proteins called cytokines.

In the normal sequence of our body's functions, cytokines help coordinate our response to viral invaders and recruit other immune cells to the site of an infection. But if all the cytokines get dumped into our blood simultaneously, too many immune cells can get recruited to the site of the infection, creating an abrupt and severe hyperinflammation that can damage organs.[16] It was a cytokine storm that ended up killing many people from COVID.

Around the same time that the sixty-five-year-old man's sample arrived at Vision Medicals, in mid-December, doctors from other local hospitals sent at least eight other patient samples to multiple Chinese genomics companies.[17] In the days that followed, one hospital in Wuhan alone sent about thirty samples to the genomic sequencing company BGI. The samples had been taken from different patients suffering from pneumonia; three were found to contain the novel coronavirus.[18] The flurry of activity was evidence that the local doctors knew they were dealing with something new and dangerous. This spike in sequencing requests itself should have been a worrisome signal. It's an unusual order pattern, to see so many doctors suddenly sending off samples to genetic sequencing labs. These kinds of signals can be a tipoff that something unfamiliar is spreading in a community. In this case, the local providers were all in search of the same thing: an answer to the question, What was making their patients critically ill?

One of the first results came back on December 27, on the sample drawn from the sixty-five-year-old deliveryman. Scientists sequenced a novel coronavirus in his lung fluid. The new virus had nearly 29,000 nucleotide bases that held its genetic instruction set, and it looked frighteningly similar to SARS-1, which killed nearly eight hundred people globally in 2002 and 2003. But the virus wasn't SARS-1.[19] It was something completely novel. That information alone should have been terrifying. A new respiratory virus capable of making people severely ill was spreading in Wuhan. Yet local officials initially withheld information from the central government in Beijing about the outbreak.[20] An employee of one Chinese genomics company said that on January 1 an official at the Hubei Provincial Health Commission instructed the company to stop testing samples related to the Wuhan outbreak and destroy all of the lab's remaining specimens. The impulse to contain the troubling results eventually spread to officials in Beijing. Two days later, on January 3, the country's top health authority, the National Health Commission, ordered the local genomics labs not to publish any sequencing information related to the novel coronavirus.[21] News of the outbreak would be shared through the postings of anxious doctors who took to social media to exchange the sequencing information.

One of these doctors was Ai Fen, the head of the emergency department at Wuhan Central Hospital. On December 30, she received a copy of another sequence report, generated on a different patient who had been admitted to her hospital with matching symptoms. She circled one phrase from the document, "SARS coronavirus," and posted it to a WeChat group shared by the doctors in her department.[22] The report was circulated widely among local doctors, many of whom came forward with more cases. More samples were sent for sequencing. More evidence emerged that a novel coronavirus was spreading in the city. More results were posted online.

By this point, there was a spike in patients arriving at local hospitals in severe respiratory distress. Ai Fen said she noticed a stream of patients in the weeks before officials finally confirmed the presence of the new virus.[23] Local doctors started to assume that the virus was a common link and that it was spreading in the community. One of the local doctors alarmed by the unfolding events was Li Wenliang, an ophthalmologist at Wuhan Central Hospital. He wrote to a group of his medical school classmates on WeChat that "7 confirmed cases of SARS were reported . . . from Huanan Seafood Market." He also shared with the same group the report and a CAT scan image from one of the patients. Screenshots of his text messages were shared on the popular Chinese microblogging website Weibo, where the hashtag "Wuhan SARS" started to trend. The chat messages were picked up by global media outlets, sparking some of the first serious reports that a troubling outbreak of a potentially novel respiratory virus might be under way in China.[24]

It also triggered some of the first overt actions by the Chinese government to quash news of the outbreak. The hashtag "Wuhan SARS" was quickly suppressed on the Weibo platform by Chinese officials, and the original post put Li in the crosshairs of local authorities, who were still working to curb the news flow. On January 3, police from the Wuhan Public Security Bureau interrogated Li and censured him for "publishing untrue statements."[25] He was forced to sign a letter promising not to do it again and was told that recalcitrant behavior would result in prosecution. He would later be labeled a radical by local government officials for disseminating the report that Ai Fen had shared. In all, eight people were detained for spreading "false news" about the case over the Internet. Many of them were local doctors caring for the stricken patients.

"No creating rumors, no believing rumors, no spreading rumors," local authorities and state media posted on the eight individuals' Weibo pages after their arrest.[26] "Rumor is a kind of virus that is worth more of our attention than real viruses sometimes," the state-run *People's Daily* wrote on its website.[27] ProPublica would later track more than ten thousand Chinese government-linked influence accounts on Twitter that had previously targeted political dissidents and the Hong Kong protests, but now began to train their attention on reports related to the outbreak.[28] A subsequent study by the Citizen Lab at the University of Toronto found that the Chinese government began censoring postings about the virus beginning December 31, when the first news of the outbreak started to circulate. Between January 18 and May 14, Citizen Lab found 2,174 keywords related to COVID that were suppressed on WeChat.[29] Many of the deleted posts, they found, were criticizing the Chinese government officials for how they handled the initial response.

On December 30, around the same time that some of the first definitive findings from sequencing were being shared with local doctors, the Wuhan Health Commission finally notified the region's hospitals that a "pneumonia of unclear cause" was spreading in the city. Providers were ordered to report suspicious cases to authorities.[30] That warning also made its way onto the Internet. Global awareness of the outbreak was now growing.

As a signatory to international public health treaties, China has agreed to inform world health authorities if it suspected that it was host to an outbreak of a dangerous new pathogen. Nations set up public databases to share information with an international network of research groups that monitor outbreaks. Yet certainly by mid-December, local officials became aware of an outbreak, and started to zero in on the culprit.[31] It became clearer that the episode met the bar for international reporting. These obligations were part of the revised International Health Regulations put in place in 2005, in part as a response to global concerns that China had put other nations in danger by withholding information about the SARS-1 outbreak for too long. Countries promised to be more transparent in the setting of future outbreaks. The new regulations required all 196 countries party to the agreement to improve systems for detecting and reporting dangerous pathogens.[32] The agreement also widened the circle of viruses that require immediate reporting to world health authorities.[33]

One of the central features of this agreement is a stipulation that countries invest in systems to detect outbreaks and report events that may constitute a potential Public Health Emergency of International Concern, or PHEIC as it has come to be known. The CDC explains that under the global compact, a PHEIC is declared by the WHO when two of four criteria are met:

- Is the public health impact of the event serious?
- Is the event unusual or unexpected?
- Is there a significant risk of international spread?
- Is there a significant risk of international travel or trade restrictions?

Once a nation identifies an event that appears to meet these criteria, within forty-eight hours it must assess the public health risks posed by the event. If it's judged to be notifiable under the International Health Regulations, it must be reported to the WHO within twenty-four hours.[34] Some diseases must always be reported. These include smallpox, poliovirus, a completely novel strain of influenza, and SARS-1. The mandatory reporting of these pathogens is meant to cover circumstances where a dangerous virus might emerge. All of the pathogens enumerated in the health regulations constitute perilous and unexpected events. The list isn't meant to be exhaustive but illustrative. By any calculus, the virus spreading in Wuhan met the spirit of these reporting guardrails and information about it needed to be shared.[35]

This international framework was adopted as a way to give the global community an early warning of a potentially dangerous new pathogen, so countries could pool resources and take steps to counter the emerging threat. Yet China would be slow to honor these conventions. At a local and then at a central level, it withheld pertinent information about the virus for weeks. China reported on it only after the outbreak became obvious. Chinese officials were praised for posting the sequencing information in early January, but even that disclosure was belated. State labs in China were later reported to have that data weeks earlier—even before many of the local doctors in Wuhan started to sequence the virus for themselves.[36]

As of December 28, with the genetic sequence already in their hands, Wuhan health officials would confirm only that twenty-seven people had been admitted to area hospitals with matching symptoms of pneumonia from an unidentified infection. Seven were in critical condition, while the others were "stable" with "controllable" symptoms, according to a media release issued by the Wuhan Municipal Health Commission on December 31.[37] By early January, with a growing number of victims who were spread across different hospitals, stricken patients were now being transferred to a single facility, Jinyintan Hospital, that specialized in handling infectious diseases. The hospital received its first transfer from a local general hospital on December 27. During the call to Jinyintan to receive that seriously ill patient, Jinyintan's vice president, Dr. Huang Chaolin, was told that a genetic testing company had detected a novel coronavirus in the patient's respiratory samples. Over the next few weeks, Jinyintan admitted more than one hundred patients believed to be suffering from the virus, filling its intensive care unit to capacity.[38]

On January 1, 2020, officials closed the food market that was thought to be a source of the outbreak.[39] For now, the market was the only purported link between the patients. Initially, local officials posited that an unknown animal, presumably butchered and sold in the market, had been infected and must have transferred a virus to the human hosts. It was a convenient theory because it would mean that the outbreak was likely contained to the market, and the virus had merely jumped from a single zoonotic source and wasn't spreading between people. But the account was based on inference alone, and it was unraveling.

Many of the early cases that were identified had some connection to the market. But not all of them. Pictures circulated around the world of Chinese health authorities, clad in biohazard suits, fumigating the market and collecting biological samples. Environmental sampling of the Wuhan market, conducted on January 1 and January 12, found evidence of the virus on door handles, the market's floor, and in sewage; but the animals that were tested turned up negative. Initially, China revealed only that positive samples had been collected, and not the evidence that animals had also been tested and were found to be negative. If this critical information had been disclosed sooner, about the negative results on tests drawn from animals, it would have cast earlier doubt on the belief that the market was the focal point for the outbreak and that the virus was spread through a zoonotic source.[40] When the Chinese government shared the initial data on the first twenty-seven human cases with Robert Redfield of the US CDC, he immediately spotted evidence of person-to-person spread. It was in plain sight.

The genetic sequence of the novel coronavirus was finally posted to a public server by a researcher in Shanghai, Dr. Zhang Yongzhen, who was believed to be acting on his own, defying specific directions from Chinese officials not to post the information. Zhang didn't post the sequence directly, instead relaying permission to an Australian colleague, Dr. Edward Holmes, a virologist at the University of Sydney, to post the information to the website Virological.org, an online portal where scientists shared sequence information.[41] Holmes then shared a link to the information on Twitter, and it quickly ricocheted around the world.[42] Zhang's lab was shut down for "rectification" by the Shanghai Municipal Health Commission on January 12, hours after Zhang had directed Holmes's team to post the sequence. It looked like retribution. It was later revealed that Zhang had finished sequencing the virus on January 5 and had sent the information through official channels to the Wuhan Institute of Virology two days before China officially declared that the outbreak of pneumonia cases was being caused by a novel coronavirus. When he examined the sequence, Zhang immediately recognized that it was a novel coronavirus and bore similarities to SARS-1. Zhang had released the sequence after growing frustrated that Chinese officials hadn't issued more forceful warnings.[43] When the data were posted to Virological.org, Dr. Vineet Menachery, a leading virologist and immunologist with the University of Texas

Medical Branch and an expert in how viruses interact with human hosts, observed on Twitter that the new virus appeared to be in the same general family as SARS-1—an alarming finding.[44] We would learn later that the Chinese authorities already possessed that information for weeks.

Until COVID, there were six species of coronaviruses that were known to infect humans. Four of these caused a largely benign illness in healthy people that, for most, amounted to little more than a common cold. In rare cases, these coronavirus strains could cause pneumonia. Two of the species originated from bats (alpha coronaviruses) and two from rodents (beta coronaviruses). Together these four viruses account for as many as 30 percent of upper respiratory infections in adults.[45] In more recent years, two new and more fearsome human coronaviruses had emerged: SARS-1 and MERS. The new virus spreading in Wuhan appeared to be a seventh entrant to this species.

Immediately after Zhang Youngzhen's team released its sequence data, China's National Health Commission finally announced that it would officially share the new virus's genomic information with the WHO.[46] The Associated Press later reported that "even then, China stalled for at least two weeks more on providing the WHO with detailed data on patients and cases, according to recordings of internal meetings held by the UN health agency."[47] Chinese physicians and public health officials had acted with conviction, and often with courage, to inform the world of the outbreak. But through official government channels, information wasn't shared the way international conventions prescribed. Even as Chinese officials reported on the sequence, they withheld perhaps the most ominous finding: that the new virus was spreading widely between people.

For other nations, getting detailed information to inform the global response proved difficult. A key step was getting early access to samples of the virus, so international researchers could better evaluate its features and start the process of developing therapeutics, vaccines, and diagnostic tests. The director of the WHO,

Dr. Tedros Adhanom Ghebreyesus, would announce in January that China had committed to share the viral samples, but Chinese officials would never do so.[48] Andrew Bremberg, the US permanent representative to the UN and other international organizations in Geneva, would urge Tedros, and other WHO officials, to use their influence with China to press for release of the samples. Privately, and recognizing the importance of getting the samples, the WHO would push the Chinese government to make them available. But Tedros refused to call on China to do so publicly, which might have forced the Chinese government's hand. He argued that, as a technical matter, China wasn't obligated to release the samples under the International Health Regulations, so he said it wouldn't be fair to pressure them publicly to do so. Indeed, the international health regulations didn't specifically enumerate SARS-CoV-2, largely because the virus didn't exist at the time these regulations were drafted. But it included SARS-1, and it clearly intended to call for the sharing of samples when a dangerous new pathogen would emerge.

At the same time, the WHO continued to publicly praise China for its purported openness. To take one of many examples, at a January 30 press conference, Tedros celebrated "China's commitment to transparency and to supporting other countries. . . . In many ways, China is actually setting a new standard for outbreak response, and it's not an exaggeration."[49] He would make similarly effusive statements on other occasions, despite the wrangling that continued behind the scenes. A year later, at a media event held a week before the end of President Trump's term, Secretary Azar reflected on the opening days of the pandemic, saying that his team was "pressing for the Chinese government to send us viral isolates from patients there. And China has still, one year later, failed to provide the first-generation viral isolates."[50] Access to those samples at the outset could have helped the world prepare. And without the source strains, it would be impossible to determine with any certainty the virus's origin, and whether the market was the focal point of the initial outbreak or, as seems more likely, one stop along its advance.[51]

The refusal to share samples was a break with long-standing practices but in keeping with recent experience. Historically, when a novel strain of influenza emerged, China shared the strains with the WHO. Then, in 2018, China abruptly declined to share samples of a strain of H7N9 bird flu that had first emerged in eastern China in 2013 and then resurfaced in 2017, despite repeated requests from the WHO and promises Chinese officials had made, as part of the International Health Regulations and similar side agreements, to share samples of novel pathogens. In particular, when it came to strains of a novel influenza, nations had committed to send the samples to certain designated research centers "in a timely manner."[52] *The New York Times* reported that scientists at the US Department of Agriculture "had such difficulty obtaining flu samples from China that they have stopped requesting them altogether." The article had been placed with the assistance of officials in the White House, who were frustrated by China's actions, and wanted to hold Chinese officials to publicly account for these policy decisions. At least four research institutions received the influenza strain only after cases eventually emerged in Taiwan and Hong Kong and those countries shared the samples.[53]

According to Secretary Azar, the US government first learned about the COVID outbreak on December 30, not through official channels but by monitoring local media as well as through a notification they received from Taiwan's economic and cultural office in the United States.[54] Surprisingly little precise reporting was filtering up to health officials during the initial weeks of the outbreak, based on Azar's account and the public reflections of other American health officials. White House staff told me that in December and early January, they had little hard information to help them calculate the full scope of the threat. For most of January, the President's Daily Brief contained little to no information on the novel coronavirus that couldn't be found in open-source reporting. *The Washington Post* would report that the volume of warnings increased "toward the end of January and into early February," citing officials familiar with the reports that were being shared. "By then, a majority of the intelligence reporting included in daily briefing papers and digests from the Office of the Director of National Intelligence and the CIA was about COVID-19," the *Post* reported, citing officials who had read the intelligence reports.[55]

Evidence later emerged that Hubei Province was dealing with what it believed was a major outbreak of flu that began on December 2.[56] Hubei was reporting twenty times the normal number of flu cases, a surge that may have started in November, putting an enormous burden on the local healthcare system. It now seems possible that some of those cases could have been SARS-CoV-2 making its first entry into the local population. Robert Kadlec would later say that the US uncovered evidence that China was starting to hoard medical supplies like masks and gowns by early December.[57] The anecdote was offered to suggest that China had a run on these products well before the US was aware of the impending risk, or at the very least, China was worried that demand for medical goods could start to rise and was stockpiling the material.

Without access to stored blood and respiratory samples from some of the earlier patients who were presenting with flu-like symptoms in November, it's impossible to know for certain if what was believed to be an epidemic of influenza was also some of the initial cases of the novel coronavirus making its first entry into the city. A year later, as part of its investigation of the virus's origin, a WHO team sent to China sought samples collected as far back as the fall of 2019, but the investigators were told by Chinese officials that the Chinese government lacked the necessary permission from patients to test samples, many of which were held in blood banks. It seemed disingenuous for Chinese officials to invoke far-reaching patient privacy protections in this instance, with such critical public health questions on the line.[58] Leaked internal documents would later show that in November, routine tests being carried out on the patients suspected of having flu were returning a higher-than-normal rate of indeterminate results.[59] Many of the Chinese patients who were thought to have complicated cases of influenza were instead sick with some other, unidentified respiratory illness.

A modeling study done later by researchers at the University of California, San Diego, and at the University of Arizona, found that the first cases of the novel coronavirus probably began spreading in Hubei Province in early November but could have been transmitted at low levels as early as October 2019.[60] *The South China Morning Post* would report on Chinese government data, that traced the first known cases in Hubei to November 17. "Chinese authorities have so far identified at least 266 people who were infected [in 2019], all of whom came under medical surveillance at some point. Some of the cases were likely backdated after health authorities had tested specimens taken from suspected patients," the newspaper wrote. "Of the first nine cases to be reported in November—four men and five women—none has been confirmed as being 'patient zero.' They were all aged between 39 and 79, but it is unknown how many were residents of Wuhan."[61] If China had shared some of the early patient samples from November, it could help reveal—or put to rest—conjecture that the virus may have been spreading widely much earlier than December 2019. The WHO team also sought wastewater samples from central China to see if the virus could be detected in sewage from late 2019 and were told the samples had been discarded.[62] In the absence of a more complete investigation, speculation would persist that the virus might have begun spreading in the fall.

The slow flow of information about COVID had echoes in the earlier experience with SARS-1 in 2002. When that outbreak first emerged, Beijing learned at some point that a novel virus was the culprit but continued to blame the cluster of infections on a bacterium.[63] Two months later, when an outbreak of the infection appeared in Canada, a Canadian lab promptly sequenced the pathogen for itself, firmly establishing that it was a novel coronavirus. It shared this information with the world. Only then did the Chinese government release data on the full scope of its epidemic, announcing that Beijing had 346 cases—a ninefold jump from the 37 cases government officials had previously disclosed.[64]

When the novel coronavirus first emerged, similar challenges hampered the early response. Without access to the viral samples, it made it more difficult for other nations to fashion diagnostic tests and to begin work on drugs and vaccines. Countries could synthesize the whole virus in the lab, but that took months and had to be done in special, high-security facilities equipped to handle dangerous pathogens. It was not until much later, on March 3, that Swiss scientists became the first group to accomplish the feat, synthesizing the virus at a high-security lab in Mittelhäusern.[65] By then, many countries already had access to samples of the live virus because it was spreading uncontrollably around the world. I was told by a member of the White House staff that at least one early sample that the US secured possession of came from one of China's neighbors. China's hesitancy had given rise to a global barter system for early samples of SARS-CoV-2.

The lack of information also put medical providers at increased risk. If you look at the medical workers in China who initially fell ill, few from emergency and respiratory departments were sick, Ai Fen later observed. It's likely that doctors in these settings had the mind-set to protect themselves. "The most seriously ill have been those from the departments on the periphery," she said.[66] These doctors were more likely to be unaware of the virus and naive about its spread and didn't have the presence of mind to wear protective gear and take respiratory precautions.

The work of many Chinese doctors and researchers early in the outbreak, to distribute information on the novel pathogen and post sequence reports and other bottom-line data, ultimately helped inform physicians around the globe about the virus. In the moment, these were heroic acts. Many Chinese scientists and physicians took risks, and their efforts saved lives. Yet, ad hoc reporting by individual doctors was not a substitute for more coordinated disclosure by the government, which didn't happen until much later in the crisis. Even on the one-year anniversary of the lockdown of Wuhan, Chinese officials tried to soften news about the government's response to the pandemic, stripping terms like "first anniversary" and "whistleblower" from Chinese websites like Weibo.[67] Li Wenliang, the ophthalmologist at Wuhan Central Hospital, who had been threatened by the local police for disclosing information about the outbreak, eventually returned to his clinical work and was himself infected with SARS-CoV-2, probably in early January while caring for a glaucoma patient who was asymptomatic but infected with SARS-CoV-2.[68] Li died of COVID on February 7. In all, more than two hundred medical staff at his hospital contracted COVID, and three of his colleagues died from the disease in March. Some of them were not made aware early enough of the virus's risks, and they were not warned to take precautions.[69]

Li Wenliang's death shocked the public around the world and galvanized criticism inside China to the government's handling of the crisis, but it didn't have any discernable impact on China's sharing of information. I was told by members of the White House staff that during the same week that Li died, the Chinese government started to clash privately with officials at the WHO, trying to head off the declaration of a PHEIC, which could give other nations political cover to apply travel restrictions in an effort to keep an outbreak contained, even though such travel measures continued to be heavily discouraged by the WHO.[70] China wanted to avoid those burdens.[71] Its officials also wanted to avoid any public suggestion that China was losing control of the outbreak. However, the situation in China had been worsening for weeks. On January 20, the Chinese Xinhua News Agency had reported that the total number of coronavirus infections in the country had climbed to 217, including 14 cases in Guangdong and 5 in Beijing. Shortly afterward, China's National Health Commission confirmed the first case in

Shanghai.[72] Chinese health officials reported the next day that six people in Wuhan had already died from the illness.[73] China's grim epidemic was under way.

Chinese officials would say that they were in regular contact with the WHO and sharing medical information with the global body. As a technical matter, this was true. But the private emails between WHO officials and their Chinese counterparts would tell a different story. The WHO would submit long lists of questions to Chinese officials, related to the scope and severity of the epidemic. In return, the Chinese government would provide achingly incomplete replies. This wrangling would continue, with the WHO resending the same questions and receiving modified replies that didn't fully address the key questions. All the while, the WHO—afraid to jostle Chinese officials, afraid to confront them, afraid to lose its limited access—would maintain that it was in constant dialogue with the Chinese government. As a technical matter, they were in dialogue. But the interchange was not productive.[74]

Twice the Chinese delegation in Geneva tried to block the declaration of a PHEIC. The first time was on January 22, 2020, and they succeeded. The second time, on January 30, the effort failed. This was right after WHO director Tedros returned from China where he was meeting with senior Chinese officials, including President Xi Jinping, as part of an effort to show engagement and support for China's efforts. Before leaving China, Tedros called his team in Geneva and told them to establish a new emergency committee to reconsider the declaration of the PHEIC. On January 29, the morning before the declaration was reconsidered, Ambassador Bremberg met with his Chinese counterpart in an effort to convince China to support the measure. Bremberg argued that a declaration of a PHEIC would not be perceived as a condemnation of China's effort to contain COVID. However, quite the opposite would be true, Bremberg said, if China continued to oppose the WHO effort. Blocking the measure, he said, would be seen as an attempt to restrain global action that could help other nations prepare for onward spread. Bremberg felt confident that he had convinced the Chinese ambassador, but he wasn't sure Beijing would sign off. That's where it was left. China continued to fight the declaration, although this time it failed. Privately, Tedros used the episode, and his eventual support for the declaration, to try to counter accusations of being soft on China.

Yet in practice the declaration would have limited impact. The idea of such a declaration was to trigger international action to contain the spread and avert, or at least temper, a wider pandemic. But many nations had not updated their domestic laws governing public health emergencies since 2005, when the world was confronting the aftermath of SARS-1 and was increasingly focused on the risk of pandemic flu. Most countries still had laws in place that tied their local action not to the declaration of a PHEIC, but to the declaration of a pandemic by the WHO—a higher standard that had not yet been met in the WHO's estimation. Indeed, the WHO wouldn't declare the novel coronavirus to be a pandemic until five weeks later, on March 11, well after it was clear to anyone watching the news that a global pandemic was already under way. "We are deeply concerned both by the alarming levels of spread and severity, and by the alarming levels of inaction," the director general said at a briefing held to announce the belated declaration of a pandemic.[75]

By that point, the ground was shifting. A day earlier, on March 10, President Xi visited Wuhan to mark the culmination of a harsh seventy-six-day lockdown of that city that had largely succeeded in arresting the epidemic there. Researchers from the University of Southampton in England, using mobile-phone location data from Chinese internet firm Baidu, estimated that if China had implemented its strict measures in early January, it would have reduced the epidemic's victims to just 5 percent of the eventual total.[76] That's a small enough outbreak that Chinese officials might have been able to fully contain the spread inside China. But from January 5 to January 17, hundreds of patients were appearing in

China's hospitals, not just in Wuhan but across the country.[77] The containment window, to avert wider transmission, had slammed shut. The contagion was ending in China, but its global spread was just beginning in Europe and America.

I spent March 10 on Capitol Hill, briefing members of Congress. In the morning I addressed the entire Republican House Conference. The Republican leadership, especially Representative Liz Cheney, was trying to build support for a COVID relief package that was coming before Congress. Cheney was deeply concerned about the threat that COVID posed and had invited me to brief the caucus that morning. She would continue to seek my perspective throughout the crisis and asked me to brief the Republican caucus on two more occasions later that spring, by telephone, after the US epidemic was under way. At the March 10 meeting, I delivered a downbeat assessment of where we stood, predicting that the virus would soon begin to spread widely in the US, and we would be forced to turn to mitigation to slow its spread. I told the members that the relief package, which would become the Coronavirus Aid, Relief, and Economic Security (CARES) Act, would be crucial to help bridge Americans through the crisis and enable the actions we would need to take to slow the spread of the virus. In the afternoon, I briefed Democratic members of the House Progressive Caucus. Shuttling between the two meetings, from a room filled with some of the most conservative members of Congress, to a meeting with some of Congress's most liberal representatives, the one constant was a palpable sense of anxiety. There was, however, little recognition that the danger that many feared was on the way had already arrived. That morning, most of the Republican caucus, nearly two hundred members, crowded into a single conference room in the basement of the Capitol Hill Visitors Center. During the meeting, one of the few tangible signs of the looming danger was the food line: the meals were ready to eat, grab and go. Two staff members, stationed behind a table, dispensed coffee into disposable cups, all to avoid people touching the same utensils.

The pandemic was under way. And it would be made worse by the distortion and suppression of information, in many different settings, at different times and levels: at first balefully, through the efforts of some government officials in China—especially local authorities in Hubei Province; and then later, in the US, amidst wrangling over how forcefully to confront the virus.

Had the Chinese government been more forthright at the outset, there was a chance that the virus could have been contained there. Later, in the US, the effort to find effective tactics and therapies to stop the virus would also suffer from a lack of reliable information as we struggled to establish a truth standard about what would be most effective, and to agree on a reliable set of actions for containing spread. In a public health crisis, reliable information is a vital currency for decision making. The spark of transmission was lit amid the suppression and distortion of key facts.

We could and should have known and done more in the US. But it's important to remember that in December 2019 and for the first few weeks of January 2020, the Chinese government didn't share vital information that could have mobilized an earlier response in China and alerted the world to the full scope of the threat much sooner. In the US, we made plenty of mistakes of our own, so it's arguable how much the earlier warnings would have helped us galvanize a more effective response. But global conventions, many implemented after SARS-1, were supposed to keep us fully informed of these risks, and they were disregarded. This should change the future course for how we deal with these threats. We'll need to rely much more on our own tools for gathering information about novel pathogens and menacing outbreaks. We can no longer depend largely on global cooperation and the competency and transparency of other nations. COVID wasn't caused deliberately, but it was enabled and nurtured by the intentional quashing of information.

CHAPTER 3

PANDEMICS AS NATIONAL SECURITY THREATS

We needed to get boots on the ground.

By late January, US officials were mounting a full-court press for China to let staff from the CDC enter Wuhan. Chinese officials initially balked, then promised access, stalled, refused, and finally relented after it was almost too late to do the CDC much good. China's willingness to sidestep some of the key commitments made as part of global agreements that were implemented after SARS-1 was felt viscerally at the CDC—which couldn't get permission from the Chinese government to let any of the dozen CDC staff, permanently stationed in Beijing, to visit Wuhan.[1]

The CDC had been trying to get access to Wuhan for more than a month, with no success. On December 31, Dr. Anne Schuchat, the CDC's top career official, was one of the first officials at the agency to spot the threat. She read a brief report in a scientific bulletin that tracks outbreaks, a description of four cases of "unexplained pneumonia" in Wuhan, and emailed a group of CDC colleagues asking if "any of your folks know more about the 'unknown pneumonia' in Wuhan." Within hours, an email reply came from Dr. Nancy Messonnier, the director of the CDC's National Center for Immunization and Respiratory Diseases, who wrote that she learned there were actually now twenty-seven patients with pneumonia. The characterization of the cases, she said, "raises concern about SARS."[2]

The next day, January 1, CDC Director Robert Redfield emailed his Chinese counterpart, Dr. George Fu Gao, a virologist and immunologist who had served as director of the Chinese Center for Disease Control and Prevention since 2017. After receiving no response, later that day Redfield called Gao to press for more information. By January 3 the two had talked multiple times about the outbreak. The following day, on January 4, Redfield sent Gao another email, again entreating for more information on the situation in Wuhan and requesting that the US CDC staff be given access to the hot zone. "I would like to offer CDC technical experts in laboratory and epidemiology of respiratory infectious diseases to assist you and China CDC in identification of this unknown and possibly novel pathogen," Redfield wrote. Gao was emphatic that there was no person-to-person transmission and no evidence of spread within hospitals. Gao's working theory was that the virus had been spread by contact with an animal, still unidentified, at the Huanan market. All the early cases seemed to be tied to that market. But Gao had sent Redfield a list of the first twenty-seven cases that the Chinese CDC had identified, and Redfield noticed that among them were three clusters where multiple family members were affected—a husband and a wife, or a child and a parent.[3] It seemed implausible to Redfield that multiple members of three different families had all contracted the virus from the same zoonotic exposure. Redfield told Gao he was extremely worried that this was evidence of human-to-human transmission, urging Gao to look aggressively through local medical admissions for people with matching respiratory symptoms who didn't identify the food market as a common point of contact.

Two days later, Redfield sent yet another note, this time attaching a formal letter offering support from the US CDC.[4] Gao called Redfield back and cried during the phone call, telling Redfield that they might be too late to stop a larger epidemic.[5] Gao had broadened his search for cases, and based on Redfield's suggestion, expanded the "case definition" he was using to include people with matching respiratory symptoms but who didn't necessarily report having any contact with the food market suspected as a source of the outbreak. Now, Gao was finding many more cases that seemed to be caused by the same pathogen. The infection appeared widespread.

Redfield believed Gao and other health authorities in Beijing were unaware of the outbreak until late December. Prior to that time, US officials believe the information was being held by provincial officials, and perhaps the Chinese military, but not shared with central health officials in Beijing. If the US had been able to get its own CDC staff on the ground when they made their first request, Redfield believes the US would have uncovered the existence of asymptomatic spread. That information could have changed the structure of the American response.

It wasn't just Gao who was giving Redfield and US health officials the cold shoulder. As the scope of the outbreak grew, information sharing contracted. Across different levels of the US and Chinese CDCs, where there had previously been regular contact among lower-level officials, the Chinese CDC went dark on their US counterparts.[6] And it wasn't just the US CDC that was having problems getting access to information in China. Gao was having his own problems inside his own country. His investigators had been rebuffed by local authorities in Wuhan. "They didn't show them all the cases," said Dr. Ray Yip, the US CDC's former China liaison, of the Wuhan authorities. "They had a couple of cases of hospital workers infected by then, and that's obviously human-to-human, how else did they get it?" Yip said.[7]

In China, the country's CDC was an "afterthought," as one member of the White House staff described the Chinese agency to me. The US had helped fund its development and trained many of its key staff, building an institution in China molded in the US CDC's own image, on the premise that in a crisis, it would operate much like America's own CDC. It was not a wasted investment. Over the years, the Chinese CDC had been an important collaborator to the US CDC and similar public health authorities. But at the outset of this crisis, the Chinese CDC had been largely sidelined. The US government had a false sense of assurance about the information it would receive from the Chinese agencies and the role that China's public health institutions in Beijing would play in evaluating and responding to the outbreak in Wuhan. Gao would not be allowed to visit Wuhan until January 19, when he was part of a team sent from Beijing to tour local hospitals and then to report back to the National Health Commission and China's cabinet on January 19 and 20. It's believed that the description shared by this fact-finding team, of the deteriorating situation in Wuhan, helped prompt authorities in Beijing to place the city under a lockdown on January 23.

It turned out that the Chinese Communist Party didn't turn to its CDC to lead the response; they turned instead to the military. More worrying, Chinese officials didn't feel a strong imperative to keep their own public health agencies informed of events. Through December, information on the outbreak was widely shared with the medical branches of the Chinese military, but not with China's CDC, according to US health officials I spoke with who were in contact with their Chinese counterparts. One illuminating fact: the Chinese authorities sent the head of their biowarfare program to Wuhan to oversee the response. In the US, we were misled by a belief that China's effort would conform to the type of response that many US public health officials would have envisioned, with public health institutions taking the lead. China's CDC was largely sidelined.

By the end of January, Secretary Azar believed he had an agreement from the Chinese government to allow experts from the US CDC into the region, and Azar briefed President Trump on the development. On January 27, the president tweeted, "We are in very close communication with China concerning the virus. Very few cases reported in USA, but strongly on watch. We have offered China and President Xi any help that is necessary. Our experts are extraordinary!"[8]

To some on Trump's political team, the president's tweets heaping praise on China's initially sluggish cooperation were a source of angst and confusion. Was the president trying to prematurely proclaim a breakthrough? Prod it along through some public sweet-talking? Or was he guided by a desire to cooperate with Xi long enough to secure a trade agreement that was under negotiation at the time? Most would agree that the public praise probably sprang from some combination of factors. In fairness, by this point the Chinese government and President Xi had already ordered the lockdown in Wuhan, a forceful action to try to contain the virus there.[9]

Most of all, based on conversations I had with the president and his staff over this time period, I believe Trump felt he got more with sugar than with salt when it came to the Chinese leaders. In one February meeting, Trump told associates that if he was tougher on Xi, the Chinese would be even less forthcoming with information.[10] The president would make a similar comment to me in a phone call. Secretary Azar spelled this out more clearly a year later, when he said, "If you want to get access and cooperation, sometimes a bit of public praise gets you further than hitting them over the head publicly. So, we press them very hard and firmly in private, but offered muted reinforcement in public, at least while we still thought we might be able to secure their full cooperation and compliance with the international health regulations."[11]

The president called me later that summer, just before announcing his intention to pull the US out of the WHO as a way to punish the organization for what the White House perceived as the WHO's sluggish response and its indulging of the Chinese government.[12] I tried to convince the president not to take the action. My position was that taking such a step could be damaging as both a public health matter and as a political one. Among my arguments, I said that the virus was going to flare in the Southern Hemisphere and then would return to the US in the fall as a confluent epidemic that would likely engulf the entire nation. I said that critics would rightly argue we weakened the WHO at a key moment, when the organization was most needed to help address epidemics in regions of the world that would continue to seed America with new infections. Pulling out of the WHO now could contribute to the conditions that would fuel that fall epidemic.

I offered the president an alternative that I believed could achieve his political goal but keep the US in the WHO. If we wanted to send a strong message to the Chinese government, I said, there were better ways to convey our displeasure. I suggested that he move to include Taiwan in the upcoming World Health Assembly. Taiwan had been a critical partner to the US during the pandemic, and its exclusion from the WHO's Health Assembly, to accommodate China, struck me as an unfortunate and ultimately self-defeating way to treat a country that had been a faithful ally in our shared efforts to combat the virus. The president had concerns with my idea, saying that Chinese officials would see it as a harsh reproach, and it would elicit a backlash. Trump seemed cautious about not pushing the Chinese government too hard, too fast. Yet, despite the public accord, Chinese officials continued to rebuff many of our entreaties. Sugar or salt, some of the obstinacy was the same.

The fact that information didn't flow fluidly from Wuhan to Beijing, and that Chinese health officials appear to have been unaware of the initial outbreak, is instructive. We depend on other nations to inform us of these kinds of risks, but we reflexively assume that they have political and public health systems in place that are able and willing to surface this information internally. In China, it appears that by the time fuller information finally reached the central authorities in Beijing, and after being subjected to some further fiddling, weeks had already passed since the initial outbreak.

It wasn't the first time that global health authorities were deliberately kept in the dark by local Chinese officials, who were perhaps acting independently to obscure information that they believed would reflect poorly on them. During the SARS-1 outbreak in 2003, local Chinese authorities had removed seventy sick patients from two hospitals in order to conceal these SARS cases from a delegation that was visiting from the WHO.[13] The Chinese officials put the patients into vans while WHO officials toured the hospitals, and they drove the infected individuals around the city to pass time. Some patients were checked into hotels until the WHO inspectors had left the country.[14]

In a crisis like COVID, information is not foresight. It takes exceptionally brave leadership to act decisively, before the full scale of a threat becomes obvious. By the time the situation is evident, it can be too late to avoid a catastrophic outcome. But when the information itself is incomplete, or deliberately obscured, it can make timely and effective action even harder to obtain.

Poor information set the US back, preventing the nation from pursuing the actions it could and should have taken. A general refrain of critics is that the president refused to take COVID seriously, but it's more complicated. Later on, the White House began to believe that the costs of the mitigation that was required to keep the virus in check outweighed the benefits. They started to discount the threat. But early on, the president was convinced enough of the seriousness of the risks that he supported a historic forty-five-day national shutdown of nonessential activity to slow the spread. If we had more complete information about the scope of the threat in January, it's possible that the actions we eventually took by the spring could have been moved up by a week or two, or that we could have been convinced to invest in a more coordinated national response before it became too late. We were at a point in time when days mattered and many of the actions that we eventually took came too late.

Better information about the COVID outbreak could have been obtained in November and certainly by December. We could have had sources on the ground to gather samples of the virus so we could conduct our own analysis to discern its features. We could have gathered sequence information from some of the initial infections. We could have gleaned more information from doctors who might have been willing to tip us off that a major outbreak of a respiratory pathogen was under way. Yet even in January, based on my conversations with White House staff, they didn't seem to have actionable reporting from China to allow them to understand fully the threat that the outbreak posed to America. The Chinese government's reluctance to share early information, and at times their outright refusal, proved that we cannot rely solely on passive reporting for our safety.

In short, COVID proved that we cannot trust as our only backstop the international conventions, cooperation, and obligations that govern our global public health order. These institutions need to be strengthened. But they cannot be our only means for being informed about new risks. We need more active efforts to gather information when there are signs of troubling outbreaks. We will have to rely more on our tools of national security, including our intelligence services.

There's an understandable uneasiness about migrating the tools of national security into the work of public health, but COVID showed that a pandemic poses a grave national security risk. The political pressure to improve our surveillance, using all of our capabilities, is simply too great. A pandemic with a respiratory pathogen proved to be an asymmetric risk to the United States. COVID hurt us a lot more than it ended up hurting many other nations. The US proved uniquely challenged in applying and maintaining respiratory precautions for its citizens at work and recreation.

Every wasted day was a missed chance to alter our fate. Implementing a public health response requires the measuring of risks and probabilities.[15] SARS-CoV-2 was a highly contagious and pathogenic virus. And by late January, the probability was already high that the novel coronavirus would seriously threaten America. As microbiologist and Nobel laureate Dr. Joshua Lederberg once observed, "The microbe that felled one child in a distant continent yesterday can reach yours today and seed a global pandemic tomorrow."[16]

Eventually, US experts were included as part of a delegation admitted into China that also included specialists from six other nations.[17] By then, it was largely moot. COVID had escaped China. If there was anything to be learned about the spread of the virus and its clinical course, China was no longer the only incubator. On January 13, the first known case outside the Chinese mainland would be identified in Thailand. Two days later, the patient who would become the first case to be identified in America got on a flight from Wuhan that was bound for Seattle. He was carrying the novel virus on his breath. On January 20, the first case was found in South Korea. COVID had broken out. The seeds of a global contagion were now firmly planted.

CHAPTER 4

THE OUTBREAK WE DIDN'T WANT TO SEE

A thirty-five-year-old man showed up at an urgent care clinic in Snohomish County, Washington, on January 19 with a persistent cough and a fever of 99 degrees. He had returned to the United States four days earlier, after traveling to visit his family in Wuhan, and saw a health alert from the CDC about the novel coronavirus outbreak in China. Because of his symptoms and his travel, he decided to see a doctor.[1]

A rapid flu test came back negative, and his recent travel history made his doctors immediately concerned that it could be coronavirus. His doctor alerted the Washington State Health Department, which in turn contacted the CDC's Emergency Operations Center, the only place in America where a patient could get tested for the novel coronavirus. At that point, the CDC was tightly rationing who received a test, because it had the capacity only to run a small number of tests a day in its lab in Atlanta. The man's travel history meant that he met the CDC's tightfisted testing criteria. After his doctors collected two specimens from his nasopharynx, he was sent home to wait in isolation, with his quarantine closely monitored by the local health department.

On January 20, the CDC confirmed that he had the virus. He was transported from his home to Providence Regional Medical

Center in Everett, Washington, where he was admitted to a special two-bed biocontainment ward that had been established to handle Ebola patients.[2] For a time, he was the only known person in the US infected with the novel coronavirus. He would also be the first coronavirus patient to receive the experimental drug remdesivir.[3] A total of fifty people with whom he had close contact were also tested and put into quarantine, but none of them ended up developing the infection. For now, health officials thought they had managed to contain the country's first known introduction of the virus that had shut down the city of Wuhan.

However, back in Washington, DC, some White House officials had growing concerns that we wouldn't be able to contain it for much longer. A week later, in an Oval Office Meeting on January 28, National Security Adviser Robert O'Brien told President Trump, "This will be the biggest national security threat you face in your presidency. . . . This is going to be the roughest thing you face."[4] Matthew Pottinger, O'Brien's deputy, had reached the same conclusion. Two days earlier, Pottinger had sent an email to White House staff asking whether the US should immediately halt all travel from China. "The situation in China is clearly out of hand," he wrote.

Pottinger had an undergraduate degree in Chinese studies and was fluent in Mandarin. He had worked as a journalist for seven years reporting from China for *The Wall Street Journal,* before he joined the United States Marines as a military intelligence officer. He served all four national security advisers during the Trump administration, eventually being tapped as deputy national security adviser by O'Brien in September 2019.[5] Pottinger was well placed on this issue: he still had sources on the ground in China with whom he stayed in touch, and his wife is a virologist who had worked on HIV for many years in the CDC's global branch. Moreover, Pottinger's older brother was an infectious disease doctor at the University of Washington.[6]

Pottinger had covered the 2003 SARS-1 epidemic for *The Wall Street Journal* and had seen firsthand the Chinese government's effort to conceal its spread. In fact, it was his reporting that broke the story—despite China's initial denial—that there were cases of SARS-1 spreading in the northern parts of the country.[7] Even though SARS-1 was largely confined to a handful of regions, it had wrought havoc in some of the world's financial capitals. It was proof of the kind of severe economic and social dislocation that a dangerous epidemic could cause. The new coronavirus looked as if it could be an order of magnitude worse. It had Pottinger very worried.

During the opening weeks of the COVID outbreak, Pottinger was in touch with doctors he still knew inside China, and the firsthand accounts he was getting were radically different from what the White House was hearing from Beijing and the WHO. The Chinese doctors told him that the virus was circulating in the community and was already spreading in several provinces. One doctor believed that at least half the cases in Wuhan were people who were entirely asymptomatic. This was a respiratory epidemic driven by invisible cases, the doctor told Pottinger. Forget 2003, the doctor said. This was 1918. Pottinger went into the Oval Office and told the president everything he had heard from his Chinese sources. There were more than 23,000 people arriving each day from China, Pottinger said; the US needed to tap the brakes.[8] Two days later Trump announced that he would suspend most flights arriving in the US from China. By this point, these measures might slow the pace of introductions in the US and buy us some time, but the novel coronavirus had already arrived. The travel restrictions wouldn't prevent a US epidemic.

Early on, Pottinger also raised concerns about potential shortages of medical supplies like masks. But not everyone was similarly convinced of the dangers. As the pandemic later took its grip on the nation, he was one of the rare White House officials who routinely wore a mask to work, despite persistent scorn by colleagues, who labeled him an alarmist. Pottinger had two older relatives living with him at home and was so worried about the lax precautions being observed inside the White House that he had his office moved from the West Wing to the Eisenhower Executive Office Building, an historic and ornate office building located next to the White House that housed many of the staff who supported the president. It was Pottinger's own act of social distancing. He was worried that he would catch the virus in the White House and bring it home with him.

Pottinger was also worried about continuity of government. He and O'Brien had decided that the two of them would remain physically separated even as they worked together. They would avoid face-to-face meetings, even when they were both inside the White House complex, to reduce the risk that they both could come down with the virus at the same time. O'Brien also backed Pottinger's risk assessment and issued an order requiring the staff of the National Security Council to wear masks. Pottinger worked with a Taiwanese official to secure half a million masks for the US, taking 3,600 for the staff of the NSC and the White House medical unit and donating the rest to the US Strategic National Stockpile.[9] Some in the White House were angered by Pottinger's action, worried about the optics of the White House staff acquiring surgical masks at a time when US officials were still discouraging the American public from buying and using their own masks. There was a shortage of medical masks in hospitals. But Pottinger's concern about the casual procedures at the White House would prove prescient; in the coming months the White House would be host to several outbreaks.

Pottinger knew that the risks were being compounded by our inability to test people for the virus. We had no reliable way of knowing how widely it might already be spreading. The US had the capacity to identify only a small number of infections, those with an obvious link to China that would qualify a patient to be tested under the CDC's narrow guidelines. The lack of testing created a false sense of security. If you look for cases only among people coming from Wuhan, you will find cases only among people coming from Wuhan. That's precisely what was happening.

Four days after the Washington State man was identified, a Chicago woman was hospitalized for pneumonia and became the second person in the US to be diagnosed with COVID.[10] One day later, in California, a third case was uncovered. All three had recently traveled to Wuhan, and each recovered. The Chicago woman's husband became sick with COVID one week after she fell ill—the first documented case of person-to-person transmission of SARS-CoV-2 in the US. In Chicago, local health officials identified 372 people that the couple had been in contact with, and of those, 347 underwent active monitoring, including 195 healthcare personnel. No further cases were identified as a result of their illnesses.[11] In total, public health officials were monitoring sixty-three additional individuals in twenty-two states at that time. All were suspected of having COVID. Almost all had symptoms of respiratory infections, and all had some connection to Wuhan. Most were in some form of quarantine. Almost all of them would test negative for the coronavirus.[12]

Back in Washington, officials thought they were finding and isolating the virus and preventing new outbreaks.[13] Robert Redfield told the White House that the CDC's efforts were, so far, containing further spread. In fact, by January 25, the virus was very likely already being transmitted to ten new people every day in California alone, according to estimates made later by researchers at Northeastern University's Network Science Institute.[14] But during the critical months of January and February, nobody knew just how widely the virus was already spreading.

So public health officials were alarmed on February 28, when a second case emerged in Seattle, in a person who had no physical connection to the first Seattle-area patient, the one who had been diagnosed on January 20 after traveling from Wuhan. Using data that mapped the precise genetic sequence of the two coronavirus infections, scientists eventually found that the two viral strains, separated by more than a month, differed only by two small mutations. They looked connected.

Dr. Trevor Bedford, an associate professor in the Vaccine and Infectious Disease Division at the Fred Hutchinson Cancer Research Center in Seattle, was advancing the use of sequencing as a way to track the migration of coronavirus across the world. It was a relatively new field of science called genetic epidemiology, or "gen epi."[15] By sequencing the virus's RNA, he could trace the subtle ways its genetic code mutated as it passed from person to person, through successive generations of replication. Using the trail left by these small changes, he could track individual virus strains as they moved around the world. Coronavirus mutates more slowly than many other RNA viruses, so as it replicates and spreads, it does not collect as many changes. Still, it mutates about once every two weeks along a transmission chain. By looking at minor differences between the two Washington State strains—the one diagnosed on January 20, and then the second diagnosed on February 28—Bedford concluded that the two cases might have been linked.

This was the first time that sequencing data was widely used in a crisis to extend the traditional approach to epidemiology. Bedford helped pioneer the modern integration of these two disciplines. What he and others uncovered meant that COVID might have been spreading in Washington for at least six weeks. Somehow, the first traveler could have managed to infect someone who escaped detection and went on to transmit the virus to others. Bedford's genomic analysis, posted online on February 29, suggested that hundreds of people in Washington State might be already infected given the amount of time that had elapsed between the two cases.[16] By this calculus, the region was already facing a substantial outbreak, and nobody knew it.

This finding would be debated, and Bedford advanced later analysis that raised doubt about whether the two cases were actually connected, but even the possibility that coronavirus had been silently spreading in Seattle for almost six weeks was a dangerous turn of events. In fact, even if the two cases were not linked, we now know the virus had been spreading in American communities since at least January. It was hard to know, however, just how widespread the infection was. During a critical six-week period over January and February, the Department of Health and Human Services had run into a series of challenges and committed blunders in its effort to roll out a diagnostic test that could screen for the coronavirus, leaving the country dangerously blind to its spread. Without a way to reliably test patients who had symptoms suggestive of SARS-CoV-2 infection, US doctors had no way to diagnose the sick or take measures to contain the spread.

One of the doctors on the front lines of fighting COVID, and trying to find ways to identify its spread, was Helen Chu, an infectious-disease specialist at the University of Washington School of Medicine. For months, Chu and her colleagues had been collecting nasal swabs from Seattle area residents, as part of the Seattle Flu Study, a novel project that they had launched with the support of the Gates Ventures, the venture capital arm of a vast public health investment enterprise maintained by Bill Gates and Melinda French Gates. (Bedford was also part of this effort.) Starting January 1, the research team had amassed more than 2,500 samples from people who had voluntarily enrolled in the study. The participants agreed to send in self-swabbed samples if they developed any signs or symptoms of flu. The goal was to see if this community-based approach could be an effective tool for identifying and observing the transmission of influenza. The researchers wanted to see if they could use this framework to develop a tripwire that would identify when flu was starting to spread—a "sentinel surveillance" program to screen for when the influenza virus was infiltrating a community. They were testing a new model for infectious disease surveillance.[17]

But Chu knew that the tools she had for screening participants for influenza, as well as the samples she had collected, could help determine whether coronavirus was also circulating in the Seattle area. After all, the swabs were from people who were reporting symptoms of a respiratory illness. If some of the participants who were sending in samples didn't have flu, perhaps they had COVID instead?

Chu's flu test relied on a process called polymerase chain reaction, or PCR. Using this approach, labs copy or "amplify" small segments of a virus's genetic material. The process uses primers and probes that are highly specific to the genetic sequences found in the particular virus that's being tested for.[18] These primers and probes will allow the test to render a positive result only if the probes are able to detect a precise snippet of RNA that's also found in the virus. This is how PCR is used to detect the presence of a particular virus and differentiate SARS-CoV-2 from more common coronaviruses. The technology can also determine how much of the virus is contained in a particular specimen, by quantifying the amount of RNA in a sample.

The key ingredient is heat. Throughout the process, RNA from the target virus is converted to cDNA (a single-stranded version of the double-helix DNA molecule), which is repeatedly warmed and cooled. With the help of special reagents, this thermal cycling triggers a chemical reaction that allows the primers to bind to the target stretch of cDNA and copy it. With each heating and cooling cycle, the number of copies of cDNA can be doubled. These DNA segments are complementary to RNA segments that are found in the virus. In just a few hours, and after thirty or forty cycles, there can be a billion or more copies of cDNA resulting from the original target piece of RNA. In a short time, PCR makes it possible to produce copious amounts of these complementary stretches of viral RNA that you're looking for, even starting with a very small amount of the actual virus.

The number of cycles needed to find a detectable level of virus is referred to as the cycle threshold, or Ct. Measuring this value can help determine how infectious a person is. If the machine needs fewer cycles in order to detect the viral RNA, it can mean there's more of the virus contained in the sample. The more virus someone is harboring, then in many cases, the more contagious they're likely to be. The number of cycles it takes to accumulate a detectable amount of coronavirus RNA is sometimes included in the results sent to doctors, but not always, and many providers don't use this information to assess a patient's condition. They treat the test result as a binary outcome—either someone has coronavirus RNA present in their swab, and they're deemed to be infected, or the RNA cannot be found, and they're judged to be in the clear.

Another wrinkle is that the PCR machines are so sensitive, sometimes they can detect fragments of dead virus that have already been attacked by a patient's immune system and chopped into pieces. These dead fragments can linger for weeks after the patient has already beaten a viral infection. Even though the dead viral fragments can no longer cause disease, they can still register a positive result on the PCR machines. It's partly for this reason that the CDC eventually decided that they would no longer declare a patient to have an active infection based on the findings from the PCR tests alone. The CDC started to rely on a clinical criterion that also considered the number of days since a person's first COVID symptoms began.[19]

These limitations have stirred some criticism of PCR as a diagnostic tool.[20] Often, if a patient has any detectable virus by PCR, they're judged to be positive for coronavirus and are treated as if they're infectious. Sometimes labs will adjust the cycle thresholds needed to call a test positive and exclude results that are reached only by repeatedly cycling the machines, say forty or more times.[21] The challenge is that a high Ct value can mean that the patient has largely cleared the infection and is no longer contagious, but it can also mean that the person is newly infected, and the virus has just started to take its grip on them and hasn't had time to replicate itself widely.[22] So counting the number of cycles isn't always a reliable way to judge whether a person is indeed contagious.

Chu and her team had the right equipment to begin testing their samples for coronavirus. As a technical matter, it was straightforward. They would have to develop primers that were highly specific to stretches of RNA that are only found in the novel coronavirus's genome, strung among its thirty thousand base letters.[23] The research team was able to develop these primers and probes in January. A bigger challenge was to prove that the test was precise. Without a "positive control," an actual sample of the coronavirus, they couldn't fully verify that their test was able to accurately identify its RNA. To prove the performance characteristics of a diagnostic, researchers need access to samples of the coronavirus RNA to assess the new test. They could use inactivated copies of the actual virus, or synthetic genomic material for this purpose. But at this stage, Chu and her colleagues had neither. Only the CDC had those samples. Without them, there was no way to prove for certain that the homespun primers would bind to the actual virus. But the team at the University of Washington had a lot of experience in genomics and in developing tests for viruses and they were confident that their new COVID diagnostic tool was reliable.

There are a number of factors that combine to determine how valid a test is, but two of the most important parameters are the test's "sensitivity" and its "specificity."[24] A test's sensitivity is also called its true positive rate. It's a measure of the proportion of positive test results that are correctly identified. In other words, it's the percentage of people who have SARS-CoV-2 infection who are correctly diagnosed as being infected with the virus. A test's specificity is also called its true negative rate. It's a measure of the proportion of people who don't have the infection who are correctly identified as being virus free. A test with a low specificity would generate a higher number of false positives, saying that you were infected with the virus when you really weren't.[25] A test with a low sensitivity would generate a higher number of false negatives, saying that you don't have the infection when you really do.

To put this in practical terms, a test with a sensitivity of 80 percent may sound reliable, but in certain settings, it's not good enough to be practical. It means the test could miss one in five actual infections. If people are counting on that result to know whether they can go to work or visit with elderly relatives, a high rate of false negatives can mean many people will engage in risky activities, where they can spread the virus to vulnerable contacts, and not know that they're infected. Likewise, a test with a specificity of 99 percent may similarly sound accurate. But again, its reliability may depend on how it's being used. Given that most people you test in a general population are not going to have SARS-CoV-2, a test with a 99 percent specificity means that a very high percentage of the people testing positive may be falsely positive. Consider this: let's say one in one hundred people you test actually has SARS-CoV-2 infection. That means the prevalence of the infection is 1 percent in the general population (which was the approximate prevalence during the epidemic in winter 2020). And let's say that a test has a 99 percent specificity, so for one hundred people you test, you will generate one false positive. That means that if you screen one hundred random people with such a diagnostic, you will get two positive results, on average. One of those positive results will be a person who is infected with the virus, and the other will be a false positive. So 50 percent of the positive results you generate will be false positives.

A lot of factors can affect a test's sensitivity and specificity. In a circumstance like COVID, where you are testing patients for a dangerous pathogen and trying to use testing as a way to control an outbreak, the reliability of the test is a critical factor. The FDA and the CDC asserted regulatory oversight for assuring its accuracy. This meant that if the scientists wanted to use their test on patients, they needed permission from the CDC and perhaps the FDA as well. The CDC has jurisdiction over tests that are being used for public health purposes as part of a pandemic response, and the FDA regulates tests that are used to diagnose individual patients for a disease. The Seattle researchers were in discussions with the CDC and Washington State health officials for weeks, seeking the ability to begin testing the samples they were receiving, to see if SARS-CoV-2 had arrived in the area. The debate with the CDC was initially over whether the University of Washington researchers would be allowed to use the test they had developed, or whether they would have to send their samples to the CDC so the agency could do the testing in its reference lab in Atlanta. Chu and her colleagues finally got permission to go ahead and do the testing themselves in the University of Washington lab, so long as the testing was done for research purposes. This meant that if they found a positive sample, they wouldn't be able to report the result to the patient. But that wasn't practical. Chu and her colleagues knew that if they identified someone with an active SARS-CoV-2 infection they would want to share the results with the participant in order to help doctors identify cases and start tracking and tracing infected patients as a way to contain further spread.

Sharing the results with patients, however, would brand their test a medical diagnostic and in turn make it subject to FDA regulation. So they also sought permission from the FDA and the CDC to use the test for diagnosing patients.[26] But federal officials repeatedly rejected this idea, even as weeks went by and the pandemic worsened.[27] The CDC would only let the team test the samples as part of a research project. By February 25, Chu and her team decided to go forward and start using the test for "research use only," agreeing to the constraints under which federal regulators wanted them to operate.

They began by testing the samples collected most recently, and they got their first hit on February 27, on a nasal swab that had been sent to them just seventy-two hours earlier from a local high school student.[28] Because the sample was so recent, because the teenager might still be contagious, and because it involved a child, they felt an obligation to notify the patient and local health authorities. The case was independently confirmed by the Washington State Health Department the morning of February 28.[29] The school was promptly closed as a precaution. That same day, Chu's colleague Trevor Bedford joined the team to sequence the genome of the virus that had infected the boy; this was the sequencing that led Bedford to conclude that this new infection might be from the same viral lineage as the man diagnosed on January 20, who had been identified as the first case to be found in the US.[30] If the two cases were linked, it meant that the virus had been spreading in the area for more than a month. Bedford and his research team immediately posted the sequence publicly to an online portal initially conceived to enable the global sharing of sequence information on influenza strains, called the Global Initiative on Sharing all Influenza Data (GISAID).[31]

Chu's effort showed how easy it might have been to authorize academic and commercial labs to start processing COVID tests on suspected cases and how hard it was for these labs to push forward on their own. The Seattle team notified this study participant without the CDC's agreement, and in response the CDC claimed that Chu and her team didn't have the appropriate patient consents in place and hadn't properly validated their test for diagnosing the participants. Once they notified the participant, as a regulatory matter, the test was no longer being used just for research, which made their effort subject to federal oversight. The CDC forced Chu and her team to pause what they were doing.

It didn't matter that the local ethics committee told them to report the results to the healthcare providers. And it didn't matter that the CDC had tried and, so far, failed to distribute its own diagnostic, leaving the nation dangerously devoid of a way to test for the novel coronavirus. To be sure, federal agencies like the FDA and the CDC were legally charged with ensuring the reliability of tests that were going to be used to screen patients for a novel pathogen like SARS-CoV-2, and the public health emergency made it even more critical that such tests be reliable. There were ways that the CDC could have given high-quality labs an easier path to advance tests in the setting of the crisis. But they jealously guarded this turf. When the team's lab reported its first positive hit, shortly thereafter the CDC told the lab to stop testing their samples.

Weeks went by before federal officials finally gave Chu and her colleagues a partial green light to resume their testing. Officials said the lab could test cases and report results on future samples that they collected, so long as the lab used a new patient consent form that disclosed that the results might be shared with health officials. But for now, they couldn't test any of the 2,500 samples that had already been collected as part of their earlier flu study.[32] It wasn't until March 4, five days later, that an institutional review board that governs the conduct of trials involving patients determined that Chu could also proceed with testing the samples that had been collected in January and February as part of the flu study, with the results reported to public health authorities and patients.[33] By then, a lot of time had passed from when those swabs were first collected. If patients who were self-reporting symptoms of flu instead had the novel coronavirus, these infected individuals could have been circulating in the community for weeks.

Chu and her team tested samples dating back to January 1, again starting with the most recent samples. The first positive hits they got were on the samples collected in late February. In all, SARS-CoV-2 would be detected in 1.1 percent of the specimens that they screened, a whopping number considering that most health officials still maintained that the novel coronavirus wasn't yet circulating in the community. Many of these patients had reported respiratory symptoms and thought they might have the flu. Now, it turned out, some had COVID. The vast majority had reported mild illness and were treated at home. For most of them, it was probably too late to trace their contacts. So much time had passed from their initial infections that, if new cases were spawned by their illnesses, the subsequent patients were generations removed from the index cases. Two of the infected patients were children. Seven of the twenty-five patients who were identified with COVID reported being sick enough to seek medical care.[34]

It was a lot of positive hits to be getting on such a small and unselected pool of patients. If you were watching the evening news, the US was supposed to have just a dozen infections nationally. HHS officials were telling the White House staff that there was no widespread community transmission. The people in Chu's study were not recent travelers from China or Italy. They were drawn from the general community. The novel coronavirus was spreading in Seattle.

The FDA would soon adopt a new strategy that would open the door for academic labs to begin testing patients and reporting the results back to their physicians. But in the meantime, Chu and her colleagues had run into some of the byzantine rules that limited the conduct of testing. The CDC exerted tight control over which labs could screen for COVID. And for now, the only facility in America that was authorized to do the testing was a single CDC lab in Atlanta.

It meant that doctors around the country had to forward samples to the CDC for testing. Physicians would first have to call the CDC and convince the agency to agree to take a sample for evaluation. It was a cumbersome process that limited the flow of specimens. By that point, there were far more people in America with unusual respiratory symptoms whose doctors wanted to test them, but the CDC's criteria still required that patients have all the signs and symptoms of COVID along with a history of travel to Wuhan or known exposure to someone who was already infected. This created a situation similar to when China's CDC required some contact with the Wuhan food market to be considered as a possible COVID case. It was only after Robert Redfield had suggested to George Gao that the Chinese CDC broaden its criteria that Gao revealed that the outbreak was probably out of control. Now, nearly two months later, the US CDC's requirements seriously narrowed who could qualify to be tested in America.[35] On March 1, the CDC's official tally of COVID cases rose from fifteen to seventy-five. Models developed later by researchers at Northeastern University show that by this date, the US probably had 28,000 infections.[36]

We had lost control. Without a widely available diagnostic test, we missed the chance to use case-based interventions—the ability to diagnose the sick, trace their contracts, and place people who had been exposed into quarantine—as a way to limit spread. The window for preventing the virus from gaining a foothold had closed. The US would now be dependent on mitigation as its primary strategy for slowing the spread. This approach relied on closing nonessential businesses, canceling school, and convincing people to stay at home. The idea was that transmission would sputter out, with the virus unable to find new people to infect.

The idea of adopting mitigation as a way to slow the spread of a virus was first conceived as part of pandemic plans crafted fifteen years earlier, when the proximate fear was of a pandemic strain of bird flu. The 2005 plan envisioned that mitigation tactics would be used in a targeted fashion. Widespread diagnostic testing was to be employed as a way to uncover cases and target the strictest measures only to those places where community transmission was already under way. Like cased-based interventions, the mitigation also hinged on the ability to widely deploy reliable tests, which we never had in time.

We now know that some regions had relatively little SARS-CoV-2 spread and could still use case-based interventions—testing, tracing, and isolating the sick—as a means to contain the epidemic. In other places, like New York City, where the virus was already widespread, closing public venues, restricting gatherings, and asking people to stay home were all critical steps for slowing the epidemic and alleviating the burden on hospitals. But without a widely deployed test, there was no way of knowing where the virus was spreading and where it wasn't already epidemic.

There was no way to target our interventions.

With effective screening, with the kind of "sentinel surveillance" effort that Chu was piloting—where providers could randomly test a representative sample of patients with respiratory illnesses, looking for those who might be harboring SARS-CoV-2—we'd have been able to identify the outbreaks in US cities more quickly, and then focus our containment measures to these hotspots. Over the longer term, to guard against the next pandemic, Chu's idea is to create a biospecimen repository, where certain respiratory samples would be routinely held for a few months after patients underwent routine testing in outpatient settings. If these specimens were customarily collected, stored, and linked to information in patient health records, then if a new pathogen started to spread, health officials would be able to go back and look for traces of a novel virus in the community to see where and how it might be spreading. Such a repository would thus serve as a banked tripwire to uncover a new outbreak.

Right now, respiratory specimens are typically held for only a week at a time. If you get tested for flu, the lab will hold your swab for a week after your results come back and then throw it away. The reason is simple: space. Labs run out of freezer space, and for the RNA to be retained on a nasal swab, it needs to be held in ultracold temperatures, typically –80 degrees Fahrenheit.

We wouldn't need to store every sample for longer periods of time, just a representative pool, perhaps drawn from higher-risk social compartments where new pathogens are most likely to first emerge. These would be settings where there are clusters of patients who are more likely to develop serious symptoms. That might include younger patients in the case of flu, or more vulnerable patients like older individuals or those living in nursing homes. When a new pathogen first emerges, it's likely to first be identified in another nation before it arrives in the US. As with COVID, we'd have some advance warning, enough to know what we should start looking back for.

To make such a system robust, we'll need a mechanism in place to test specimens for a novel pathogen even if we don't have the explicit permission from the patients who gave those samples. Rules governing informed consent should be amended to address a declared public health emergency. We need to safeguard privacy concerns and make sure results are shared confidentially. But the imperative is to rapidly identify emerging outbreaks with novel pathogens. Most patients who were motivated enough to go to the doctor and get themselves swabbed for an obvious illness will want to know whether they may have been infected with a dangerous new respiratory virus that couldn't be identified at the time that they gave their samples.

With earlier warning of COVID's presence in America, we might have had time to intervene with targeted steps to slow the spread before the epidemic exploded. We could have delayed the US epidemic, and perhaps reduced its scope. China would use testing programs, practiced at a massive scale, to snuff out spread in regions outside Wuhan. In the province of Guangdong, health officials set up clinics where they screened more than 300,000 people with mild flu-like symptoms, identifying about 450 positive cases.[37]

Such a sentinel surveillance program had been proposed by healthcare officials in the Trump administration, but it never got off the ground. On February 13, Secretary Azar told the Senate Finance Committee, "The CDC has begun working with health departments in five cities to use its flu surveillance network to begin testing individuals with flu-like symptoms for the coronavirus. Many questions about the virus remain, and this effort will help see whether there is broader spread than we have been able to detect so far."[38] Shortly after, a CDC action plan dated February 14—the day following Azar's testimony—was circulated inside HHS, laying out the details for the five-city surveillance program. Drafted by the CDC, the "Surveillance Enhancement Plan" outlined what it called a phased expansion, where the sentinel program would be expanded to all fifty states within two months. Within six months, routine screening would be added to samples collected by the large commercial labs, including LabCorp and Quest. The plan noted that the system would hinge on the availability of commercially available tests for COVID-19. The planners clearly thought they had more time than they did. Six months was a long time to wait to have testing services available in the nation's commercial labs. No part of this surveillance plan for sentinel cases ever got started.[39] The CDC was never able to field the diagnostic test that they promised.

When the idea for sentinel surveillance was first proposed at a meeting of the White House Coronavirus Task Force in early February, nobody questioned the merits of the approach. Instead, the task force members worried it might be too late. Health officials acknowledged that they needed to have in place a surveillance system for coronavirus similar to the one they had for flu. They conceded that the effort was overdue, and they didn't know how much spread was already under way, according to White House staff members who attended the meeting. "It was a sign of how much trouble we were in," one attendee told me. "We would never catch up in time."

Until that point, the US had not recorded a single case of coronavirus infection in New York, which seemed improbable to some of the task force members. How could a city as dense and porous as New York not have a single case of the novel virus? "It was like the dog that didn't bark," this person told me. "It was a sign to a lot of us that we were way behind the curve. That's when some of us started to get even more nervous. We were never going to make it." By the time that New York City reported the first person who was positive for SARS-CoV-2 infection, the City had only conducted 32 tests on patients. It wasn't even enough testing to serve as a tripwire.[40]

The United States would never deploy enough tests to implement widespread screening, even for a rudimentary surveillance effort. In the absence of surveillance testing, health officials were forced to rely largely on a far less sensitive tool to tell whether coronavirus was starting to spread in local communities: syndromic surveillance. This approach relies on dozens of different streams of information on things such as how many people were seen in emergency rooms with unusual respiratory symptoms or the number of patients with flu-like illness who subsequently test negative for the flu. Syndromic surveillance can also include more esoteric information, such as how many times physicians are ordering complicated diagnostic tests like the Biofire Respiratory Pathogen Panel that are able to screen simultaneously for a battery of less common respiratory infections. Doctors will often order these complex tests for a patient who has an atypical pneumonia that defies a more obvious diagnosis.[41] In January, during Italy's flu season, doctors in the Lombardy region were ordering many more chest CAT scans than usual to evaluate complicated cases of pneumonia that confounded them. Many doctors thought these cases might have been complications of flu. It was really COVID making its first entry into the local population.[42] A spike in the use of complex imaging and diagnostic screening tests could be a signal that a serious respiratory illness is spreading, and it's stumping local doctors. Looking for these unusual ordering patterns by doctors is one element of syndromic surveillance.

Without a diagnostic test to actually test people for the presence of the virus, that winter, the Coronavirus Task Force relied entirely on syndromic surveillance data that showed mostly flat trend lines that were consistent with historical norms. Based on these data, task force members were told there was little risk that coronavirus could already be spreading widely in America. But a big part of the challenge was that officials at the CDC were applying a flu model to coronavirus. "People think that because you prepared for flu you were prepared for coronavirus, and it took eight to twelve weeks to realize that the two viruses behaved very differently," one senior White House official said to me in the spring. When it came to a novel coronavirus, we also lacked a foundational infrastructure for testing and developing countermeasures: we can make new flu vaccines by changing the strain we use in existing manufacturing processes. Similarly, the country has a lot of flu tests that can be adapted to a novel strain of influenza. We lacked any similar foundation to build on for a new coronavirus.

Falsely assured that SARS-CoV-2 wasn't spreading in communities in America, federal health officials were repeating a now-common refrain in their public appearances: "There is no need to panic." In reality, they didn't have the tools to know for sure. You have to admit what you do not know. Government officials often make the mistake of trying to answer every question. Sometimes, when you lack data, when you're confounded, the most reassuring message is to admit what you do not know. It builds trust, and building public trust is critical in a crisis. Public officials also make a mistake telling the public "not to panic." When people hear the phrase "There is no need to panic," the first thing that many people will ask themselves is: "Should I panic?"

When I was FDA commissioner, we uncovered a potential carcinogen contaminating a common class of blood pressure medicines. It was at trace levels, and we didn't believe it posed a serious risk. However, we put out substantive alerts regularly, updating on what we knew, on the progress of our investigation, and on the evaluation we were doing to fully quantify the risk.[43] Rather than aggravating the public concerns, our constant communications, and our efforts to clearly relay what we did and didn't know and a timetable for developing more answers, proved reassuring.

In a public health crisis, eventually, you'll need to ask the public to do hard things, and the public will have to trust the integrity of your appeals. You cannot advance policies or appeal for collective action if the public isn't on board. Once you pursue a policy, and get met with public rejection, or noncompliance, your capacity to implement future actions is eroded. If the public refuses to follow guidance in a public health crisis, if there's widespread opposition or protest, it becomes difficult or even impossible to advance additional measures, to take strong actions.

This is what happened with COVID. The federal government lost trust and credibility early, by its inability to accurately convey the true scope of the hazard. The lack of reliable information on COVID's spread, and the inability of people to access testing, degraded the integrity of the response.

Inevitably, at the outset of a crisis, you may be accused either of overreacting, if your worst projections don't materialize, or of underreacting if the situation spirals out of control. But if people reject the public health measures, we fail. Making the case for our actions, and galvanizing support for them, was essential to the success of pandemic response. Early on, the reassuring messages that the president and public health leaders were offering weren't properly informed. They simply didn't have the information they needed to conclude the risks were low. By the end of February, in the US, there were plenty of reasons for Americans to be concerned.

Some White House officials also started to grow more skeptical of the reassuring reports from HHS. Among them were Grogan and Pottinger. Both didn't understand how the CDC could be so sure that coronavirus wasn't already circulating in America given the events unfolding around the world, where the pathogen had spread furtively only to explode after cities had become saturated with it. Grogan sometimes enlisted the help of his fellow task force member Ken Cuccinelli, the acting deputy secretary of Homeland Security, to press the health officials, testing their conviction. But in task force meetings, the health experts, led by the secretary of Health and Human Services, held firm. The data they were monitoring showed no signs that the novel virus was spreading in the US. The syndromic surveillance, which was set up to monitor for the spread of flu, didn't show any concerning changes.

CHAPTER 5

LOOKING FOR SPREAD IN THE WRONG PLACES

A pivotal moment occurred at the January 21 Coronavirus Task Force meeting in the White House Situation Room. Joe Grogan was pressing hard on whether the coronavirus could be spreading without being detected. Secretary Azar turned to Anthony Fauci to address Grogan's concerns. "What would be the epidemiology to justify your question?" Fauci pointedly asked Grogan.

Fauci was technically correct; they had no epidemiological data to prove that community spread was under way. But why was that? Because of the CDC's failure to deploy diagnostic testing that would have shown community transmission was happening. Against the backdrop of the events unfolding around the world, and the cases that were starting to emerge in the US, fear among some White House staff was growing. Grogan, Pottinger, and others started to become very concerned that we could be missing a lot of infections.

It turns out we were. Without a diagnostic test to screen patients, health officials were relying instead on a specific tool that the CDC used for syndromic surveillance called the Influenza Like Illness surveillance system, or ILI. It's a network of about one hundred public health labs and more than three hundred clinical labs located in all fifty states, Puerto Rico, Guam, and the District of Columbia. The labs report to the CDC each week the total number of respiratory specimens they've tested for influenza and the number of positive specimens they found, along with demographic information on who was getting sick with flu.[1] At the peak of the flu season, the CDC collects as many as 75,000 samples a week. Typically, no more than about 30 percent will test positive for influenza. That means, in a normal season, there are about 50,000 suspected flu cases being reported to the CDC each week that test negative for influenza. These patients testing negative for flu are harboring some other respiratory infection. The CDC looks at historical patterns to make sure that the number of respiratory illnesses testing negative for flu doesn't show any unusual outliers—a sudden spike caused by a new respiratory epidemic. But with fifty thousand samples testing negative for influenza, a building epidemic could be easily hiding in plain sight.

In normal times, the system helps monitor the ebb and flow of the seasonal flu. In a pandemic, it's meant to serve as a tripwire to signal early signs of a new outbreak. By looking at this data and comparing it to historical norms, CDC officials told the task force they would know if a new respiratory illness was spreading. For the CDC, flu was the model for how they would detect COVID. They expected SARS-CoV-2 to behave essentially like influenza, and they expected their flu surveillance tools to be effective in identifying its spread. They thought they would be able to see if community transmission of the coronavirus had begun. But COVID wasn't acting like flu. And the tools they were using were inadequate to detect a burgeoning epidemic.

Some of the shortcomings were inherent to the approach. For one thing, this system for syndromic surveillance is backward looking. By the time you collect and analyze the data, it's at least a few weeks old. So, if a new virus was spreading exponentially, it could be widely diffused by the time a signal showed up on the dashboard. The other problem is that syndromic surveillance is not very sensitive. There could be thousands of cases of coronavirus infection circulating in a major US city, and the monitoring might not show any unusual trends. It could take tens of thousands of infections to register a discernible signal, especially with the majority of cases showing no or minimal symptoms. Most patients wouldn't even present to doctors for evaluation.

Syndromic surveillance also leaned on data streams that were incomplete. The CDC relied heavily on claims data from the Medicare program, but that data took about three months to be collected and reported because of the delay between the delivery of healthcare and payment for those services. Providers don't always turn in their bills on time. Later, Medicare would go back and do its own analysis of the data sets that it was feeding to the CDC in February. It was part of an after-action exercise to gauge the effectiveness of the response, and how useful the Medicare data would have been at detecting spread. Medicare analysts found that the data didn't show a signal, in the form of a spike in visits for respiratory illnesses and related diagnoses, until the end of February, which seems understandable. That's when the epidemic first started to explode. But much of the claims data from these February visits wouldn't have been made available to the CDC until March or April, because of the delay between patient encounters and the filing of billing claims. That's when the signal on the spread would have been first revealed to the CDC.

These shortcomings were aggravated by the way the coronavirus spread, through asymptomatic transmission. As many as 40 percent of infected patients didn't develop noticeable symptoms, and thus never show up on the symptom monitoring.[2] In January, it wasn't so clear that asymptomatic spread was a significant driver of infection. But by February, evidence had accumulated that individuals who had not yet developed symptoms of COVID and those who were infected but would never develop symptoms were playing a significant role in transmission.

Finally, a system that was looking for prominent respiratory symptoms as a way to detect COVID's spread was just not sensitive enough to serve as an early tripwire for community transmission. For those who were diagnosed with COVID, about 10 percent developed significant respiratory illness that required hospitalization, and only about 2 percent became critically ill.[3] That's from the pool of patients diagnosed with symptomatic disease. During the spring 2020 COVID wave, it's likely that we were diagnosing only about one out of every ten or fifteen actual infections. With more testing in place, by the summer, that proportion improved to about one in eight infections, and by the fall, one in four. Thus, even thousands of cases lurking in a large city might not be enough infection to generate an identifiable signal of disease, especially because it was late winter, and the virus was able to hide in the shadow of flu season.

Moreover, the syndromic surveillance was confounded by the advance of COVID itself. The ominous news of the epidemics in China and Italy had the American public on alert. The more coronavirus dominated the news, the more people were avoiding crowds, wiping down their airplane seats with Lysol, and dabbing their hands with generous amounts of Purell. All of these behaviors could reduce the spread of flu and other respiratory illnesses at the very time that public health officials were looking for unusual spikes in respiratory symptoms as a way to portend COVID's arrival. If flu diagnoses began to fall unexpectedly at the same time that COVID was slowly rising, then the spread of COVID could be masked by an offsetting drop in flu cases. In relying on influenza surveillance, public health officials were comparing current trends to a historical baseline that didn't reflect the current risks. If flu cases were starting to fall prematurely, past trends might no longer have been a reliable baseline.

Another challenge with relying on the influenza surveillance was that the system was looking for signs and symptoms related to respiratory pathogens. As we would soon learn, COVID was not entirely a respiratory disease. Many of its manifestations are respiratory and flu-like, but not all of them. A surveillance system looking primarily for flu-like symptoms could miss an illness with more heterogeneous effects. That's probably a lesson learned for new pathogens. Don't think you understand the pathogen you're dealing with until you've gained a lot of experience with it.

Finally, another big part of the reason why the syndromic surveillance wasn't effective was because we were having an intense and late flu season, and COVID cases could easily have been obscured by an epidemic of flu infections. The timing of the introduction of SARS-CoV-2, arriving in the US just as we were experiencing a flu epidemic, was absolutely optimal for keeping it hidden. For all of these reasons, any ability to pick up a signal could come at a point when we were so far into an outbreak of COVID that any hope of containment would have already failed.

"The public health people on the task force were saying that there was no spread, or we'd see it in the ILI," one White House official who sat in on the task force meetings told me. "It wasn't until the week of March 6 that you saw a clear separation, and that was retrospective that you saw the separation. Until then, we were never told the ILI was picking up anything weird ever, we were told they were using the ILI network and it's not detecting coronavirus spread. . . . CDC was telling us we have a surveillance system and it's the [influenza-like illness surveillance] network and that we're in good shape and that was totally untrue." That same week, on March 5, the assistant secretary for health at HHS, Dr. Brett Giroir, told a group of reporters at a Capitol Hill press briefing that the "best estimate" of the overall mortality for COVID might be as low as one-tenth of 1 percent, which would have put it in the range of flu, to which he explicitly compared it. And it certainly wasn't higher than 1 percent, he said. The best estimates of the case fatality rate that spring put COVID's mortality higher than 1 percent. But in linking it to the death rate from flu, it was a measure of how much some health officials were still viewing COVID through the lens of influenza. By that point, others in HHS had a firmer grasp on the virus's true pathogenicity. Anthony Fauci had sent an email to a *Wall Street Journal* columnist three days earlier, on March 2, alerting them that for COVID-19, "the mortality rate is approximately 2%. For seasonal influenza it is approximately 0.1%"[4]

Compounding its error of overreliance on the ILI, the CDC clung to its use of syndromic surveillance for months, even after there was ample evidence of asymptomatic spread. Those inside the healthcare response told me that the CDC was resistant to changing their surveillance or considering its limitations. "They built up this system season after season and it worked well for flu, so they were reluctant to pull away from it," one senior HHS official told me.

But while the syndromic surveillance was not flashing red in January and February, it was not green, either. For most of February, analysis of the surveillance data showed that the proportion of people with flu-like symptoms but testing negative for influenza was at the high end of the historical range.[5] This signal was hardly reassuring. Set against the backdrop of the situation unfolding in China, and the risk that COVID would spread to America, the data were disquieting.

If there had been an awareness earlier to the large portion of SARS-CoV-2 spread that was through asymptomatic patients, and a mindfulness that syndromic surveillance would miss these cases, it could have inspired a pivot toward more widespread and aggressive testing earlier, including screening for asymptomatic people. (Of course, the US didn't have the capability to conduct widespread screening because we couldn't deploy a reliable test for SARS-CoV-2.) It was not until the late winter that the CDC finally acknowledged the possibility of using widespread testing of asymptomatic individuals as a way to contain transmission.[6]

"It was almost like CDC didn't want to believe it," one person involved in the White House response told me. Many observers assumed that the White House had prevented the CDC from recommending large-scale testing given the president's public statements about more testing producing more cases, which in turn produced more public fear and scrutiny.[7] But this aversion to testing was also organic inside the CDC. The agency clung too long to the approach it knew for flu. This caused the CDC to make other mistakes with its data. Two of the biggest failures was the CDC's overestimation of the risk of spread from contaminated surfaces, and its underestimation of the risk from asymptomatic transmission. The two errors were mutually dependent.

For a long time, without adequate testing and with no emphasis on diagnosing asymptomatic cases, the CDC was attributing a lot of the transmission it was seeing, and that it couldn't explain, not to asymptomatic infections but to "fomites"—a circumstance where someone touches a surface that had become contaminated with the respiratory secretions of an infected patient and then touches his or her own nose, eyes, or mouth. The CDC saw a whole bunch of cases where they couldn't trace patients back to a sick contact, and so the agency wrongly assumed that the transmission chain must have been lit by touching a surface that was contaminated with respiratory secretions. Instead, it was probably through direct contact with an asymptomatic person.

Despite the fact that influenza can also spread asymptomatically, the CDC didn't assume that the same would apply for the novel coronavirus. On January 28, Anthony Fauci said during a Coronavirus Task Force press conference that in "all the history of respiratory-borne viruses of any type, asymptomatic transmission has never been the driver of outbreaks. . . . Even if there's a rare asymptomatic person that might transmit, an epidemic is not driven by asymptomatic carriers."[8] Around the same time, a report had come out in a major medical journal, *The Lancet,* casting some of the first prominent doubts on this scientific dogma. It stated, "As shown in this study, it is still crucial to isolate patients and trace and quarantine contacts as early as possible because asymptomatic infection appears possible."[9] Relying on symptoms to find cases, the CDC also discounted the usefulness of PCR (polymerase chain reaction) tests, of the sort advanced by Helen Chu and her team at the University of Washington. But without the ability to diagnose some of those asymptomatic index cases, the CDC couldn't trace the true patterns of spread. Privately, the CDC still defends its influenza surveillance and the applicability of its flu-based tactics to COVID.

These failures caused the perceived risk of fomites to be overestimated, and the perceived value of better respiratory precautions, like the use of masks, to be underestimated. We put far too much emphasis on cleaning surfaces, when we should have been taking more steps to improve airflow and filtration in confined indoor spaces and to get N95 respirators to individuals at high risk of bad outcomes. Asymptomatic spread played an especially large role in transmission from, and among, younger people. The CDC really didn't get a clear picture of the extent of the asymptomatic cases and its role in transmission until universities started to test their students and uncovered the full accounting. The agency's early models on spread underestimated the scope of the pandemic because they were largely based on transmission patterns for flu. Compounding the error, the CDC wouldn't share its methodology, so those who might have been able to uncover the shortcomings and correct for the CDC's faulty assumptions had no way to fully understand how the agency made its projections.[10]

By February, I, along with others, was worried that SARS-CoV-2 could already be spreading widely in America. On February 21, I appeared on CNBC's evening news show to discuss epidemics that were now under way in South Korea, Japan, and Italy. I was set to appear after an interview with Anthony Fauci. We didn't yet have any firm evidence of local transmission in the US, and Fauci seemed secure, for now, that it was not yet spreading in American communities.

The host of the program, Wilfred Frost, posed a question to Fauci that grew out of an op-ed that I had published in *The Wall Street Journal* the day before, in which I warned that community spread of coronavirus was likely already under way in the US. I wrote:

> A mere 15 cases of the Wuhan coronavirus have been diagnosed in the U.S., according to the Centers for Disease Control and Prevention, and that number hasn't budged in a week. But the true number of cases is unknown because the U.S. is testing only those who recently arrived from China or have been in close contact with confirmed patients. Public-health authorities need to be prepared for a wider outbreak. The CDC says it will set up a pilot program in five states to screen some patients with unexplained lung infections. But that program hasn't started, so we can only hope that cases didn't get into the U.S. undetected and begin spreading. It's important not to overstate the danger: If thousands of people were infected in, say, New York, more patients would be showing up at hospitals with serious lung infections. The outbreak would be obvious. But since most people with the virus suffer only a mild illness, dozens and perhaps even hundreds of cases may be circulating undetected.[11]

Picking up that thread, Frost asked Fauci: "Some people in the last couple of days in particular have started to elevate their concerns as to whether this disease is already present in the US and spreading across people who are otherwise apparently healthy. Is that something that is impossible to know either way for sure and how worried are you about that possibility?"[12]

Fauci replied, "Well, certainly it's a possibility but it is extraordinarily unlikely and let me explain why. The reason is if there were people who were actually spreading it, you would not have them identified, isolated, and contact tracing which means you would have almost an exponential spread of an infection of which we are all looking out for. We have not seen that, so it is extremely unlikely that it is happening. However, not to be overconfident and to directly get some data, what the Centers for Disease Control and Prevention is doing is picking out six sentinel cities where they will see people who come in with symptoms of the flu, who test negative for the flu, they will then test them for the coronavirus so it would be a good indication if we are in fact inadvertently missing cases of the coronavirus infection. So the pattern of what we are seeing argues against infections that we are missing and what we are going to do preemptively will make it even more proof that that is not the case, so we are not relying on overconfidence."

Fauci had told the *USA Today* editorial board on February 15 that if testing showed that the virus had slipped into the country in places federal officials didn't know about, then "we've got a problem."[13] Of course, we now know that by February 21 local transmission was under way.[14] Fauci and others on the task force based their assessments on the best data they had. For a long time, without a way to actually test for the virus, the best data they had was the ILI surveillance.

After the appearance on CNBC, I was driving home and called a senior White House official to relate the exchange. I asked, "How can you guys be so sure that coronavirus isn't spreading here?" It seemed likely that, at that point, we had community spread of the virus but couldn't yet detect it. "I don't know what to tell you," the official replied. "We keep asking them," referring to the health officials on the task force, particularly the CDC officials. "They tell us it's not spreading."

I was very worried about the reassuring message, that it might deter doctors and nurses who could be confronting the illness in their wards from taking precautions to protect themselves because they were being told it wasn't spreading in America. In Wuhan, a lot of healthcare providers had contracted COVID early on in the outbreak because they were unaware of the virus's presence. I didn't want to see providers in the US make the same mistake because they weren't armed with good information about its spread. So, right after I hung up the phone with the White House official, I took to Twitter to send a message of my own: "U.S. officials express a lot of confidence that if #COVID19 was spreading we'd see it in our epi surveillance. We should be hopeful. But providers and front-line personnel should still have caution, awareness that community spread could become apparent in U.S. Be safe, Be alert."[15]

A press conference on February 25 is now viewed as a turning point in how federal health officials spoke about the mounting risks. Dr. Nancy Messonnier, the director of the CDC's National Center for Immunization and Respiratory Diseases, warned—for the first time—that community spread was all but inevitable. But when you unpack what was said, the actual statements still hewed closely to the prevailing narrative that the coronavirus hadn't yet arrived in our communities.

The part of the message that was widely reported, and created a stir, was when Messonnier said coronavirus would soon begin spreading at the community level, and the "disruption to everyday life might be severe." That is the quote that triggered a brisk stock market sell-off that unnerved President Trump, who was traveling abroad in India. The market reaction would be a final nudge that would cause the president to hand oversight of the COVID response to Vice President Mike Pence. But before Messonnier delivered that stark warning, there was another part of her message that didn't get as much notice. She reinforced the CDC's view that the virus hadn't yet begun to spread in the US and had been largely stopped at our border. "To date our containment strategies have been largely successful," she said at the start of the call. "As a result, we have very few cases in the United States and no spread in the community." Messonnier's briefing would occur the same day that Mardi Gras festivities were getting into full swing in New Orleans, one of a number of superspreading events that would unsuspectingly unfold that week.[16]

In the weeks before and after the press conference, Robert Redfield would also affirm the message that the threat to the nation remained low. "The American public needs to go on with their normal lives," he said.[17] In fact, SARS-CoV-2 was spreading widely.

The point here isn't that the federal health officials were wrong.

The point is that they were working with faulty tools, and from faulty data sets. Because health officials couldn't test widely for the presence of the virus, they overrelied on information that they had access to, which was based on a model for flu. This is the central failing that needs to be fixed for our future preparedness. Without a way to test for SARS-CoV-2, to diagnose community cases, to understand how COVID established itself, and to uncover the asymptomatic spread, the US had no way to know for sure if it was spreading. The country certainly had no way to know if it would reliably show up in our surveillance system for flu-like symptoms. We were more situationally blind than we assumed.

By the end of the month, the data feeds were finally starting to show troubling signs. "The flu tests were going down, but suddenly the ILI stopped going down or started to go up, we started to see some funny blips, curves that should have gone smoothly down were starting to flatten out or even rise," a senior health official working on the federal COVID response told me. "We saw something in Chicago and in Seattle, and then later something was happening in New York. . . . We saw little blips here and there, but they were initially small. Then they grew larger."[18]

Behind the scenes, by late February, there was agreement on the task force that they might soon need to shift the nation's posture and start implementing targeted mitigation in four cities—Seattle, Boston, New York, and Santa Clara, California. Secretary Alex Azar recommended calling for the equivalent of snow days in those cities, slowing activity to stem the spread of the virus and, hopefully, avoid the need for broader shutdowns. The task force had run a tabletop exercise to rehearse such an escalation. The task force members told me that, at the meeting, they had concluded that community spread of the disease in more than one city was now likely.

They prepared to meet with the president on the evening of February 26, after he returned from India, to secure his agreement to this escalation and the new focus on community mitigation. But they would later tell me that the political fallout from Messonnier's briefing caused Trump to put the brakes, for now, on further action. The Messonnier press call was seen inside the White House as a barometer of how the public might react to tougher action. If the mere hint of mitigation prompted markets to swoon, some on the White House political team argued that it would be utter carnage if they actually implemented the measures she had discussed. This was the calculus taking shape. It prompted the White House to freeze further actions for a full two weeks, while they considered their options. It was a "lost two weeks," in the words of one senior health official. The virus was reaching an epidemic proportion, although it was not yet obvious. It would soon burst out of control. The two-week pause would probably seal a more difficult fate for the nation.

By the first week in March, the system was finally flashing red, especially in New York City, where there was an unmistakable rise in influenza-like illness that couldn't be explained by people testing positive for the flu.[19] There was a 50 percent increase in the number of emergency room visits for respiratory illness. It was an ominous turning point. That's when CDC director Robert Redfield called the New York State health commissioner and said, "I think you have a problem." The ILI finally showed a spike. Lab-confirmed cases of flu were now falling sharply, but visits to providers for flu-like illness were rising even more quickly.

Some of those mounting doctor visits were from people who had COVID.

On March 6, the Coronavirus Task Force was shown a detailed presentation by Robert Redfield of data identifying the uptick in suspicious cases captured through the CDC's ILI surveillance system. Before that point, the CDC hadn't identified or reported to the task force any concerning signals coming from that data. But, as I noted, the ILI surveillance data is insensitive and backward-looking. The actual cases that drove the blip had occurred weeks earlier. Community spread was well under way. Now the CDC had the data to prove it.

We would learn later that by February, the West Coast was

heavily seeded with cases imported from China that circumvented our travel restrictions.[20] It turns out that the February 2 ban on air travel from China was quite leaky. An estimated 40,000 American residents had returned to the US from China through major airports after the ban took effect.[21] And in the month before the travel restrictions, nearly 300,000 people had traveled into the US from China.[22] Airport screening for symptoms was also porous, and there was no way to identify asymptomatic carriers. Meanwhile, the East Coast was being seeded heavily with cases coming from Italy. Wuhan and Italy had close ties. Italy was the first G-7 nation to join China's investment and infrastructure project, the Belt and Road Initiative, and many Chinese textile workers had immigrated to Prato, the historical capital of Italy's fashion business.[23] Garments would be manufactured in Wuhan, and shipped to Prato, where the final stitches were "made in Italy."[24]

A review of CDC records by the news organization ProPublica found that of the more than 750,000 travelers screened at US airports by mid-September 2020, officials found only 24 cases of COVID.[25] We now know that many hundreds of cases got through. We closed down most, but not all travel from China early on. But we let travelers continue to arrive from Europe even as the epidemic raged there. Matthew Pottinger and others had urged restricting travel from Europe much earlier than our March 11 ban that affected travelers from twenty-six European countries, but the proposal met resistance from Secretary of the Treasury Steven Mnuchin and others on the economic team, who worried about the financial impact of such a move.[26] The economists had been worried that the virus, and our actions to stop its spread, could trigger a "doom loop" in their vernacular. This is a situation where a sharp and sudden drawdown in the supply of goods and services eventually leads to a corresponding decline in demand for these economic products, which further erodes their supply.[27]

In retrospect, we could never stop the virus at our borders. At best, we could have delayed its fuller entry. Screening programs and travel bans were long known to have limited effect in preventing or even slowing epidemics.[28] One widely cited modeling study, which examined the impact of travel bans in the setting of a pandemic flu, found that implementing strict travel bans that reduced infections by 99.9 percent would delay the peak of a hypothetical US pandemic by just six weeks. This was if the restrictions were implemented early, on day thirty of a hypothetical foreign epidemic that had not yet spread discernibly to America, and the estimate was modeled from influenza, which is less contagious than SARS-CoV-2.[29] In other words, even draconian bans on travel, implemented to stop a virus that was less contagious than SARS-CoV-2, would, at best, be temporizing. It would buy some time; it wouldn't prevent the epidemic.

As for screening programs that would merely evaluate travelers for symptoms of COVID, when such an approach was implemented in Australia in response to the 2003 outbreak of SARS-1, for example, 1.84 million people were screened, 794 were quarantined, and no cases were confirmed.[30] Across Canada, China, and Singapore, thermal scanners had been used to screen more than 35 million travelers in 2003, detecting nearly 11,000 fevers but uncovering not a single case of SARS-1.[31] Still, Canada would record 250 cases of SARS-1, and Singapore 206 cases. During past pandemics, it was reported that people took fever-reducing drugs to defeat the scanners.[32] Yet the same thermal devices and techniques would be used in US airports to screen passengers in the first few months of 2020, building a false sense of security.[33]

On top of this fundamental weakness, our screening system was also plagued by countless shortcomings. The CDC's outdated data systems didn't ferry information directly to airports. The agency relied on phone calls, faxes, and spreadsheets attached to emails to alert airports as to the arrival of travelers who should be denied entry or subjected to extra screening. This system made it hard for local officials to identify and stop passengers who were arriving from hot zones. At one point in mid-February, the CDC's system went offline completely, allowing thousands of passengers to stream into the country uncontested. A lot of people got through the checkpoints. Some were carrying COVID with them. Later analysis found that of the 675,000 travelers who were subjected to enhanced health screening at airports, fewer than 15 cases of COVID were identified.[34] In the end, the number of federal personnel charged with implementing the screening who caught COVID exceeded the number of infected travelers that the efforts would intercept.[35] It would be the first of many instances where we played catch-up to the crisis.

The inability to field a reliable diagnostic, to deploy it in scale, and the overreliance on syndromic surveillance that was inherently flawed, were historic failures that left us badly at risk. If, as a later analysis has suggested, the first case in Seattle (the one diagnosed on January 20) never started a chain of transmission, and the subsequent outbreak was the result of a second case that had arrived in the city much later, then the virus hadn't yet gained a foothold in the community until February.[36] The window to act on the initial cases that we knew about in Seattle, and the ones we didn't know about in New York, San Francisco, and other cities, may have been open longer than we thought—if only we had had a way to test for the virus and isolate it. It wasn't even a hard test to design and manufacture. And still, we badly bungled its rollout.

CHAPTER 6

THE ZIKA MISADVENTURE

The problems started early.

The old playbook that governed America's pandemic response always had the CDC going first. The agency would be the first to have access to samples of a novel virus, since any dangerous new pathogen would be sent to the CDC's labs and would require special handling. The CDC would be the first agency to design a diagnostic test using the special access it had to these viral samples. It would be the first to use its new test to screen samples on behalf of other doctors around the country. Then, if there was a risk of a novel pathogen spreading more widely, requiring more testing than what the CDC's central lab could handle, the agency would be first to develop a test "kit," to be distributed to a network of about one hundred public health labs operated by the states. This would expand the nation's testing capacity. Eventually, commercial test manufacturers would get viral samples from the US government that they could use to develop their own test kits. The CDC would take charge of growing the virus and then sending it to the National Institute of Health, which would work with contractors to process it for controlled distribution to test developers.[1] The commercial manufacturers would get these kits authorized by the FDA and distribute them to a broader base of private and academic labs. It was a sequential process that would unfold over months. This is how the system was supposed to work, and the CDC tightly guarded its institutional prerogatives.

The CDC got an opening on January 11, when Chinese health authorities first published the genetic sequence of the novel coronavirus. Using this information, the CDC's Core Lab started working on a test in earnest. Designing it took seven days.[2] The CDC got a second opening on January 19, after patient zero sought care from an urgent care clinic in Seattle, four days after arriving from Wuhan. A respiratory specimen was collected from the patient and flown to the CDC.[3] The agency was able to confirm that he was positive for SARS-CoV-2, using a preliminary test fashioned from the sequence data. Now the CDC officially had a live sample of the virus to complete the development and validation of its new test. The agency also began efforts to grow the virus in larger quantities, to create a resource it could eventually share with other labs.

The CDC's ability to quickly design that initial test was laudable. In 2009, when the H1N1 swine flu first appeared in the US, the CDC had acted with similar speed, developing a diagnostic test in about ten days and getting it authorized by the FDA.[4] Each of those H1N1 flu test kits had contained enough reagents for one thousand clinical specimens.[5] But when the CDC tried in 2020 to mass-produce its test for SARS-CoV-2, and share it with labs, it would run into problems.

Other countries were also racing to develop tests for SARS-CoV-2. On January 12, using a test developed with the assistance of CDC personnel who were stationed in Thailand, that country became the first nation to confirm a case of SARS-CoV-2 outside China, in a traveler who had arrived directly from Wuhan. By the end of January, the Thais had diagnosed eleven more patients with COVID.[6] South Korea and Taiwan were each also about to field their own test.

The World Health Organization was also in the game. It didn't make a test but served as a clearinghouse for different testing protocols, with the actual development work done by independent institutions and companies. The WHO would review data and qualify certain tests for use, serving as a de facto regulator for countries that didn't have strong regulatory authorities of their own. On January 13, the WHO posted a protocol for a test designed by a German group that was similar to the CDC's design but simpler to manufacture. Several countries used this instruction set to start making their own test kits. By January 16, a German manufacturer had started producing these kits in bulk. Soon, the WHO adopted that German test inside its own laboratories. But the CDC's experts said the European test could be less accurate than the test that the CDC was developing. The German protocol, the CDC argued, might fail to detect the virus if it mutated into new variants that were distinct from the original strain.[7] The CDC stuck with its own design.

But back at the CDC, efforts to field a test for the US market started to run into trouble.

On January 24, the CDC posted the blueprint for its test. This information could have helped other sophisticated laboratories to fashion their own versions of that test to be run in their own high-complexity facilities.[8] But the CDC stated on its website and in a newsletter for public health labs that only the CDC could conduct the test. The agency considered SARS-CoV-2 a special pathogen that required careful handling and, as a consequence, the CDC exerted exclusive control over the collection and testing of viral samples. The plan was for the CDC to eventually package its new test into kits that it would manufacture itself and distribute to state public health labs. Then, after the public health labs were running the new test, if there was still testing demand that couldn't be met by these labs, commercial and academic labs could make their own laboratory developed tests and get permission from the FDA to offer these tests to patients. Eventually, if there was still a need for testing that couldn't be fully met, medical device companies would get approval from the FDA to manufacture test kits in bulk and would then distribute those test kits to the nation's labs to run on high-capacity testing machines. This sequential approach to expanding the supply of testing had been followed, with mixed results, during past outbreaks involving other novel pathogens.

Once the CDC announced that it was developing a test kit that could be deployed to other labs, commercial manufacturers who might have entered the market decided to wait. Why, some reasoned, develop a test only to have the CDC flood the market with its own kits? The CDC, it appeared, had things under control. Meanwhile, officials at the CDC thought they had time to get their own test developed and deployed to labs. They expected COVID to behave like the other two novel coronavirus strains that had emerged in recent years, SARS-1 and MERS, each of which had triggered limited outbreaks before they receded. And yet there were reasons to be skeptical, even in early January, that SARS-CoV-2 would follow this familiar trajectory. The new virus seemed to be far more transmissible than either SARS-1 or MERS. There was some indication early on that people could transmit the virus before symptoms set in, increasing the opportunity for spread. SARS-1 and MERS also made their victims seriously ill and patients became highly contagious only after symptoms set in, by which point patients would be too sick to move around. That made these viruses less likely to spread widely. COVID was different. A CDC scientist who worked on the agency's initial response to COVID would later say: "It was being treated as a MERS situation or a SARS situation. . . . At that point we thought it was going to be a limited activity."[9]

Under the pandemic playbook the CDC was following, the plan was for the FDA to grant the CDC permission to distribute its test under an Emergency Use Authorization (EUA). To meet the standard for emergency authorization, the CDC would need to prove that it's "reasonable to believe" that the test may be effective in diagnosing patients. An EUA is more flexible than the FDA's usual review standard, which requires a "reasonable assurance of safety and effectiveness" in diagnosing patients. The EUA pathway was fashioned by Congress (as part of the Project BioShield Act of 2004) to provide an efficient way to get medical countermeasures to the community in the event of a public health emergency, including a pandemic or an attack by a chemical, biological, radiological, or nuclear weapon.[10]

The CDC submitted partial validation data to the FDA on January 27, so that the FDA could start its review process while the CDC completed the rest of its development work. Four days later, on January 31, the CDC asked the FDA for enforcement discretion and permission to ship the test immediately, even as the CDC completed its application. The FDA agreed. On February 3, the CDC submitted a complete EUA request to the FDA, and the authorization was granted the next day, coinciding with the formal declaration by Secretary Azar that "there is a public health emergency that has a significant potential to affect national security or the health and security of United States citizens living abroad."[11] This declaration by HHS cleared the way for companies and labs who were developing therapeutics and diagnostic tests to file EUA requests with the FDA.[12]

The CDC asked the FDA to require, as part of its authorization, that certain conditions be met by labs that wanted to use the CDC's new test. For one thing, the CDC wanted the FDA to specify in the authorization that the test could be used only on respiratory specimens. The CDC hadn't yet demonstrated the accuracy of its test for detecting coronavirus in other body fluids like saliva. The CDC also wanted to include in its FDA authorized labeling that whenever the test generated an inconclusive or positive result, the lab performing the test would be required to send the specimen to the CDC, which would run the sample in its own lab to confirm the finding. But the CDC's central lab in Atlanta could only handle a few hundred tests a day. If COVID cases started to grow, or if a lot of inconclusive results were being generated, then the CDC's reference lab would be quickly overwhelmed. The CDC also intended to confine the use of its test to the 115 US public health labs.

Taken together, the CDC's restrictions would limit how much testing would initially be done. Those constraints would be overshadowed, however, by a tragic failure. As I'll detail later, when those tests finally shipped, they didn't work. Instead, the entire country remained tethered to the CDC's single lab for all of the nation's testing, because it was the only place in the US that had a workable test. The CDC had prevented other labs from developing their own tests and now, its own kit was defective, and the CDC's Atlanta lab was the only facility that could get the test to work. By March 1, the CDC had tested fewer than 3,600 samples. Senior HHS officials would tout this figure in a way that implied that 3,600 Americans had been tested for COVID.[13] But even that small figure overstated the agency's throughput. ("Throughput" refers to a lab's capacity to run a lot of tests in short periods of time.) Under questioning at a Senate hearing on March 3 by Senator Richard Burr, the CDC's deputy director, Dr. Anne Schuchat, revealed that many of these were multiple tests conducted on the same patients who were being serially tested for the virus, not tests on 3,600 unique patients. In fact, far fewer than 3,600 different patients had been tested for the virus.[14]

During this critical stretch of time, the country was largely blind to the spread. The US lacked a simple test that could diagnose people for SARS-CoV-2, and we would never catch up. From January 18 to March 7, the CDC tested only 3,869 specimens for the virus.[15]

The country had missed the window to field a diagnostic test and deploy it widely enough to detect the early spread, isolate the sick, and try to reduce the scope of transmission. We wouldn't have been able to avert a US epidemic, but we might have delayed its start and reduced its severity. Instead, the virus was heavily seeded across the country. The CDC had missed the game.

Zika should have been a wake-up call for the CDC and the nation. For years, Zika was a seldom-seen infection, but in late 2015 its spread was first detected along the northern coast of Brazil, raising concerns that it might reach the US. The infection had been discovered in 1947 in the Zika Forest in Uganda, where scientists were conducting routine surveillance for yellow fever. Until 2006, Zika was reported only sporadically in humans. There were no known outbreaks. Only fourteen cases of human infection had been documented worldwide. But things changed in 2007, when the virus caused the first large outbreak on the Pacific island of Yap, in the Federated States of Micronesia. That was followed by an even bigger cluster in 2013 in French Polynesia. Then, on March 2, 2015, Brazil notified the WHO of reports of an unusual illness that was characterized by fever, rash, and joint pains.[16]

At first, Zika was not suspected, and no tests for the infection were performed. Nearly seven thousand cases of the unfamiliar illness were reported in the northeastern states of Brazil from February through April.[17] Eventually, Zika was determined to be the culprit. Later, studies using sequencing data showed that the virus had probably arrived in South America in 2013, but it took time for a sufficient number of cases to accumulate before the disease exploded into public view.[18]

By 2016, its presence in Brazil was firmly established. The WHO declared that a large cluster of cases of microcephaly—a devastating condition where babies are born with tiny heads—was being caused by the infection. On February 1 of that year, the WHO declared that Zika's spread in Brazil constituted a Public Health Emergency of International Concern (PHEIC).

Zika is transmitted through the bite of a mosquito, and the infection is notoriously hard to test for. Blood tests can have difficulty detecting the virus, depending on how long you've had the infection. Diagnosing Zika is complicated further by the fact that 80 percent of those infected will never show any symptoms so they would be unlikely to seek testing. Its continued spread in the Americas seemed inevitable.[19] Inside the CDC, the need for a reliable test that could be widely deployed gained urgency. It was just a matter of time, many believed, before the infection arrived in the US.

Initially, the CDC relied on the lab at its branch in Puerto Rico to run the tests. The facility specializes in the diagnosis of dengue fever, another infection that, like Zika, is transmitted by the bite of a mosquito. Both pathogens belong to a class of viruses known as flaviviruses, which are generally spread through mosquitos or ticks. The entire class is named for the yellow fever virus, which is part of the same family of infections. The term was applied because yellow fever often causes jaundice in its victims (*flavus* means "yellow" in Latin). For most of these viruses, human beings are dead-end hosts; once a mosquito or tick transmits the infection to a person, the virus eventually dies inside its human victim. People can't transmit the infection any further; only the tick or mosquito can spread it. But Zika is different. It adapts well enough to its human victims that people can spread the infection on their own. Zika, for example, can be sexually transmitted.[20] Sexual transmission for dengue is only anecdotally reported, and direct spread of yellow fever from one person to another doesn't occur. If a person is infected with Zika and then bitten by a mosquito, he or she can also transmit the pathogen to the bug. The mosquito, in turn, can spread the virus to other mosquitoes, which can then infect more people through their bites. Zika's ability to spread in this fashion made prompt diagnosis a key to preventing an epidemic. When, as expected, the Zika virus started to spread in the US, the CDC's single testing site in Puerto Rico became overwhelmed. Testing backlogs emerged. Test results were taking weeks to reach patients, reducing their usefulness to doctors.

So the CDC's virology team started to share instructions with public health labs on how they could fashion their own Zika blood tests using PCR. Speed was critical. There was a narrow window to stop a wider epidemic. If doctors could properly diagnose victims, then they could isolate them and limit the virus's advance. And if they could keep the infection from spreading to the US mosquito population, a wider epidemic could be averted. Controlling the mosquito population was key. But equally critical was identifying the virus's human hosts.

By the end of 2015, just sixty-two symptomatic cases of Zika infection had been diagnosed in the US, but it was widely agreed that this represented just a fraction of the total infections. The scarcity of diagnostic tests was hampering the CDC's ability to uncover more cases. Because Zika infection didn't spread through inhalation, it wasn't going to race as quickly through the population as a respiratory virus, but that didn't mean that the CDC had endless time to get a working test into the hands of providers. The urgency of making testing more widely accessible was driven by the fear of the severe congenital abnormalities that the virus was known to cause. If Zika became widespread, hundreds of women could unknowingly harbor the infection during pregnancy, with potentially shattering consequences for their unborn children.

By this point, the CDC was under intense pressure to develop a diagnostic test that could end the testing backlog. In early 2016, the agency made the decision to sideline the PCR-based test that it was using, and instead manufacture a new test that was supposed to be more precise and could be more widely distributed to labs as a complete kit.[21] The new test was designed by the CDC's lab in Puerto Rico and was called the Trioplex, because it screened for Zika as well as for two other flaviviruses—dengue and chikungunya fever. These infections are spread by the same species of mosquito that carries Zika, and they cause symptoms that can mimic Zika infection.[22] By testing for all three viruses at the same time, the CDC wanted to make sure that Zika wasn't being confused with these other diseases. But the clinical utility of simultaneously screening for all three infections was controversial. These other infections were not epidemic at the time. Zika was. That should have been the focus of the CDC's attention.

The CDC scrambled to make the new test, evidenced by careless and sometimes dangerous handling of its components. In one notable incident, the CDC shipped a specimen of chikungunya virus that was supposed to be inactivated. It turned out to be alive as it went through the mail. The viral sample had gone from the CDC's high-containment facility in Fort Collins, Colorado, to a refrigerator in a CDC lab in Atlanta that used lower-level precautions because it wasn't supposed to contain dangerous, live pathogens. "We are gravely concerned that a conscious decision was made to ship this pathogen before confirmation [that it was inactivated] because of urgency," Congressman Fred Upton, chairman of the House Energy and Commerce Committee, said at the time. The CDC admitted that time pressure had allowed the lapse to occur.[23] The episode underscored the long-standing problems with quality that plagued the CDC's test development.

On March 16, 2016, the CDC finally sought authorization from the FDA for its Trioplex test. The CDC wanted to start distributing the new test to state public health labs immediately. One day later, the FDA authorized the test for emergency use. As with other tests distributed for emergency use, the FDA included language in its authorization that the CDC test might not be fully reliable. The FDA instructed providers to use the results in the context of other clinical findings that could help doctors confirm whether a patient was really infected.[24]

But the CDC was far more confident in the test, even though it would end up struggling to get it to work. The agency instructed public health labs to scrap the PCR-based "Singleplex" and instead, switch to the new Trioplex. In practice, the CDC had sidelined a protocol that was already deployed and working. And it was all done to screen for two pathogens that were not even spreading in America with a test that, it would soon be shown, wasn't as sensitive at detecting Zika as the one it replaced.

Compounding matters, the new CDC test was more difficult to run, requiring a manual that was forty-one pages long, as opposed to two pages of instructions for the Singleplex test. The Trioplex had the advantage of being packaged as a kit, with qualified primers and probes that were manufactured and provided by the CDC, along with detailed instructions for how to interpret the results. So it was supposed to be more reliable. The Singleplex was just a protocol—labs would have to assemble the tests themselves from components that they would purchase separately. But the Singleplex was working. Six days later, on March 23, the CDC director testified before Congress and said that the agency had already manufactured more than a half million Trioplex test kits. The CDC was all in on the new test.[25]

Problems surfaced almost immediately.[26] Public health labs couldn't get the new Trioplex tests to work right. The issues were so acute that the CDC's reference lab that had the most experience testing for Zika, its Fort Collins Arbovirus Diagnostic Laboratory, made the extraordinary decision not to use the CDC's own test, though state public health labs were not informed of this decision. Dr. Robert Lanciotti, the chief of the Fort Collins lab, had raised some of the sharpest concerns.[27] Lanciotti was well regarded in his field and had developed the test for diagnosing West Nile virus.[28] He analyzed the Singleplex and Trioplex side by side and found that Trioplex missed about 40 percent of Zika infections that were detected by Singleplex. An independent lab ran a similar study and came to the same conclusion. Documents from the time show that the FDA also had concerns about the new CDC test.[29] But the CDC's Emergency Operations Center continued to urge state public health labs to switch to the new Trioplex test, not telling them that the new test had generated results showing it could have less sensitivity at detecting Zika. Still, at least seven state and public health labs rejected the Trioplex test and continued to use the Singleplex.

To address the issues, the CDC modified the Trioplex to boost its sensitivity. But the new procedures relied on extracting higher amounts of genetic material from the Zika pathogen. This required a complicated set of steps and advanced laboratory equipment that was out of reach for many public health labs.[30] The Trioplex's problems were so serious that the presidents of the three major organizations representing the nation's microbiology and virology labs wrote an extraordinary letter to the CDC rebuking the agency and echoing many of Lanciotti's initial warnings.[31] And on January 12, 2017, the CDC modified its stance on the Trioplex test, informing public health labs that they could discard the non-Zika components of the Trioplex as a way to get the test working.[32] It was an eerie harbinger of the problems that would plague the CDC's test for COVID.[33]

As for Lanciotti, he saw the problems as being larger than just the challenges with the Zika test. For him, the episode exposed deep-rooted weaknesses in the way the CDC went about developing and deploying a test in a crisis. "Time will tell," Lanciotti wrote in 2016, "how effective it is and what unforeseen problems arise" with the CDC's testing approach.[34] Calling the CDC's methods "rigid," he warned about the vulnerabilities of a process that left public health labs almost completely tethered to the CDC to manufacture and supply the only test kits that would be available in a public health crisis.[35] In part for airing these concerns publicly, Lanciotti was removed from his post. Ultimately, after an investigation, the Office of Special Counsel ordered the CDC to reinstate him.[36]

The Government Accountability Office later faulted the CDC for not having a plan to supply its diagnostic test to commercial labs that could broaden the testing for Zika, and a subsequent audit would fault CDC leaders for ignoring Lanciotti's initial concerns.[37] Among the findings in a 2017 report, the GAO would write: "CDC developed the first two authorized diagnostic tests for the Zika virus and offered these tests to public health laboratories, but not to some manufacturers. Some manufacturers did not have access to the authorized CDC tests and encountered difficulty acquiring authorized tests from other manufacturers. Without a clear and transparent process for distributing CDC diagnostic tests, the agency may not be able to develop the capacity of the commercial sector to be able to meet the needs during an outbreak."[38]

The window to prevent a Zika epidemic was closing. The CDC's Emergency Operations Center "continued to promote a questionable assay with misleading communications that led laboratories to believe that it was in fact the best available test for the detection of Zika virus," Lanciotti would write two years later, in response to an investigation by the Office of the Special Counsel. "Not recommending the best available test during an epidemic of a novel pathogen, in which the potential to miss cases is evident with implications for proper clinical case management, is in fact a threat to public health."[39]

Those words were written in 2018. They could have been written in 2020 as we grappled with the damaging aftermath of the CDC's botched rollout of its initial test for COVID.

CHAPTER 7

THE CDC FAILS

The problems with the CDC's COVID test seem to have started inside the agency's Respiratory Virology Lab in Atlanta. They had sprung from a contaminated component that was meant to make the test more precise but also made the process more complicated and prone to failure. There were three main parts to test, referred to as N1, N2, and N3. Each component was a different set of primers and probes that would bind to a specific segment of the coronavirus's RNA. The first two reagents, N1 and N2, were specific to SARS-CoV-2. They targeted two regions in the virus's RNA that code for the production of a structural protein that plays a key role in the coronavirus's replication and assembly.[1] The third component, N3, was unique to the CDC test and was not included in other countries' tests, including the one developed in Germany and adopted by the WHO. The N3 component was designed to bind to any SARS-like coronavirus, including some strains that were found in bats but had never been known to have infected humans.

Stephen Lindstrom, the director of the CDC's National Center for Immunization and Respiratory Diseases at the Atlanta campus, was the official in charge of developing the new test. He had helped make the decision to add the third component, telling colleagues it would help detect SARS-CoV-2 even if the virus began to mutate.[2]

It wasn't in the original test design. I'm told that the decision to add the N3 component was made by Lindstrom and his immediate colleagues, but the agency's director, Robert Redfield, wasn't informed of the design choice until after problems emerged.[3] Fielding the COVID test was the most crucial task the CDC was facing during one of the most epochal moments in its history. What test design the CDC was using, and its potential shortcomings, struck me as a decision that the CDC director should have been informed of ahead of time. It struck me as a troubling breakdown in accountability and awareness. "They decided that the most important thing they needed to do was get no false positives, so they added a third primer pair, that wasn't in the original test design," I was told by one health official close to the process.

The CDC test, with its N3 component, was designed to be more precise. But, like the Zika test, it was also a lot more intricate. The trade-off for its precision was added complexity. By including the N3 component, the CDC had made the test more difficult to manufacture and harder to perform. Precision mattered, to a point, but in the setting of a pandemic with a dangerous virus, speed and ease of manufacture also mattered. Like the test for Zika, the CDC's new COVID test was overengineered, which in this case may have opened it up to problems. But unlike Zika, this was a respiratory virus that could spread quickly through a population. Instead of producing a test that was more accurate, the N3 component ended up becoming a point of failure.

State and local labs were eager to get the new COVID tests and start running their own samples. The CDC had assured public health labs during a January 28 conference call that "CDC's goal is to get [FDA authorization] as quickly as possible and expects the assay will be ready to deploy within two weeks, possibly sooner."[4] Labs were waiting for the CDC's test, but they were growing impatient. The day before, seeing some early challenges with the CDC test, senior staff in the FDA's device center had reached out to all of the commercial manufacturers with testing platforms installed in labs throughout the country, to find out their timeline for developing their own tests for the novel coronavirus and to ask whether there was anything the FDA could do to help. The FDA was poking around for a hedge in the event the CDC test stumbled. But many commercial manufacturers were worried that they would invest in tests that wouldn't get procured or reimbursed. One CEO told the FDA that switching over a single line of production meant not supporting a test that they were already marketing—a task that would cost millions of dollars. The US government had made no assurance that a successful test would be purchased. There were no promises of minimum buys and no promises of reimbursement. The CEOs told the FDA staff that up until that point, no other officials from HHS had been in touch even to ask if the companies would get engaged. For now, many commercial manufacturers with the capacity to develop tests were mostly waiting to see what happened with the CDC's efforts.

The first indication of problems with the CDC test appeared on February 3, when quality checks done at the agency showed that the third reagent, the N3, might have been generating false results. The CDC assumed that the setback was just a faulty lot, so they set aside the production run that included the problematic kits. But on February 6 came another red flag. A final quality check showed that the test kit could fail 33 percent of the time. For reasons that have not been fully explained, the CDC decided to ship the kits anyway.[5] They were initially sent to labs in thirty-three states.[6] On the same day, the World Health Organization directed the shipment of about 250,000 COVID test kits (manufactured by a German firm) to more than 70 labs around the world.[7]

Once the public health labs got the CDC kits, they first tried to validate them by testing sterile water, a common practice to confirm whether a test kit worked. Sterile water should have generated a negative result since it contains no coronavirus RNA. The labs were distraught to discover the samples testing positive as if the water contained live virus.[8] This was not an isolated problem, either.

Twenty-four of the first twenty-six labs that tried to run the tests couldn't get it to work. Nobody knew why.

Within two days of shipping the kits, the CDC started to hear about these problems. At first, the problems seemed isolated to the N3 component, causing the tests to register false positives. On February 9, Scott Becker, the head of the Association of Public Health Laboratories, sought permission from the CDC for the labs to use the test without the N3 component. He received no response.[9] The next day, CDC officials had a call with the FDA to report the problems. The CDC's diagnostic kit was subject to the FDA's regulation because the test was intended for making decisions about patient care. The CDC told the FDA that ten of the roughly one hundred public health labs that had received the kits were unable to get the tests working due to problems with the N3 component. At that point, there was still no clear news from the other labs. The CDC staff told the FDA officials that there may have been a manufacturing problem with N3, and that a new batch would be made to replace the bad lot. The CDC was confident this would fix the problem.

The N3 component was not required in order to make the test work, and it had not been included in the tests being used by the WHO, South Korea, China, Thailand, or any other nation running PCR tests. But instead of dropping the extra component, the CDC set out to fix it by manufacturing fresh batches. The CDC lab responsible for designing the test was offering two different reasons for its resolve to stick with the N3 component. Which version you got depended on whom you asked. One justification was that the component was necessary to make the test more precise, and it could become essential in the event that the virus mutated and evaded N1 and N2. The other justification was that the FDA wouldn't let the CDC change the design once the submission had been made to the regulatory agency. The FDA disputed the latter assertion.

Neither reason could fully explain the CDC's decision to stick with the faulty N3.

On February 11, staff from the CDC lab that had designed the test informed the directors of the state and local labs about their efforts to fix the test. "Thank you for your patience as CDC investigates reported sporadic aberrant reactivity in the N3 assay," the CDC staff wrote in an email. "After consultation and agreement with FDA, CDC is currently manufacturing and quality control testing a new N3 primer/probe set. . . . We hope to provide this replacement component as soon as possible."[10] Then, on February 12, in another phone call with the FDA, the CDC reported that twenty-six of the one hundred labs were reporting problems. By now, the CDC had already manufactured a new lot of the N3 component and said that the fresh reagent would be ready to ship within a few days.

At this point, staff at the FDA and senior officials at HHS said they were led to believe by the CDC that many of the public health labs had gotten the test to work, and therefore testing was under way in many states. This wasn't true. In reality, few of the public health labs had affirmed that they got the test to work, a fact that at least some people at the CDC apparently knew. By February 16, the FDA started to get word that the problems extended beyond the N3 component. The N1 component was also causing glitches and may also have been contaminated.

It was becoming clear that the trouble with the CDC's manufacturing process might be the result of a more systemic failure, perhaps related to contamination at the CDC site that was making the kits, and this was now affecting many parts of the agency's diagnostic test. The CDC wasn't a manufacturer. In an ideal system, the CDC would have contracted with commercial manufacturers from the outset and wouldn't have tried to make test kits on its own, even for the public health labs. The CDC eventually reached out to contract manufacturers around February 10, but agreements to enlist the help of commercial companies to make the testing components weren't completed for about ten days, due to complex issues that would arise in the contracting process. Among other contracting matters, the CDC was trying to assure protection of the agency's intellectual property around its test design.

In fairness to the CDC, the "playbook" for responding to a pandemic didn't envision that these functions would be turned over to a commercial manufacturer. But that should have been built into the process from the outset. A commercial manufacturer would have not only greater capacity to produce the test kits in massive volumes, but deeper experience and better quality controls in place. But there was no procedure in place to make the handoff to the private sector—no contracting mechanism in place and no government entity clearly empowered to turn the task over to private manufacturers. Nor did the CDC try to self-organize such an effort when it ran into problems. The view was that a combination of fidelity to past practices, parochialism, and institutional pride initially discouraged the CDC from turning to contract manufacturers, until its protracted challenges were beyond fixing. The playbook that HHS was working from also didn't envision the CDC using a commercial manufacturer. The CDC was supposed to make the first test kits, and there was no thought given to deviating from that blueprint.

On February 16, the CDC offered the first in a series of contingency plans if matters couldn't get back on track. One plan was to change the cutoff point for how much SARS-CoV-2 RNA the test would need to detect in order to register a positive result. If the CDC changed the cutoff, then a small amount of RNA, including the contaminant that was fouling N3, might no longer cause the test to produce a positive finding. But making this change would also reduce the test's sensitivity, since it would make it more likely that a respiratory sample with a small amount of viral RNA might also yield a negative result. The other plan was to drop the N3 component entirely. To the FDA, dropping N3 seemed like the better option. However, now N1 also appeared to be contaminated to some degree. Worse, the CDC still didn't have an analysis of its data to support that either change would fix the problem or that the changes wouldn't negatively affect the test's performance by reducing its sensitivity. The plan for now was to make replacement kits and the CDC would consider contingency plans only if it couldn't swap the faulty kits with a functional replacement. It was now more than a week since the issue had first surfaced. The next day, *Politico* reported that the CDC planned to distribute a second, corrected test to the labs. The report caught the FDA by surprise because the CDC hadn't yet put forward a corrective plan.[11] It was from this article that the FDA also learned, for the first time, that only three of the more than one hundred public health labs had figured out how to get the CDC test to work. Until then, based on its February 12 call with the CDC, the FDA was under the impression that more than seventy labs were able to run the test. The next day, the CDC clarified that thirty-six labs had problems with the N3 assay, and six were running into challenges with N1, while seven had gotten the test to work. The other fifty or so labs hadn't reported anything to the CDC.

Most, it turned out, either couldn't get the test to work or, hearing of the problems, didn't even try. They were waiting to receive the replacement reagents.

Inside the FDA, the staff was growing anxious. There was a palpable urgency to get a working test into the field so providers could start widely testing patients. Around this time, Peter Marks, the longtime head of the FDA's center for biologics, the group that approves vaccines, returned from an assembly in Europe at the WHO, where he had been briefed on SARS-CoV-2, and he announced at a morning meeting with the FDA's senior staff that "it could easily turn into the worst pandemic in a century." In a telebriefing on February 21 and then again on February 26, the CDC continued to tell stakeholders that it had testing under control through its own lab in Atlanta. On the first call, the agency reported "no lag time" for testing and on the second call it reported that "turnaround at

CDC is within a day. There is a little bit of shipping time. But that's the process."[12] However, the CDC continued to equivocate on a path forward for the test kits that could be used by other labs. On February 21, the agency had sent its first analysis to the FDA that was intended to support each of the two contingency plans: either to drop the N3 component entirely, or to change the specifications of the test to make it less sensitive, so that it wouldn't register false positives as easily. The CDC was taking a long time to decide which of the two options they wanted to pursue. Two days later, the CDC came back to the FDA and identified some problems with the spreadsheet of data that the CDC had provided to support the analysis, and so the agency's staff had to resubmit a corrected version. Additional revisions came on February 25 and again on February 26. Each time the CDC was correcting its own errors. It was now becoming clear that the CDC was either reluctant to move away from its original test design or having a hard time reaching a decision, or both.

In the midst of this, frustrated by the CDC's unclear process and decision making, the FDA began proactively to draft a statement to grant the CDC enforcement discretion to enable the use of the test with just the N1 and N2 components until a modification to the CDC's test could be formally authorized. The FDA wanted the public health labs who had kits to be able to start using them immediately, without the contaminated N3 component. By this point, the data showed that using just the N1 and N2 components would make the test a little less sensitive at detecting SARS-CoV-2, but it was still within an acceptable range. The public health labs already had the kits in hand. They just had to be told they could start using the tests without N3.

The way the FDA saw it, even if N1 was also contaminated, but at a lower level, then the test would end up generating a false positive on the N1 component and a true negative on N2. So, at worst, the test would produce an inconclusive result. In such a circumstance, labs would know that they couldn't trust such a conflicting result, and they'd assume something was wrong with the test. They'd seek out a confirmatory test from the CDC's own reference lab. So even if the N1 component was also faulty, it wasn't likely to put patients at risk by making doctors believe the person was positive for SARS-CoV-2 when in actuality they didn't harbor the infection.

But the CDC wanted to wait for new lots of N1 and N2 to be prepared before implementing this fix. On February 25, the CDC sent the FDA an email telling the regulators that it wouldn't allow the public health labs to run the test with just the N1 and N2 components of the previously distributed kits (the ones that were already in the hands of the public health labs). Instead, CDC officials were going to tell the labs that they should continue to wait until the CDC sent labs newly manufactured lots of N1 and N2. The FDA disagreed with this plan. That evening the FDA suggested to the CDC that labs with the kits should be permitted to use the existing N1 and N2 components immediately, which could be accomplished by removing N3 from the instructions. The next day, the FDA heard that the labs were destroying the original kits, figuring that the faulty kits were now useless. So the FDA intervened on its own, and asked the Association of Public Health Laboratories to tell its member labs not to destroy these kits; additional information would be coming shortly. The FDA then provided the CDC a proposed draft of new instructions for the public health labs, allowing them to use the existing kits without N3. It remained unclear to the FDA what the CDC would decide to do. It was the CDC's test, and it was their decision, but the FDA provided as much flexibility as it could; granting the CDC enforcement discretion to let the labs use the CDC test by dropping the troubled N3 component, and the FDA urged the CDC to move forward with this solution.

On a call with all of the public health labs later that day, the CDC promised the labs, once again, that new components were coming soon but that an interim solution had been fashioned. It was the same suggestion that the public health lab association had requested on February 9 and that the FDA had been pushing behind the scenes. The CDC had finally made up its mind: It would allow the public health labs to use the CDC kits and drop the faulty N3 component.

At this point, the public health labs were falling into one of three buckets. A few of the labs never had a problem with the CDC test and would continue testing. They may have gotten test kits that didn't suffer a contamination. For the labs that had problems with N3, they could drop that component, reverify the test with just N1 and N2, and if it worked, begin testing. But then there were still a third group of labs that had persistent problems with both N1 and N3. For these labs, the interim solution wouldn't work, because the tests would be generating inconclusive results. They had to wait for the remanufactured kits to arrive, and in the meantime send their samples to the CDC and see if the agency would agree to test them. The public health labs in New York City and New York State were in this third group, and they were effectively out of the testing business until the CDC was able to replace its faulty kits with ones that worked.

There was already good reason to believe that New York could soon become ground zero for COVID in the US. Looking at where the virus was already raging and where travelers were arriving from, it was clear that the patterns of spread would bring the virus into New York City. The New York labs were deeply frustrated, and they made their issues known publicly. The Association of Public Health Laboratories held a separate call with the FDA immediately following the February 26 national call, to seek a path forward for that last group of labs, the ones who couldn't get N1 or N3 working, which included the labs in New York. Dr. Jill Taylor, the director of the Wadsworth Center, New York State's main public health lab, said on the call that her lab's inability to test was "intolerable." New York State hadn't diagnosed a single case yet, although we now know the virus was already spreading widely there, especially in New York

City.[13] It was the same day that New York governor Andrew Cuomo, reacting to mounting concerns about the increasing global spread, set aside $40 million to fight the virus and announced a plan for possible quarantines at homes, hotels, and hospitals.[14] The New York public health officials knew they were at the precipice, and they didn't have a test to see over the cliff.

At a press conference the same day, New York City mayor Bill de Blasio said that the city was prepared to start doing its own testing with a test that it would develop itself, inside state labs. It was a clear rebuke to the CDC. On the call, Taylor had told the FDA staff that New York State wanted to start using its own laboratory developed test if the FDA would allow it. However, on the earlier call, the CDC had reiterated to the labs that new testing components for the CDC kits, presumably free of the contamination, were coming soon, an increasingly fleeting promise that continued to freeze the labs in place, waiting for the CDC's new reagents to arrive. Many labs were capable of developing their own molecular tests; however, it would take time to validate these home-brewed tests, and, hearing the CDC's assurance that help was around the corner, most labs chose to wait.

While this was going on, one of the CDC's contract manufacturers, Integrated DNA Technologies, Inc., or IDT as it was known, decided to start making its own primer and probe kits using the blueprints that the CDC had already published on January 27. This was the key component that labs needed in order to assemble their own test kits and no longer have to rely on the CDC. Initially, IDT began accepting orders for a version of that panel that could be used for research use only, and not for testing and reporting results to patients.[15] However, using the primers and probes, sophisticated labs could now assemble their own tests using the CDC's original protocol. And IDT had made a reliable set of these components that could be mass-produced. With the IDT kits, labs would now have all of the components they needed to start testing patient samples.

IDT was primarily a supplier of custom nucleic acids for the research community, but it also had a contract with the CDC to help the agency manufacture reagents that the CDC would distribute through channels like the International Reagent Resource—a consortium established by the CDC years earlier to provide genomic components that scientists use to carry out basic research and to make diagnostic tests.[16] While IDT wasn't intending to become a contract manufacturer for labs that were now trying to assemble their own tests for COVID, the company found itself in that role. And the FDA, which had been prodding the CDC to turn for help to one of its contract manufacturers, seeing the testing shortage growing more acute, now stepped up its efforts. The CDC and the FDA contacted IDT on February 25 to discuss IDT's ability to ship more kits that could fall under the CDC's testing protocol. IDT shipped its primer and probe kits to the CDC that same day to undergo closer evaluation, and four days later, on February 29, the CDC agreed to perform quality checks on the IDT kits, to support their use by labs that wanted to use these kits to make their own tests based on the CDC's protocol.

After the CDC reviewed IDT's kits and certified their quality for release, the FDA announced during a March 2 webinar that any lab that already had a kit from IDT's initial lot—the first run of kits that were initially intended for research use only—could now use those same kits to start testing patients. The IDT kits had now been qualified by the CDC and authorized by the FDA for use in clinical practice.

The CDC had been willing to put the IDT primer and probe kits under its existing FDA authorization for just the public health labs, and now the FDA asked the CDC to broaden it to allow the kits to be distributed to other labs. Initially, the CDC didn't want to take responsibility for kits that would be distributed under the CDC's emergency use authorization but would be run by commercial and academic labs that the CDC didn't directly engage with. The agency felt it couldn't supervise how the tests would be performed if they were being used outside of the public health labs.

However, the FDA decided to extend the authorization enabling the use of the IDT kits to all laboratories, not just the public health labs, even though the CDC hadn't requested the expansion. The FDA now urged the CDC to ship the IDT kits to clinical labs outside the CDC's public health network. This finally happened on March 4, with shipments to labs in Washington State and Northern California, where the testing shortages and the clinical needs were especially acute.

The IDT kits would fill the void that had been left by the CDC's sequential failures with its own kits.[17] By mid-March, IDT had a sufficient number of primer and probe kits, authorized by FDA, to make it possible to manufacture about 10 million tests that followed the CDC's protocol. Most people assumed that the testing hole was filled when the CDC finally got its own test working. In reality, it was IDT kits that supplied the market. The contract manufacturer shipped kits in such massive quantities that they became the backbone of the nation's testing supply for weeks. The episode showed the merit in turning to a contract manufacturer, and the early opportunity that was lost when the CDC decided to do these things in-house.

The whole experience was a massive failure. So, what happened to the CDC's own test?

The labs learned much later that the third and extra reagent that CDC included, the N3, was contaminated with coronavirus RNA. But N1 was also contaminated, although at a lower level. At some point in the manufacturing process, someone had probably walked from a dirty room, where scientists used RNA, into a clean one where they assembled the CDC kits, and carried the sticky strands of genomic material with them. Then some pieces of the RNA had gotten into both the N1 and the N3 components. This meant that two of the three primers that were designed to tell whether the test was detecting SARS-CoV-2 were themselves contaminated with strands of coronavirus genomic material. As a consequence, when the test registered a positive result, it wasn't detecting COVID. Instead, it was cross-reacting with itself, detecting the bits of RNA that had accidentally seeped into the system's reagents.

A later federal review would conclude that the CDC had failed to exert proper quality control over the kits in spite of "anomalies" that were discovered during the manufacturing process and were known to the CDC staff in charge of the program.[18] The CDC took a lot of risk by insisting on making the synthetic genomic material for the positive control component of the test kits in the same building where the agency would be assembling the completed test kits. Performing all these tasks in the same facility violated careful lab practices.

Manufacturing the synthetic SARS-CoV-2 RNA that would serve as a "positive control" should have been outsourced to a lab that was disconnected from the CDC facility that was assembling the completed kits. It was critical that the two processes be kept far apart. RNA can be hard to work with. It attaches to surfaces and spreads easily.[19] The risks of doing all these things in the same facility are well known in the lab industry and a common source of problems. As one research paper warned: "Even when dedicated rooms and directional movement is employed, false-positives can still occur. In these circumstances, the source of contamination can often be linked to the extraction of other positive samples in the same room at the same time."[20] But the CDC had decided not to outsource the production of these "primes," as they were called, because the agency worried that using a contractor would add too much time to the process. It was estimated that partnering with another lab or manufacturer (like IDT) to make the primes at the outset would have added about ten days onto the process. Given the urgency of getting a test to patients, and to save those ten days, the CDC decided to do everything inside its own lab complex, but mistakes made in the process caused the agency to lose six critical weeks.

Like the setback that the CDC had with its test for Zika four years earlier, the agency may have created the conditions for failure by overengineering its test for COVID and then being wedded to that more complicated design even after problems arose. The complex test design necessitated a more complicated manufacturing process, creating more opportunity for error. Years earlier, the CDC had taken similar risks with its test for MERS, only to suffer a low-level contamination that ruined some of the MERS test kits the agency had manufactured and already shipped to public health labs. That time, the CDC also decided to make segments of MERS genomic material (that would serve as the positive control) in the same facility where the agency was also manufacturing other test components. The segments of synthetic genomic material managed to seep into the MERS test kits even though the various constituents were being made on different floors of the CDC's building.[21] The agency's problems with its test for MERS should have been a lesson on just how easy it was for RNA to spread inside its manufacturing facility.

A year later, the FDA staff said that they believed the contamination of the COVID test kits originated in the CDC's respiratory virology lab. That was where the CDC was not only assembling the test kits for public health labs, but also where the CDC was doing some of its own clinical work, processing patient samples that doctors had sent into the agency for testing. Those two functions—assembling test kits and screening patient samples—should also have been performed in completely separate facilities. The coronavirus RNA may have jumped from one of the patient samples that the CDC was processing and into the manufacturing process for the kits it was assembling in the same facility. Making matters worse, the CDC later acknowledged that it hadn't put in place proper controls and standard operating procedures to fully prevent inadvertent contamination. In a February 19 email, a scientist at CDC's core lab, Dr. Nicky Sulaiman, told Lindstrom that another lab employee "had agreed to also decontaminate all . . . areas and instruments to be used in further manufacturing of the reagents," after the problems first arose.[22] The email suggested that the CDC knew that the lab might be contaminated.[23]

When the CDC made the first batches of test kits, the respiratory lab was probably still clean. So the first runs of those kits were assembled without any contamination. The CDC retained this first batch of kits for its own use, to process patient samples that were being sent to the agency by doctors. But then, when the CDC assembled later batches of kits, to be shipped to the public health labs, a contamination occurred. When the agency realized the setback, and went to make new batches of kits, the contamination persisted. This scenario probably explains why the CDC was able to continue to test patient samples in its own lab without any problems, but outside labs weren't able to get the CDC test kits to work for them. By keeping that first run of test kits, the CDC had unknowingly retained for itself the only tests that were free from contamination.

The inability to firmly establish where the contamination occurred underscored the loose controls in place. Some close observers would never be comfortable ascribing the initial contamination just to the virology lab. A federal health official told the *Wall Street Journal* that the protracted problems suggested that the complications might not have been N1, or N3, but a more fundamental flaw with how the test was designed.[24] One CDC official told me that the third primer was "self-dimerizing," meaning it cross-reacted with itself.[25] This means that the primer basically bound to itself rather than to the target RNA in the SARS-CoV-2 virus. However, this theory wouldn't explain the problems with N1, nor does it account for why other developers who used the CDC design to manufacture their own tests didn't have the same problems.

The CDC issued its own internal report, which also identified the virology lab as one potential source of the contamination. But the CDC report stopped short of concluding that this lab was the cause of the problems. The report's uncertainty underscored the challenges with the agency's controls. The CDC's after-action report couldn't settle on where the troubles originated. HHS wouldn't release that report to the public, and it was never widely shared outside the agency. In that report, the CDC seemed to agree that the virology lab was not following its own protocols.

CDC leaders insisted that the test never registered a "false positive"; instead, agency officials told HHS and the White House that the test was registering an "indeterminate" result. That seemed like a distinction without much of a practical difference and appeared to many inside the administration, who heard the CDC's argument, as a defensive posture that reflected a measure of the agency's indifference to the scope of the failure. The bottom line was that when the test was being used to screen sterile water, it shouldn't have found any coronavirus RNA. The fact that it was registering a positive result showed that the test couldn't be trusted.

An after-action report by HHS would find that "time pressure to ship test kits out quickly—and before [quality control] had been conducted on them—might have compromised" the ability to "identify certain anomalies in data and realize the possibility of contamination."[26] The review also found that the CDC's lab practices "may have been insufficient to prevent the risk of contamination."[27] While this was going on, the public health labs continued to struggle with a decision of their own. Should they pursue their own lab developed tests and seek authorization from the FDA to use them on patient samples—a process that could have taken weeks? Or should they wait for the CDC to fix its own kits? With the CDC saying each week that the problems would be fixed in another week, most labs decided to wait, and more time was lost. The CDC had tethered the nation's public health labs to its faulty products for more than a month.

Many academic labs and commercial manufacturers were also, understandably, waiting to see what happened with the CDC test. They didn't immediately try to fashion their own tests or test kits. The CDC was controlling the space. Moreover, once the CDC obtained SARS-CoV-2 viral samples from the infections that emerged in Seattle, it initially wouldn't share them with labs that were equipped to develop their own diagnostic kits. The agency didn't view it as a part of its mission to assist these labs.[28] It would be weeks before commercial manufacturers could get access to the samples they needed, and they'd mostly have to go around the CDC. One large commercial lab would obtain samples from a subsidiary in South Korea.

Inside HHS, Secretary Azar and his team continued to hold a belief that the CDC was going to manufacture kits that would be widely shared, something the agency had never done before. The focus on the CDC thrust the agency into a challenging position for which it wasn't prepared. The CDC had never manufactured kits for use beyond the public health labs. Even in 2009, when the CDC quickly fielded a test for H1N1 swine flu (making a flu test was a much more straightforward endeavor), the agency supplied these tests only to the public health labs and only in quantities sufficient to satisfy the limited needs of those labs.[29]

In the case of SARS-CoV-2, even if the CDC had distributed its kits flawlessly, the US would always have needed a way to manufacture massive quantities of tests for the rest of the clinical market. The process for getting those tests could have started at any point. Commercial manufacturers had to get in the game. Certainly, when it became clear that the CDC was having problems, that process should have gotten under way immediately. The shortcomings weren't all on the CDC. The leadership at HHS should have stepped in much earlier.

By the time HHS gave up on the CDC's test and turned to the commercial and academic labs by the end of February, the window for slowing the epidemic had probably closed. By this point, it would take too much time to ramp up the testing capacity of independent labs. We would have needed these labs to get into the market much earlier. It takes weeks, and often a few months, for labs to develop and fully validate a test for a novel pathogen. It requires the development of specialized reagents, and the creation of protocols for fitting these tools into precise kits for testing. This is how you ensure that a test is generating accurate and reproducible results. To identify and contain spread, we needed a reliable and widely available test in February and certainly by March. We didn't have one. The SARS-CoV-2 virus was now everywhere.

The manufacturing problems at the CDC were compounded by another challenge. This time, it was a policy obstacle. It materialized on February 4, when Secretary Azar invoked authorities under section 564 of the Federal Food, Drug and Cosmetic Act, a necessary step that followed his earlier declaration of a public health emergency.[30] The FDA provision gave HHS emergency powers to advance medical products to combat SARS-CoV-2. Now, products intended to address the crisis would be able to come to market under an emergency use authorization. However, there was also now a Catch-22 when it came to getting more tests into the market.[31] Usually, high-complexity labs operated by companies like Quest or LabCorp, or by academic institutions, are free to fashion their own laboratory developed tests, known as LDTs, for detecting a viral pathogen. These LDTs are subject to the FDA's oversight, but for most of these tests, the agency has exercised "enforcement discretion" and has chosen not to apply regulatory requirements—so long as the LDTs are run inside the sophisticated labs that assembled them and vouch for their quality and precision, and the tests are not turned into kits and resold to other labs.

But when it came to COVID testing, things changed once the public health emergency was declared and section 564 was invoked. Now SARS-CoV-2 became a special pathogen, and any test for it subject to FDA oversight, whether it was an LDT developed by, and used inside, a sophisticated lab, or a test that was packaged into a kit and widely distributed to multiple labs.

This was the intent of Congress and the general framework set forth by the Pandemic All Hazards Preparedness Act, a law passed in 2006. The policy was not new; this approach had governed tests used in the setting of public health crises for years. It was well understood that LDTs that screened for a novel virus would be subject to FDA review once that pathogen became the subject of a public health emergency, and the same policy had been in place during the Zika pandemic and H1N1 swine flu.[32] In fact, COVID would be the seventh time that LDTs would become subject to active FDA oversight during a public health emergency. The public health rationale for having the FDA regulate these tests was straightforward: having accurate tests would be an important part of stemming a new contagion, so Congress wanted the FDA to make sure these tests would be reliable.[33] Not all labs provide the same level of oversight, and variability in the quality of LDTs has created public health problems. When labs would eventually start marketing their own lab developed tests for COVID, the FDA would find "design or validation problems" with 82 of the first 125 EUA requests and "several have been denied authorization," according to Dr. Jeffrey Shuren, the head of the FDA's device center.[34] The scope of the problems surprised the staff at the FDA, especially given how well established PCR-based testing was in the market. So, under the law, now that section 564 had been invoked, clearing the way for the FDA to grant emergency-use authorization for medical products aimed at addressing the pandemic, laboratory developed tests would also need to seek such an authorization. It was a speed bump to getting these tests quickly on the market. And the leadership at HHS was surprised by the conundrum, unfamiliar with the policy, or the implications that once the declaration was made under section 564, the lab developed tests would come under the FDA's oversight. Nobody envisioned the problems that would occur with the CDC kits, how fast the pandemic would spread, and the need to get the commercial and academic labs quickly into the testing game to fill the void.

By late February, the CDC's delays were so acute that Scott Becker at the Association of Public Health Laboratories and Dr. Grace Kubin, the lab director at the Texas Department of State Health Services and past president of the public lab association, sent an extraordinary letter to FDA commissioner Dr. Stephen

Hahn on February 24, asking him to grant enforcement discretion that would let the public health labs "create and implement" their own laboratory developed tests. "We are now many weeks into the response with still no diagnostic test available," they warned.[35] That same day, I posted a lengthy thread to Twitter outlining specific steps that the FDA could take to get lab developed tests into the market, using an established playbook. I wrote, "FDA could adopt a tailored approach, recognizing the expertise of academic labs that develop and run highly complex and novel tests every day, and address the expectations for test design, validation, and performance to help labs pursue EUAs."[36] To create such a pathway, I said, the FDA could outline the criteria that labs would need to meet in order to offer laboratory developed tests for emergency use. Labs could then start using their tests with patients so long as they could self-certify that they met those standards. Eventually, labs would submit EUA requests to the FDA proving their tests' performance, but in the meantime, the FDA could allow the tests to be used to fill a critical shortage.

Ultimately, a new framework was advanced. The FDA issued a guidance document on February 29 that would bake in the basic elements of that policy prescription.[37] But in the first week after the new pathway went into effect, only six clinical labs came forward to say that they had validated their own tests and would begin using them. Developing tests takes a lot of lead time. It can take even longer to get a testing service to scale. LabCorp and Quest both announced their testing services on March 5, which then started a few days later. Initially, the two commercial labs were able to run a combined total of just 2,500 patient samples a day.[38] For their part, the public health labs—which also started to run their own lab developed tests—were offering another four thousand tests a day across about one hundred sites. It was a lot—but still only a fraction of what was needed.

The conundrum exposed how fragmented our national testing enterprise is. America is unique in our reliance on lab developed tests for a large portion of our diagnostic testing. In most countries, labs run mass-produced test kits that are made by major device manufacturers. In the US, in many instances, labs assemble their own tests using basic starting materials. Part of this stems from the way labs are paid. Sometimes the margins are better when labs assemble their own tests rather than run kits that they buy from commercial manufacturers. So the LDTs are an important part of the total testing supply in the US. Yet even the most sophisticated lab could run, at best, a few hundred tests a day if it was using a test that it had developed in-house. In a public health emergency, we would need to deploy mass-marketed kits that could run on the large testing platforms such as the LightCycler machine made by Roche, or Hologic with its similar platform called Panther. Instead of running a few hundred tests a day, these kits and machines would give labs the capacity to screen thousands of patient samples each day. This was the only way to quickly ramp up the massive amount of testing that would be needed to keep up with a fast-moving pandemic. The problem is that nobody ever planned for such an event.

European and Asian countries rely much more on these kits and high-capacity machines, rather than the more bespoke tests that US labs develop in-house and can't be run at the same volume. But with the CDC struggling to deploy kits that could run on these high-throughput testing instruments and with no other kit manufacturers in the market, it became an acute problem when the labs were frozen out of the game. Absent the mass-produced kits, the LDTs were the only other option.

By March, there was an ominous refrain that COVID was spreading more widely than we knew, and we needed to get more tests, any tests, into the market. In New York, Mayor Bill de Blasio and his health commissioner publicly pleaded with Commissioner Hahn to allow the state and not the FDA to certify any New York lab that met the state's already-stringent regulatory criteria. It was a way to fast track the city's ability to start using its own laboratory developed test. On March 5, de Blasio would write on Twitter: "We need the FDA to fast track approval for testing methods developed by private institutions. And we need it NOW."[39] The next day, Dr. Raul Perea-Henze, the city's deputy mayor for health and human services, wrote a letter to Hahn: "The importance of NYC having additional testing kits and expanding our testing capacity cannot be overstated," he said. That same day, de Blasio would use Twitter to urge again that "we need the FDA to approve faster, more efficient testing. The faster the results, the faster we can limit the spread."[40] Three days later, with no apparent movement, the mayor would again take to Twitter, writing, "To the federal government: we need you to do the simplest thing in the world and have the FDA approve automated COVID-19 tests. We're bringing back hundreds of results. We could be bringing back thousands. The FDA could do it with the stroke of a pen TODAY."[41]

I understood what de Blasio was trying to do. There were times when I was the FDA commissioner that I would use Twitter as a way to draw attention to actions that were frustrating our public health goals, and I saw Twitter as a way to instigate steps by others that were important to us for achieving our mission. I put out a series of tweets calling on Internet companies like Facebook to start cracking down on ads for opioids.[42] I had used Twitter repeatedly to call on vaping companies like Juul to take specific actions to curtail youth abuse of their products. One tweet that helped support a specific action was during an ongoing outbreak of *E. coli* in romaine lettuce. I called on the produce growers association to implement better tracking and tracing of their products, including changes to their packaging that would make it easier for regulators to isolate which growing region and farm was the source of any products that were found to be contaminated.[43] Up until that point, the industry was reluctant to implement packaging that could allow produce to be easily traced to a specific farm, which made it harder for the FDA to isolate the source of an outbreak and implement more targeted recalls.[44] I believed my messages on Twitter, urging the labeling changes, helped break a deadlock.[45] Twitter was a nontraditional communications tool for the FDA. I was the first FDA commissioner to use the platform, the first commissioner even to have an account.

On March 13, New York governor Andrew Cuomo spoke to President Trump and Vice President Pence about the state's ongoing testing needs, "and they finally approved New York's request made earlier in the week to be able to approve any lab in the state to do COVID testing—a real breakthrough that practically took the FDA out of the lab-approval equation for New York," Cuomo would write in his autobiographical account of his role in managing his state's crisis.[46] Cuomo said he had also appealed to Pence for help in securing permission from Commissioner Hahn. The FDA move was indeed significant; it gave New York almost sole authority to authorize its own labs to conduct COVID testing. But I think Cuomo was wrong to assert that his phone call with Trump finally unlocked the action. The FDA's career staff had already finalized the first step in that plan on March 12, giving New York's Wadsworth lab permission to authorize tests made by the twenty-eight other labs that held New York State lab permits. The FDA staff's decision would give New York State sole authority to regulate laboratory developed tests that were advanced by the sophisticated labs in its own state. Basically, New York could act as its own FDA in authorizing new tests. A few weeks earlier, the FDA had already enabled the first of these labs to begin using tests they had developed themselves.[47] Now New York State would be in the driver's seat.

I was in the Oval Office with President Trump during that phone call with Cuomo. I had been asked to meet with the president and other members of the task force that day to discuss the US response to the crisis. The governor was calling the president to thank him for helping New York City secure a field hospital that was going to be constructed in the Javits Convention Center. The issue of testing didn't come up, probably because the FDA's career staff had already worked out all the issues on their own, interfacing directly with New York State health authorities. The FDA ultimately expanded this opportunity to other states in a March 16 guidance, outlining the steps states could take to earn a similar power to stand in for the FDA in authorizing their own labs.[48] In the end, only eight other states would opt to do this. Testing would ramp up very slowly.

The availability of tests wasn't the only obstacle. The CDC's guidance on who could qualify to be tested for SARS-CoV-2 was deliberately parsimonious, designed to make sure that demand for testing would match the meager supply of available test kits. In a February 19 presentation to state health officials, the CDC defined who they thought ought to be tested: "You had to have had close contact with someone confirmed to have COVID-19, or to have traveled from China and then had respiratory symptoms and a fever at the same time."[49] Yet, by then, health officials were well aware that patients could be infected with SARS-CoV-2 but not mount a fever.[50] Three weeks earlier, NIAID director Anthony Fauci had cited the asymptomatic spread as one reason why he had withdrawn his earlier opposition to bans on travel from affected regions and decided to back the restriction on certain travel from China that took effect on January 31.[51] But the CDC maintained its overly narrow testing criteria well into March. "By early March, it was evident that the virus should have been spreading in the U.S.," the cofounders of *The Atlantic*'s COVID Tracking Project wrote a year later. "Yet the CDC's stringency about who could be tested and the lack of clear testing data meant many federal leaders simply didn't acknowledge that reality."[52]

The WHO also maintained a narrow case definition at the outset of the pandemic and was more explicit about their rationale. It needed to conserve scarce testing resources. But recognizing that the virus was spreading in communities, the WHO updated its guidance at the end of January, broadening its criteria for who should get tested. The CDC, by contrast, would stick with its tight restrictions much longer. That narrow framework achieved its ostensible purpose. HHS officials would boast that they were able to process all of the tests that were sent to the CDC. This was technically accurate. Of course, the claim didn't account for all the requests for testing that the CDC simply refused to accept or the providers who decided not to make an appeal because they already knew that it wasn't going to meet the CDC's high bar. The agency took deliberate steps to enforce guidelines that would make sure it didn't receive more samples than its single lab could handle. In late March, the CDC went so far as to edit an article that was slated for publication in a science journal, to remove a passage inserted by a Washington State public health official that called for widespread testing at senior assisted-living facilities. That statement encouraged more testing than the CDC was prepared to allow or was able to handle at the time. The edits to the article were viewed as a measure of the steps that the CDC was willing to take to constrain demand for tests to the small number of samples that the agency was able to process. "I would be careful promoting widespread testing," the CDC editor noted, according to drafts of the manuscript.[53]

Clinicians and local health officials would later say that they often had to press CDC officials for days to get the agency to accept a sample from a patient that doctors suspected of having COVID. The first case of community spread diagnosed in Solano County, California, on February 26 would ultimately lead the CDC to rewrite its testing guidance, because it showed the public health consequences of the overly narrow testing policy. By the time this patient was tested and diagnosed with COVID, more than two hundred workers at two hospitals had been in contact with the infected patient—and thus had risked exposure to SARS-CoV-2.[54] At one of the hospitals, so many healthcare workers had to be placed into quarantine that the facility had to temporarily close its intensive care unit because it ran out of available staff.[55] The CDC's criteria for granting a test were so strict that the agency reportedly refused to test a nurse who fell ill after treating a COVID patient.[56] The implications of this constrained testing policy were far-reaching. The CDC's narrow criteria pretty much removed any chance for using testing and tracing to identify the community spread that was building in cities like New York. The testing approach was not rooted in clinical considerations. It was born of necessity. The CDC couldn't field a diagnostic test that could meet the needs of providers.

There were other obstacles to getting more tests manufactured and deployed into the field. The commercial, public health, and academic labs also didn't have access to the genomic material or inactivated viral samples they needed as positive controls to use for validating tests that they could develop on their own. If laboratory developed tests were going to fill the void, then someone had to get the labs the components they needed.[57] And as for the large manufacturers and big labs, they were unsure if they'd be reimbursed for making and running their own LDTs. They had been stiffed by government funding agencies after Zika, and they hadn't forgotten.[58] The CDC was also exerting its intellectual property rights over its test design. If labs wanted to develop their own tests, they would need to copy the CDC's design, which the agency said was proprietary. Negotiating a right of reference to the CDC's recipe added additional complexities and delays, especially for the commercial manufacturers that wanted to be able to market kits to laboratories. Taken together, these layers of obstacles and intricacies froze out of the market solutions that could have filled some of the void left by the CDC's own stumbles.

The CDC shouldn't have been charged with making test kits in the first place. The agency just doesn't have the requisite know-how to take on a major manufacturing challenge. It's not an agency structured to do things at this kind of scale or to respond to a major crisis. It's a high-science organization that does epidemiological investigations and careful research. This was a massive operational challenge. From the very outset, we should have turned to a contract manufacturer. Developing a test that can be run in your own lab is a very different skill than developing a kit that can be run inside anyone's lab. The commercial device makers have the requisite expertise to manufacture test kits that can be run in different labs and on different platforms. Some flexibility needs to be built into the design of kits to allow them to be easily fitted on different kinds of testing equipment. This is a development and manufacturing skill that the CDC simply doesn't have. But it's precisely the work in which the large diagnostic kit makers have deep experience. Their representatives are in labs every day to understand which testing machines are present, and how kits can be made to work on the largest complement of these platforms. The CDC had no similar ties to the academic and commercial labs for which it was expected to suddenly provide test kits. Its culture didn't easily allow for the kind of cross-industry collaboration that had to occur. It's not what the agency did. It wasn't the task it was designed to do. It's not what it knew.

The CDC's single biggest breakdown in its response to COVID was the abortive rollout of its COVID test. In their defense, officials at the CDC pointed out to me that developing tests for commercial and academic labs was never their job. It wasn't what they had ever done before. Their experience was in working directly with state public health labs. They had done it successfully in 2009 in response to H1N1 swine flu. But once the agency ran into challenges with their development of that initial test for SARS-CoV-2, for weeks after, there still wasn't another test being pursued in the US other than the effort that was under way at the CDC. Seeing the CDC's challenges, the leadership at HHS needed to mount a concerted effort to develop an alternative, pulling together other public health agencies with relevant tools such as FDA, NIH, and the Biomedical Advanced Research and Development Authority. There was no plan B. It was almost as if once the CDC tried—and failed—to develop a test for COVID, the focus shifted exclusively to remedying the CDC test even though the agency was never meant to, and never would be able to, supply the entire market with testing. When Secretary Azar asked the CDC why the agency wasn't shipping its test kits to private hospitals, CDC officials had to tell him that the agency never provided test kits to the private sector, only to state labs. But in this case, the CDC was expected, for the first time, to supply test kits directly to commercial and academic labs.

Into the late spring, CDC officials still found themselves explaining to the leadership at HHS that it wasn't the CDC's historical role to work with commercial and academic labs, or to help industry develop test kits. Every year the agency develops a test kit for flu, but it doesn't send this kit to hospitals or commercial labs—only to state public health labs. Instead, the needs of hospitals and private labs for flu tests are met by commercial manufacturers.

We had repeated openings on testing; chances to start turning things around before the crisis spiraled out of control. One such opportunity came on March 4, when the entire lab industry gathered in Washington, DC, for the annual meeting of the American Clinical Laboratory Association. The trade association conference straddled a turning point when the pandemic was becoming firmly rooted, and there was a palpable sense that the US would be next to fall to COVID. Now, in the ballroom of the Grand Hyatt Washington were the chief executives of all of the major commercial labs. It was probably the last conference that these CEOs would attend in person for more than a year. Standing behind a glass podium and flanked by four potted trees, Dr. Stephen Hahn of the FDA would deliver a twenty-minute speech of which he would devote a mere five hundred words to the pandemic. He reminded the lab executives of their regulatory obligations if they wanted to develop tests for COVID and mentioned some accommodations that the FDA had mapped out to ease that process. It was largely a standard recitation of what was already on the FDA website. It was a missed opportunity. By that point, we needed a stronger call to action.[59]

I saw the annual meeting of the lab industry as a critical turning point. It was an opportunity to rally the commercial labs and test makers around the effort. The lack of a clear call to action was emblematic of a larger organizational problem. No single agency or person really owned the problem. And HHS leaders never convened the department's operating divisions, including the FDA, the NIH, and the CDC, into an organized effort to tackle the key challenges we faced.

There was plenty of precedent for the FDA commissioner stepping in to quarterback an industrywide response in a moment of public health crisis. One came in 2004, when I was working as a senior advisor to FDA commissioner Dr. Mark McClellan.

There was pressure on the US government to allow the intellectual property protecting drugs for the treatment of HIV/AIDS, which had been developed by western manufacturers, to be appropriated by generic drug companies, mostly located in India. These generic firms had promised to expand treatment options in Africa by manufacturing less costly versions of the pills.

The entire public health community was committed to increasing the number of HIV patients in Africa who could be offered treatment. But some of the knockoff drugs were of suspect quality, and those of us working at the FDA at that time worried about the prospect of using American funds to buy millions of doses of what, we feared, could be inferior drugs.[60] We believed that all patients deserved the same high-quality treatment options available to Americans.

So we led an effort to bring the drug industry together with the US government, whereby drug makers would allow their patented medicines to be developed into cheaper combination pills that would be easier to distribute and use in the more austere settings found in many public health clinics in Africa. The FDA, in turn, would put these new products through a process where the combination medicines could be reviewed; if they met the agency's standards for approval, they would receive "tentative" FDA approval. Since the drugs still had patents blocking their commercialization in the US, they couldn't win full FDA approval. But this provisional status would be sufficient to ensure the high quality of these copies and would in turn pave the way for the US government to legally purchase and distribute the drugs in Africa.

As part of this effort, the FDA agreed to publish a guidance that would outline all the circumstances where the FDA believed there was already sufficient clinical data to support the use of different drugs in combination. Any drug maker who could successfully combine different medicines to make one of these combination drugs could apply for tentative FDA approval just by demonstrating that the new combination pill was chemically the same as each of its individual components and would deliver the same amount of medicine. Drug makers wouldn't need to also undertake the time and cost of developing clinical data to prove that these combinations were effective, since the FDA had already specified that in advance.

There were other aspects to our overall plan to promote greater access to HIV/AIDS drugs in Africa. As another part of the effort, I asked three drug makers to join a hastily arranged meeting at the FDA, where we sought the cooperation of Merck, Gilead, and Bristol Meyers Squibb to pool the intellectual property they each owned for the components of what we believed could be an ideal once-daily pill for the treatment of HIV. We promised the companies an efficient development and review process if they would join together to develop the pill, and they agreed. The drug they would make was Atripla, one of the most important HIV medications to be launched in the last twenty years.

Taken together, the process we set in motion became a key part of the President's Emergency Plan for AIDS Relief or PEPFAR, and I believe that combined FDA effort helped save millions of lives. PEPFAR was a singular public health achievement.

Sixteen years later, with COVID spreading across America, I urged White House officials to meet with the lab industry executives while they were in town. Dr. Deborah Birx, who served as the White House coronavirus response coordinator, was urging the president and vice president to do the same thing, and Vice President Pence did meet with the chief executives of the major lab companies, though Hahn did not attend.[61] The very morning of Hahn's speech, I did what I could to encourage more concerted, urgent action, especially with the entire lab industry assembled in Washington at the moment that the epidemic was about to explode. So, following a television appearance on CNBC, where I had already raised the issue of getting more COVID tests into the marketplace, I sent a message on Twitter from the backseat of a car on the West Side Highway in New York. I wanted to make the stakes abundantly clear:

"On #coronavirus testing: The nation's big clinical labs meet in Washington today for a convention and their CEOs are in town meeting with federal elected leaders. This is their moment. They ought to step up. They may be judged by what they now do."[62]

CHAPTER 8

NOT ENOUGH TESTS AND NOT ENOUGH LABS

There were two main technologies available to test for SARS-CoV-2: antigen-based tests (which measure whether you have whole particles of live virus in your body, called a virion) and molecular tests that rely on PCR (and measure the presence of the virus's RNA). Each has pros and cons. Besides their lower cost and greater convenience, antigen-based tests can offer some clinical advantages over molecular tests, which are otherwise considered the gold standard for diagnosing infection.

Once you've been infected for eight or ten days, you might not have live virus in your body anymore. The immune system has already attacked and destroyed whatever virus you had. What's left, at that point, is often dead virus, RNA fragments, that can no longer grow in viral cultures, or infect someone, even though you continue to shed these viral particles.[1]

Sometimes the PCR tests can still detect genetic fragments, and return a positive result, even though you no longer have active virus in your system, and you're therefore no longer contagious. By contrast, antigen-based tests are likely to return a positive result only when there's whole, live virus still in your respiratory secretions. The general criticism of antigen tests is that they may not have the same sensitivity as PCR, meaning they can miss positive cases. But a good antigen test can have high specificity, meaning that when it detects virus particles it's more likely to be picking up live SARS-CoV-2, and you're more likely to have an active infection.

The PCR tests sometimes try to correct for this challenge by quantifying the amount of RNA found in a sample. When the RNA level is very low, it may be an indication that the person is recovering from the infection and is no longer shedding live virus. Therefore, the patient is no longer contagious.[2] Or it may be because their infection is early, and the virus hasn't had time to replicate in high numbers and reach its full saturation. The antigen tests can take longer to develop and may not be available at the outset of a crisis involving a novel pathogen. But the tests can often be produced more cheaply and in larger quantities than molecular tests like PCR tests. The trade-off is that antigen tests are generally less sensitive. They may be more likely to miss positive cases. The key is fitting the right test to the right public health purpose.

What you need is an all-of-the-above approach that sets in motion a deliberate effort to build the four layers of testing needed in a crisis. First are laboratory developed tests that can be deployed quickly to allow you to get initial tests into the field. Next are manufactured test kits that are based on PCR and can be run on high-throughput testing instruments and can allow high-volume testing at clinical laboratories. Third are point of care molecular tests that many doctors already have in their offices and can be updated to test for SARS-CoV-2. Finally, inexpensive antigen tests can also be used at the point of care or deployed to schools and workplaces, and eventually can be used as at-home tests. These are the four layers of a resilient system. A plan to develop each of these testing platforms should have been under way from the start. The part of the system that got short shrift for a long time was the non-PCR tests. We rely far too much on lab testing in this country. Government payers like Medicare favor lab-based tests over point-of-care diagnostics because they're accustomed to paying for services. This has created a healthcare system that is biased against tests that can be put into the hands of consumers, which is precisely what's needed in a pandemic.

The first antigen test directed against COVID was a lateral flow test developed by Quidel and authorized by the FDA on May 9.[3] Similar to the way a home pregnancy test works, a lateral flow test for a virus uses antibodies bound to a nitrocellulose (paper) membrane. The antibodies are designed to bind to specific regions on a particular virus.[4] When a sample is added (in this case, the swab from someone's nose) the sample will flow along the test paper and then onto an absorbent pad where the antibodies are bound. If the virus is present in the sample, the antibodies will bind to the viral target, and together, the antibodies and the bound virus particles will continue to move along the test strip until the bound complex reaches a readable line. Other lateral flow tests would later be authorized by the FDA, including one by Abbott called BinaxNOW that's about the size of a credit card and initially sold for five dollars.[5] Upon its approval on August 26, 2020, it was made available in massive quantities owing to investments Abbott had made in manufacturing. A patient can swab himself or herself, add drops of some reagent fluid on the paper from a small bottle, and then fold the special paper over the nasal swab. A readable result appears in about fifteen minutes.[6] This test was adopted widely in the fall of 2020 and was also being used to screen people entering the White House. By April 2021, it was finally sold directly to consumers, without a doctor's prescription, in a pack of two tests for about twenty-five dollars.[7]

Each type of diagnostic test has an important role when we take the time to understand the context in which each can be optimally used. A comprehensive approach to pandemic planning will take into consideration the different kinds of tests, how they should be best used, and when they can be made available after a new pathogen emerges. We didn't have a deliberate strategy to deploy these different layers of testing. Every company stepped into this market at different times, with little meaningful coordination. A real pandemic playbook would have mapped out these decisions in advance and seeded the development of these different testing layers.

Some government entity or official should have been charged with advancing these different options in a coordinated fashion. We should have prioritized the development of the point-of-care tests early on, and supported their production with US government funding, and guaranteed purchasing and reimbursement. Such an effort would have needed to be led by the secretary of Health and Human Services and would have required collaboration across the CDC, the FDA, and other federal agencies like BARDA and the Centers for Medicare & Medicaid Services. No one agency, or public health official, owned this challenge. We needed an all-of-the-above approach to get the four layers of testing into the market. We needed a coordinated response that deployed the right test to the right patient at the right time.

When antigen tests finally reached the market in large quantities, the federal government initially cornered the entire supply, buying more than 150 million of the BinaxNOW tests by September (spending $760 million) with no detailed plan on how to use them.[8] They shipped most of these tests into nursing homes, following a strategy to focus resources on preventing infection in the vulnerable elderly. But nursing homes found the tests a poor fit for their needs.

Many nursing homes worried that the antigen-based tests were not sensitive enough to screen an asymptomatic population and the tests would miss positive cases—a problem in a nursing home, where there are many vulnerable residents. If nursing homes used the tests to screen their staff, and then missed workers who might have asymptomatic infection and be able to spread the virus, the consequences could be grave. It was well established that many outbreaks in nursing homes were triggered by infected staff members who didn't know they were shedding the virus.

State and local health departments expressed similar concerns. A testing approach in these facilities needs to be as airtight as possible, since the risk of a single introduction of virus can be catastrophic. Thus many of the antigen tests went unused. A federal survey found that about 30 percent of thirteen thousand facilities that were provided the rapid tests hadn't used the equipment. During a period of time when regulations from the Medicare program required nursing homes to do routine testing if there was an outbreak in adjacent communities, hundreds of facilities didn't use the antigen tests they were given.[9] They stayed with PCR.

Because of the antigen tests' potentially lower sensitivity, the FDA initially authorized them to be used only to diagnose SARS-CoV-2 infection in people already showing symptoms, because that's what the data submitted to the FDA supported at the time. Individuals who are already symptomatic generally have a higher viral load in their secretions and are more likely to register a true positive result on these tests. By contrast, patients who aren't yet showing symptoms, or may never show symptoms, may have less virus in their secretions, so they may be more likely to test negative even though they are still harboring the infection.[10]

Despite these limitations, a lot of people still sought to use the antigen tests to screen largely asymptomatic populations at nursing homes, including some of the federal officials in charge of expanding the national supply of tests. Manufacturers weren't seeking emergency use authorizations for screening in asymptomatic populations, so the FDA didn't have a lot of data to support the reliability of the antigen tests in these settings.[11] The companies making these tests were worried that their technologies wouldn't perform as well under these circumstances. They didn't want to generate a whole bunch of data showing that the tests weren't that good at detecting the virus in people who weren't showing symptoms. In many cases, they didn't pursue these trials.

So, when used to test asymptomatic populations as a way to find and isolate those who could spread infection, the tests needed to be used with their limitations in mind. That said, they could still be effective tools for detecting infection in the general population, when used properly. The key was using them in the right settings—generally in lower-risk settings and with people who were at lower cumulative risk of bad outcomes from the virus. It was also important that the tests be used as part of a program that required regular, repeat testing of the same population. Modeling showed that frequent, repeated antigen testing could help prevent outbreaks in schools and other congregate settings like jails.[12] By repeatedly testing the same population, it increased the odds that a test would detect infection in a population. A test that might miss an infection on one day would pick it up when the same people were retested a day or two later.

This kind of serial testing is not as feasible using PCR tests, which are costly and hard to administer. In most cases, tests that used PCR required that samples were sent off to specialized labs. This meant that results could take a few days to be returned. That limits the usefulness of PCR tests for screening a population at work or in a school, where the goal was rapid detection to prevent the infection from being introduced into these congregate settings in the first place.[13]

Since the antigen tests were cheap and easy to use, they could have been added as another layer of protection to help prevent outbreaks, in addition to other mitigation that was already being taken, like masks, distancing, and symptom screens. But a nursing home wasn't the sort of lower-risk environment to which an antigen test was appropriately suited. The federal government had sent the antigen tests to the one place where they shouldn't have been used. With so many frail residents, it was a setting where you needed something that was as close to foolproof as possible.[14] You needed a zero-fail testing solution. And it was well known that the antigen tests could generate a fair number of false positives and false negatives.

The problems surfaced immediately. Multiple reports circulated of situations where nursing homes tested a patient with an antigen test, got a positive result, and then moved the patient into a COVID-only ward, believing the patient was infected. Then, they would find out on a subsequent PCR test that was run to confirm the result that the initial finding from the antigen test was a false positive. In the meantime, the patient had become infected with SARS-CoV-2 after being placed alongside COVID patients. Nevada's Health Department ordered nursing homes to stop using all antigen tests after reviewing results from 3,725 tests and finding too many false positives.[15] These were situations where patients tested positive on the antigen test but didn't have SARS-CoV-2.[16] Federal health officials deployed the antigen test to the one venue—nursing homes—where the generally lower performance of these tests was perhaps most poorly suited.[17] Not only did we have a shortage of tests; we didn't use the tests we had in an appropriate manner. It was another measure of the disorganized state of testing in the US and the lack of a comprehensive national strategy.

In October, a few months after federal health officials started to distribute the tests widely to nursing homes, the American Health Care Association, which represents for-profit nursing facilities, sent a memo to its members warning of the problems. "Due to the lower sensitivity and specificity of these test devices, not all state public health departments allow for their use, and many have certain requirements in place for using these tests appropriately," it said.[18] Yet federal health officials continued to promote the tests for nursing homes. "The nation needs more, not less testing, especially in congregate settings," Dr. Brett Giroir, the assistant secretary for health at HHS, said in a letter to state health officials on October 8. The ban on antigen tests that Nevada had imposed was "based on speculation. It may cost lives," Giroir wrote. "A false positive should not be a concern," he said. "It's a reality of the test."[19]

In the setting of a pandemic, precision matters. But for certain applications, speed, low cost, and ease of use can matter just as much. Some testing can often be better than no testing at all. If the test is going to be used to do serial screening of the same asymptomatic and, ideally, lower-risk population, the ability of an antigen test to detect infection will rise with the repeated screening. Used right, and used in a setting where the consequence of a false positive is lower, the antigen tests could be an important additional layer for reducing risk in congregate settings.

The following year, the FDA would create a streamlined path for antigen tests to be authorized for this sort of serial screening of asymptomatic populations. The agency would allow tests that had already proven that they have strong performance in testing symptomatic patients to get an emergency authorization for asymptomatic screening, so long as they were being used in a serial testing program such as repeat testing to screen a population at a school or workplace. Post-authorization, the test developer could then submit real-world evidence, characterizing the performance of the test when used for asymptomatic testing, to shore up the initial authorization.[20]

A good lateral flow test will not provide the same accuracy as a lab-based PCR test, but it can be much more accessible. For the purposes of screening a large population, the lab-based tests are not practical. We learned this when the labs became overwhelmed in the spring and fall. Unless you worked for the White House or a professional sports league, it was impossible to get access to PCR testing to conduct serial screening, as a way to prevent outbreaks in your school or workplace. We need an all-of-the-above approach when it comes to testing in a crisis, and we need to fit the right test to each purpose. But we'd never have enough of any test during the first year of the pandemic, and we didn't use the tests that we had properly.

For most of that first year, we would depend principally on molecular tests that relied on PCR. However, here again the lack of coordination across federal agencies and a lack of leadership from HHS hampered our response. On March 6, while touring the CDC's headquarters in Atlanta, President Trump told reporters that "anybody that wants a test can get a test."[21] A day earlier, the CDC had provided clearance for IDT to manufacture and distribute kits based on the CDC's basic testing blueprint for PCR tests to be done in laboratories. But kits were only one part of the equation. Tests are authorized for use on specific testing equipment, or "platforms," where the test's performance has been validated. The issue is that not every platform is found in every lab—a large commercial lab often has a variety of platforms on hand, whereas a small lab may have just one. And if a test is authorized for use on a platform that a lab doesn't have, it can't use the test.

So an equally critical issue, in addition to having enough test kits, is the installed base of lab equipment that exists in US labs and having the right kits to run on the largest complement of machines. The nation's installed base of testing equipment is a hodgepodge of testing instruments. Every lab decides what testing machines it wants to buy, and it ends up creating a patchwork of equipment. In the future, the US needs to support the establishment of a strategic network of labs with a uniform base of instruments that can be used to process the largest possible pool of test kits. Then, when an outbreak occurs, we can develop tests that will run across the largest number of installed testing machines. It was too late to take such a strategic approach for COVID.[22] Because the CDC mostly worked with the public health labs, it didn't know what equipment was in commercial and academic labs, or how to make the most of those machines.

This obstacle became apparent in early March, when COVID test kits started to finally flood the market, but access to testing wasn't increasing. On March 7, FDA commissioner Stephen Hahn said that he would soon be able to deploy 2.1 million COVID test kits into the market.[23] However, the number of test kits was not the most important metric. If only one lab was equipped to run those kits, and it could process no more than one hundred samples a day (the limit for many public health labs), then your capacity was one hundred tests a day. To track the actual availability of testing to patients, I launched a project at the think tank the American Enterprise Institute, to collect data from labs across the nation on how many tests they were able to process each day. The first update was posted on March 9. We calculated the daily COVID testing capacity in the US to be only eight thousand tests a day, most of it from the public health labs.[24] Labs all over the country started to report their capacity to us, and for a period of time, our running total was widely cited by federal officials and the media.[25]

These issues came into sharper view at a congressional hearing held later that week. Representative Mark Pocan, a Wisconsin Democrat, would pick up on our data in a series of questions he posed to Commissioner Hahn at a hearing held to review the agency's budget. "So, help me to make sure I understand this," Pocan said, "you said 989,000 people can get tested with the tests that are out there right now, but as I understand it, the processing, we can only process about 16,000 a day right now . . . it appears not truthful if we say there's 989,000 people that can get tested, but you can only process 16,000 a day, that's a real problem."[26]

Pocan was right. Touting the number of kits that had been manufactured only told part of the story. It obscured the broader challenges. The key was getting commercially manufactured kits into the market in mass quantities and making sure the kits were authorized to run on the widest set of installed testing instruments. A year later, Brett Giroir, who had left HHS following the presidential election, told CNN's medical correspondent Dr. Sanjay Gupta, "When we said there were millions of tests available, there weren't, right? There were components of the test available, but not the full meal deal."[27]

Testing kits mattered. But what also mattered was ensuring that the big labs could draw on their sizable capacity to process lots of specimens with the high-throughput machines that could run hundreds of tests at a time. Up until that point, most of the testing volume was being supplied by low-volume, lab developed tests, and the

IDT kits that were mostly going to the public health labs, even the largest of which didn't have enough testing machines and personnel to process more than a few hundred tests a day. The big commercial labs, with their large instrument sets and greater capacity to process samples, had to get into this market to reach the kind of testing volume we needed, but they couldn't yet turn on their systems. They got into the market late and, absent direction and coordination from the HHS secretary's office, they were still pursuing FDA approvals to start using their high-volume testing machines.

That, more than anything else, was holding back testing supply in the nation.

HHS could have jump-started this market in January or February and paid commercial manufacturers to make kits in large volumes, prioritized the clearance of tests that could run on the widest array of instruments, and guaranteed reimbursement to the commercial labs for the tests by agreeing to buy a certain volume of testing in advance. However, HHS was still looking for the CDC to solve the problem. It seemed like once the CDC had stepped forward and tried to develop a test, the HHS leadership, including Secretary Azar, would continue to focus on the CDC as the fulcrum of this challenge when, in practice, the CDC was not going to be able to supply the market even if it hadn't run into problems. The only way to solve the testing shortage was to get the big commercial labs fully engaged. The professional staff in the FDA's device center were independently working to validate the high-throughput instruments, and they had privately reached out to some of the commercial labs and test manufacturers, to try to coax them into the market, but the FDA's effort was not part of a coordinated strategy. Some of the agency's career staff would try to seize the initiative.

Getting the high-throughput molecular machines authorized to test for SARS-CoV-2 using PCR would come late in the game. By March 13 the first manufacturer, Roche Diagnostics, got FDA authorization for one of the high-volume Cobas PCR platforms that were the backbones of many academic and commercial labs.[28]

About 122 of the Roche testing machines were installed across US labs.[29] Other companies, including Thermo Fisher, Abbott, and Hologic, would also get EUAs for their own platforms in the weeks that followed, often authorized within a day of submitting a request to the FDA.

Authorizing the new tests was a critical step. But for now, it was too little, too late. The epidemic was already well under way. The lack of accessible testing would contribute to the excessive spread of SARS-CoV-2 and to the disproportionate burden borne by Americans who worked in lower-wage jobs, where they came into contact with the virus but had no way to use testing to avoid unknowingly bringing COVID into their homes.

On March 15, Governor Jay Inslee of Washington State announced an order to close all restaurants, entertainment, and recreation facilities.[30]

San Francisco issued a stay-at-home order the next day.

On March 19, California's governor, Gavin Newsom, issued a statewide order directing residents to stay at home except for essential activities.[31]

New York and Illinois followed with stay-at-home orders on March 20.

We had no way to diagnose actual infections, let alone intervene to prevent those who were contagious from passing the virus on to others. So we had to apply our measures to mitigate spread of SARS-CoV-2 to the whole population, since we couldn't identify those who were carrying the infection. The plan that we had drilled for always envisioned the CDC taking the central role in developing a test and following a very staged process in deploying it across the nation. This system could work with a slow-moving outbreak. It was no match for a fast-moving pandemic.

CHAPTER 9

SHORTAGE AFTER SHORTAGE

Getting the test kits and the machines to run them on was only one part of the challenge. It also took specially equipped sites and trained personnel to collect the samples. You needed to collect sputum. That meant the nation's installed base of blood specimen collection sites were of no use. You couldn't send patients into the local LabCorp or Quest. The states would have to stand up brand-new sites that were capable of collecting thousands of respiratory samples a day.

By late March, it was clear to senior White House staff that America's testing capacity was insufficient. It was a public health and political liability. Testing supply was increasing across the country, driven largely by growth in the capacity of the large commercial labs, but cumulatively, we still weren't conducting enough tests to keep up with demand as the epidemic surged. The inability to get tests performed was creating backlogs at hospitals. Sick patients were sitting in hospital isolation beds waiting for the results of a COVID test to determine if they had to remain in special COVID units, or if they could be sent to a general medical bed or transferred to other facilities. Hospitals were already short on beds and supplies, and the inability to clear patients who didn't have COVID was slowing discharges. The hospital labs were simply overwhelmed.

So, the decision was made by the White House Coronavirus Task Force to direct states and commercial labs to restrict testing to hospitalized patients and healthcare workers.[1] However, this created new problems. It meant an end to community testing. By rationing testing and reducing any chance of diagnosing mild and moderate cases that didn't rise to the level of needing hospitalization, the directive frustrated the ability to control spread.[2] People were much more likely to self-isolate and take other steps to reduce the likelihood that they might spread the infection if they actually had a positive diagnosis. Now, people in the community who had symptoms of COVID were told just to assume they were infected and self-isolate at home. At one point in March, the testing problems were so acute that one White House official had asked the National Institute of Allergy and Infectious Diseases whether the Centers for AIDS Research, a network of seventeen NIH-funded AIDS research centers that are located at major medical centers, could lend some of their screening capacity that's normally focused on testing for HIV. But the NIH wouldn't be able to spare the capacity.

By now, it wasn't just the diagnostic kits and high-capacity testing instruments that were in short supply. To perform a PCR test, labs also need specialized reagents that facilitate the process and are consumed each time a sample is analyzed. These reagents include the primers that bind to the specific stretch of RNA that you're trying to copy, like the N1, N2, and N3 reagents that caused problems in January and February. They also include special enzymes that help copy the genetic material that you're amplifying (called polymerase), and a buffering solution that keeps the process stable. Many of these reagents are manufactured outside the US, and when COVID hit, there was a global run on these supplies. "Current challenges that need to be resolved include uneven testing capacity and supplies throughout the US, both between and within regions, significant delays in reporting results (4–11 days), and national supply chain constraints, such as PPE, swabs, and certain testing reagents," the White House wrote in a draft report on how to create a national testing system. Those supplies should have been sitting in a strategic stockpile, ready to be accessed once the emergency arrived.[3] (That White House report was never released.)

On March 5, the same day that the IDT kits based on the CDC protocol were finally authorized by the FDA, North Carolina's secretary of health, Dr. Mandy Cohen, sent a letter to Secretary Azar—one of many similar letters he would receive. "I'm writing to alert you to my significant concern that there are inadequate laboratory supplies available to state laboratories to test for COVID-19," she wrote. "As you know, testing for COVID-19 is a complex procedure. The laboratory supplies we have on-hand from the CDC are only adequate to perform the necessary steps to test approximately 150 individuals. While we appreciate the additional supplies en route to us, they will not meet the increased number of individuals that will be identified for COVID-19 testing." At this point, North Carolina had diagnosed two cases of COVID in the state, under circumstances that suggested that community spread could be under way.

The frailties that plagued this supply chain were exemplified most vividly by a humble tool: the swabs used to get a biological sample of the virus from a patient's nose. Most of these nasopharyngeal swabs were made by a single plant in northern Italy—a country that was experiencing its own devastating epidemic. The manufacturer, Copan Diagnostics, which was based in Italy's hard-hit Lombardy region, was allowed to continue production even while most other work in the region had stopped. Making swabs for PCR testing isn't so straightforward. Doctors prefer certain kinds of "flocked" swabs for collecting samples from the nasopharynx, because they're easier to use and more likely to get a good specimen. To work right, the swab tips need to be spun from polyester, not cotton. Cotton may interfere with PCR testing and the proper extraction of the viral RNA, so labs use polyester swabs to collect patient samples when they are testing for nucleic acids. Copan's manufacturing plant in Italy could produce 720,000 of these specialized polyester swabs in a day and as many as 100 million in a year.[4] That was a lot of swabs— if there had not been a pandemic going on. But there was now a global run on the Italian swabs, and in the US, we were running out of them. The swabs became the bottleneck.

On Sunday, March 15, the Trump administration said that by the end of the week it would expand testing by supporting a national network of two thousand high-volume testing sites, each able to process up to four thousand samples a day. At that point, they had enough swabs to conduct about eight thousand tests a day, well short of what they needed. During a conference call that night with about three hundred representatives of commercial, academic, and state labs to announce the plan, Brett Giroir at HHS said that the Trump administration was expecting "speed bumps" in the availability of test supplies as testing expanded, specifically citing shortages of swabs and other equipment.[5] At one point the FDA, along with BARDA and the Department of Defense, stepped in to arrange a military transport out of Aviano Air Force Base in Italy, just to get shipments of swabs from Italy to the US.

To boost supplies, initially, the US government leaned on Puritan Medical Products, the sole American producer of the polyester swabs. The company, located in Guilford, Maine, was part of a one-hundred-year-old family-owned business, Hardwood Products Company, that also manufactured disposable wood products including ice pop sticks, toothpicks, and tongue depressors.[6] Puritan was the medical arm of the business that made plastic swabs used to collect specimens for a broad range of clinical tests. The company's grant of authority from the FDA to manufacture swabs dates back more than fifty years.

Before the pandemic, the US market consumed about one million nasopharyngeal swabs a year, mostly for medical testing in hospitals. Producing the swabs was a sleepy but stable business. Puritan and Copan manufactured almost all of the nasopharyngeal swabs that were used globally, with Copan the bigger of the two companies, controlling about 60 percent of the market. According to Bloomberg News, "Both companies have patented their swabs and repeatedly sued one another for design infringement. Rival manufacturers have had no interest in entering a static market for a relatively low-profit-margin product dominated by two litigious incumbents."[7]

As the COVID pandemic got under way, federal health officials reached out to Puritan in mid-March, and Puritan was able to boost production to make 10 million swabs a month by April.[8] By the end of July, with federal assistance, the company opened a second manufacturing facility, and increased production to 25 to 30 million swabs per month. By the end of 2020, and with the help of a $51.2 million grant from the Cares Act, it had opened another plant and was producing a total of 50 million swabs per month.[9] Puritan was now supplying most of the US market.

Federal officials then turned their attention to companies that manufactured other kinds of swabs to see if they could repurpose their operations to start making the specialized instruments. This effort ran into problems and revealed how hard it was to retool the production of medical supplies in the middle of a crisis. Some of the health officials who were involved in procuring the swabs were not aware that they needed to be made from polyester, not cotton. And so there was concern at the health department in Thurston County, Washington, when one of the first shipments of these repurposed swabs arrived, because the swabs were packaged in large boxes marked "Comforts for Baby Cotton Swabs" and were filled with thousands of loose swabs that looked like Q-tips.[10] To ensure sterility, the swabs should have been individually wrapped in sterile packaging. Another shipment I saw, which went to health officials in Illinois, was packaged in boxes marked "180 swabs" and loosely packed in cartons marked with the cotton logo. The cartons read "Cotton Swabs, crafted with care for maximum softness," alongside a logo of a bee sleeping on a clump of cotton, with the words "Comforts for Baby" encircling the bee.

In the end, it turned out that all these swabs were made from polyester.[11] HHS had tapped U.S. Cotton LLC—best known for its Q-tips—to manufacture them, and the company had modified 7 of its 180 machines to switch from making cotton to spun-polyester swabs, but hadn't changed the imprints on its packaging to reflect the new contents.[12] HHS had given the okay to ship these swabs with the wrong labeling to expedite access. But the cotton markings created confusion at the testing sites. One health official, when the swabs arrived, sent a picture of the cotton logo on the box and quipped, "Maybe we can use them to dab our tears." A letter had been shoved into the huge box with the "baby Q tips" that confirmed that they were in fact polyester. "The packaging and labelling being utilized . . . does not accurately reflect its contents," the letter read. By 2021, Puritan had used a $75 million investment from the Department of Defense to renovate yet another factory, and by the spring it was making another 100 million swabs a month.[13] Puritan had cornered the US market. It took a year, but swab supply was finally keeping up with demand.

The problems with the federal supply chain were not confined to swabs. Thousands of containers full of vials, tubes, and other fluid receptacles were shipped to states to be used for collecting and storing patient samples until they could arrive at a lab for processing. Many tubes came loosely packed in huge bins, mislabeled, and in the wrong kind of sterile packaging. State officials couldn't verify some of the contents and had to discard entire shipments.

One of the things I learned at the FDA, watching other critical medical products go into shortage, is that it's often the lowest-margin constituent in a complex supply chain that's most vulnerable to shortages. Often the only way to produce such a part profitably is to manufacture it at very large scale, which means it's likely to be made by a small number of big, consolidated suppliers. To reduce costs, these manufacturers are often located outside the US, where labor costs are lower. Moreover, because it's a low-margin product, that manufacturing facility is probably operated cheaply, and lacks investment and upkeep, leaving it more prone to disruptions. It's also for these reasons that, when problems arise, they usually arise early in production, when the low-margin materials first get baked into the manufacturing process.

When it comes to the manufacture of drugs, these low-margin components are often the starting chemicals. Many of these chemicals are made in China, and more specifically, Hubei Province. This is why there was concern about the risk of drug shortages when the epidemic first emerged in China. I testified on this issue on February 12 before the Senate Committee on Homeland Security and Governmental Affairs, in a hearing on Protecting the US from Global Pandemics, right before the US epidemic got under way. In that hearing, I told the Senate that "China's chemical industry, which accounts for 40 percent of global chemical industry revenue, provides a large number of ingredients for drug products. It's these starting materials—where in many cases China is the exclusive source of the chemical ingredients used for the manufacture of a drug product—that create choke points in the global supply chain for critical medicines."[14]

When it came to testing, the shortages of key components would persist for months. In July, a survey conducted by the Association of Public Health Laboratories found that 20 percent of public health labs would run out of at least one item required to do their tests within a week without steady resupply.[15] In the case of testing systems, the low-margin components that went into shortage were the reagents used to run the tests, and the swabs used to collect samples.

All of the cheapest and least complex parts of the testing supply chain would become the bottleneck to expanding test access. By Christmas 2020, a full year after the pandemic started, the FDA would formally add pipette tips to its list of medical devices in shortage.[16] Pipettes are devices used to draw a small amount of liquid and transfer it from one place to another with disposable plastic cone-like tips. Labs use pipettes to extract samples of viral RNA and move them onto the machines that test for the presence of the genomic material and evaluate its sequence. They also help transfer the different chemicals and biological material for these testing processes. So, when new variants of the virus emerged in the US, and we pursued a large-scale effort to sequence more SARS-CoV-2 samples in order to detect the new mutations, demand for pipette tips surged and their limited supply became a key rate-limiting factor in efforts to expand sequencing.

Utah was one of the states that was stymied by the shortages. It had set a goal to sequence 10 percent of all its COVID cases, but as late as February 2021, they were sequencing 3 percent. Increasing the testing rate required robotics, and the robots used lots of pipette tips.[17] Dr. Kelly Oakeson, the chief scientist in the Utah Public Health Laboratory, said they could do more sequencing, but they didn't have enough pipette tips to process the samples.[18]

By April 2021, more than a year after the pandemic began, labs in fourteen states reported having less than a month's supply of pipette tips on hand, according to a survey by the Association of Public Health Laboratories. The problems were so acute, the public health labs called on federal regulators to enact rules to prioritize the delivery of pipette tips for newborn screening programs, to make sure some of the nation's most critical lab tests would be able to continue.[19] We were back where we started.[20]

It wasn't just reagents, pipette tips, and swabs that were in short supply. In 2006, Congress appropriated money to add personal protective gear to the Strategic National Stockpile, funding the inclusion of about 50 million surgical masks and 100 million N95 respirators to that supply.[21] In late February 2020, Secretary Azar testified to the Senate Appropriations Committee that the national stockpile held more than 30 million N95 respirators. It was a small number, and far less than anticipated. But in actuality, the situation was even worse. A day later, Azar would clarify that the stockpile actually held 30 million surgical masks, the kind that tie behind the head. Not the N95 respirators that can more fully protect against aerosol spread. The stockpile had only 12 million N95 respirators, along with another 5 million that were expired and no longer certified for their integrity.[22] Many of these N95 respirators that were once in the stockpile were used during the H1N1 swine flu pandemic more than ten years earlier, and never replenished.[23]

The stockpile was a running sore throughout the crisis. In April, the federal government started distributing ventilators from the stockpile to backstop hard-hit cities. When the machines arrived, they didn't work. The *New York Times* reported that HHS had assured governors that the stockpile was holding about ten thousand ventilators in reserve and had another ten thousand on order.[24] But the government had allowed a key contract for maintaining the machines to lapse, and more than two thousand of the ventilators on hand were not functional.[25] Similarly, when masks were shipped from the stockpile and arrived at hospitals, there were reports that healthcare personnel found them covered in mold, and the rubber straps had become dried and brittle.[26]

The idea of the Strategic National Stockpile was to prepare for biomedical contingencies that could pose an existential threat to the nation. But much of the emphasis had been on using the stockpile to counter the risk of biological agents that might be used as weapons. A lot of the stockpile's resources had been carefully curated to counter specific pathogens—anthrax, smallpox, and especially bird flu. A lot of emphasis was put on developing and stockpiling drugs and vaccines to counter these individual threats. Less emphasis was placed on building broad capabilities that would be needed to respond to a pandemic with a virus we didn't anticipate, which would create a run on items like masks, ventilators, and testing supplies. The pathogen-by-pathogen focus for deciding what to stockpile meant that there were few targeted resources for countering viruses that didn't make the cut. Coronaviruses weren't seen as a pathogen that could seriously threaten us.

The question of what risks the stockpile should be focused on addressing was a longtime source of internal wrangling, with some at the CDC wanting to put more emphasis on naturally occurring threats like pandemics, and others at HHS arguing that more of its focus should be on countering bioweapons and investing in the countermeasures we would need if biological agents were used in a conflict. Historically, the stockpile was under the supervision of the CDC. During the first two years of the Trump administration, the Office of the Assistant Secretary for Preparedness and Response (ASPR) was agitating for control over the stockpile, and to focus more of its mission on the threat posed from bioweapons.[27] The first secretary of Health and Human Services, Tom Price, had put off the entreaties, recognizing that it was a big decision and would need more careful analysis, and he declined to move control of the stockpile from the CDC to the ASPR. The issue was also sitting with the president's homeland security adviser, Tom Bossert, who was being advised to take a briefing on the matter before any final decision was made, given the potential significance of such a reorganization and its broader implications. But when Price was fired by Trump, and Deputy Secretary Eric Hargan briefly became acting secretary, Hargan agreed to move the stockpile from the CDC and put it under the ASPR. The protracted battle over where the stockpile would reside, and over the role it should play, shifted attention away from shoring up its basic elements like masks and ventilators, while people fought over its bureaucratic administration. Once the stockpile was moved under the ASPR, its primary mission would predictably evolve, with more emphasis put on the risks posed by biological weapons and less on pandemics.[28] The torturous wrangling, and the decision to dislodge the stockpile from the CDC and begin the cumbersome process of moving its supervision to the ASPR, diverted attention away from its core activities just as COVID was about to sweep over the nation.

Even a properly equipped stockpile, however, has limited usefulness. Certain products are hard to store and hard to distribute. In a protracted crisis, even a well-stocked inventory can be depleted.

Maintaining a pandemic-scale stockpile can also be cost-prohibitive because of the sheer amount of equipment and countermeasures you'd need to hold in order to adequately cover the population. To guarantee a sufficient supply, we'll need the capacity to expedite production of these items and ensure that it's manufactured domestically, so we'll have more control over the supply chain in a crisis.

This is true especially when it comes to the low-margin commodities that are most likely to be used in large quantities in a crisis and also have fragile supply chains. That means maintaining manufacturing sites in the US, where construction and operating costs are higher than in lower-wage regions of the world. Ideally, we'd overbuild these facilities to make sure that there's some excess capacity in the event we need to increase output during a crisis. This will be a key part of creating a hot base for pandemic preparedness.[29]

The stockpile should be reoriented away from the idea of creating warehoused inventories of the products we anticipate needing in a crisis and toward a system of vendor-managed inventory by the federal government. Under this approach, the purchases done by federal agencies like the Department of Defense, the Department of Veterans Affairs, or the Indian Health Service, can be placed through the stockpile and sourced using domestic manufacturing sites that the federal government preferentially contracts with as a way to support US-based production of these materials. It might cost more, but the money would help guarantee that we have a domestic source for critical supplies. The cost would become a part of our national expenditure for security and public health preparedness. Health agencies could buy all their masks through the stockpile, which would, in turn, purchase them from domestic manufacturers. The stockpile, at any one time, might hold a year's worth of supply and then pull from this rotating inventory to satisfy the collective demand of all of these healthcare agencies. The inventory can be kept at government warehouses or held on location at the manufacturing sites. Take the example of masks. Buyers would receive masks drawn from this stash. As soon as some N95 respirators leave the warehouses to support a hospital's immediate needs, a new supply comes into stockpile, where it will sit for twelve months before being distributed later. In this way, the US will always have a twelve-month inventory on hand that can be forward-deployed, all at once if necessary, in the event of a crisis. We would also have a domestic manufacturing base capable of replenishing that supply.

We need to view the maintenance of this kind of capability as a key part of our national security.[30] In the realm of defense, we sometimes overpay for the production of high-tech military equipment in order to maintain their domestic manufacture. Submarines and jet fighters can be built and assembled more cheaply in another market. But we recognize the importance of making sure that we can control their manufacture in US plants and retain the ability to repair them in domestic production facilities. We can't trust that plants in other markets will be open to us in the event of a global crisis.

Maintaining these capacities inside the US could come at a relatively small cost. Consider the domestic supply for surgical masks and N95 respirators. The entire US market for medical-grade N95 respirators is about $150 million annually.[31] Nationally, before COVID, US healthcare personnel were using about 50 million medical-grade N95 respirators each year. Requiring half that market to be manufactured in the US, even at a 50 percent premium, would have cost about $40 million; a large sum, but such an investment could help secure a significant amount of capacity for domestic manufacturing of N95 respirators. Supplying this entire market, even at crisis levels, could be achieved through production at US-based facilities at an attainable cost, especially when viewed against the strategic imperative of maintaining the domestic capability to produce critical components. In 2019, the entire global market for N95 respirators was roughly $800 million to $1.1 billion.[32] Yet much of the personal protective equipment we used came from China, which held on to its supplies once COVID hit. Estimates were that China was the source of 72 percent of the surgical masks and 54 percent of the medical gowns imported to the US when COVID arrived. According to HHS, 70 percent of the N95 respirators used in the US were made overseas.[33]

Once COVID struck, demand surged. For one ninety-day stretch in the spring of 2020, we used about 140 million N95 respirators.[34] In March, some US hospitals were using 1,700 percent more N95 respirators, 500 percent more isolation gowns, 860 percent more face masks, and 300 percent more surgical masks, according to one estimate.[35] A lot of money was spent trying to quickly establish domestic manufacturing of all these items. In the Coronavirus Aid, Relief, and Economic Security Act, or CARES Act, $1 billion was set aside for this effort. But the lingering question is how to maintain this excess capacity once the pandemic is over. In past crises, or in preparation for them, we built the domestic capacities we thought we'd need, only to see them atrophy once the crises failed to arrive, and attention shifted elsewhere. This was the case with our preparations in the early 2000s for a feared pandemic involving a bird flu. The need for the domestic manufacture of critical supplies, and the need for policy approaches that provide for sustained capacities, will be a recurring theme in how we prepare for the next pandemic. As I'll get into later, these same challenges would surface when it came to the manufacture of adequate supplies of drugs and vaccines.

It may seem like swabs should be easy to make and should be interchangeable, but it's not that straightforward. The key to the performance of a test is its accuracy and the reproducibility of its results. That precision comes from exerting careful control over the different aspects of the testing system. This includes exactness in how viral samples are collected and transferred to the test. One swab may collect a sample differently than the next. Some swabs can be used only with certain kinds of samples, or for certain kinds of tests. Small deviations in how a sample is collected can affect the quality of the test and, in turn, its precision. How seemingly minor components affect the overall reliability of a test shows how complex these supply chains can be.

The accuracy of a test needs to be confirmed by using a proven menu of testing components according to an exacting blueprint. This is how labs that adopt a test can ensure that it will produce the same results every time. The way a test is performed is carefully tailored based on its components. A test that is validated to work on one kind of testing instrument with a certain set of reagents, pipette tips, and vials can be highly sensitive to deviations—even changes to something as seemingly simple as a plastic well used to hold the sample or a polyester swab used to collect it. This is another reason why it was important to have FDA oversight for these tests. One way to stretch the supply of these components, in addition to boosting manufacturing, is to design test kits from the outset in a way where the available components can be used interchangeably. Building this kind of resiliency into testing systems should be a part of our pandemic planning.

In the spring of 2020, COVID overwhelmed our testing infrastructure. When the CDC's test was authorized, many of the materials needed to perform it were in short supply. To increase capacity, the FDA's professional staff evaluated the performance data of several testing components to determine if certain materials and instruments could be used interchangeably. While this wasn't the agency's normal mission, the FDA was quietly serving as a clearinghouse and publishing FAQs to help labs mix and match among scarce materials. A coordinated national testing plan, had one existed, could have supported this kind of work as part of a larger effort to oversee allocation of test supplies in a crisis, with testing systems designed for maximal interchangeability and validated in different configurations to maximize the use of the available supplies and components. Mixing and matching should have been designed into the system.

Furthermore, there was no federal coordination for distributing the test kits or supplies to states. It became a fracas, with individual states vying for resources in a global marketplace. Governors traded tips on how to get reagents. Governor Larry Hogan of Maryland, whose COVID advisory board I served on, sourced thousands of test kits from South Korea, using personal connections though his wife, a Korean American.[36] I spoke to Governor Charlie Baker of Massachusetts about his effort to get a private jet to bring in supplies under the peering eyes of federal authorities who, he feared, would confiscate the goods he had sourced from China. Other governors banded together to make their own coordinated purchases, outside the reach of the federal government, which was bidding against them for the same tests and supplies.[37] By August, seven governors (three Republicans and four Democrats) would form their own purchasing compact in collaboration with the Rockefeller Foundation.[38] The first-of-its-kind effort was conceived and advanced with the help of the foundation's president, Dr. Rajiv Shah, who had previously served as the administrator of the US Agency for International Development (USAID).[39]

In Massachusetts, I was part of an advisory group to Governor Charlie Baker. Testing was at the center of his efforts to address the epidemic from the very beginning and helped Massachusetts get better control over the fall and winter surge of infection. Early in the crisis, Baker had turned to his state's academic labs, especially the Broad Institute of MIT and Harvard, to help fill the testing void in the spring. The state's SARS-CoV-2 testing went from 3,000 a day in March to about 100,000 a day at the beginning of the fall. This was primarily a result of increasing testing capacity through laboratory efficiencies and automation, as well as better lab operations and the smart adoption of technology. On a per capita basis, Massachusetts's testing capacity was the best in the nation most days. The state had many free testing sites, and people were encouraged to use them.[40] Massachusetts even pioneered the testing of sewage wastewater as a tripwire to give policymakers a way to identify early when and where community spread was getting under way in a certain area. (People shed the virus in their feces.) The data gleaned from testing raw sewage allowed the state's public health officials to measure how prevalent the virus was in a community and better target interventions.[41]

Even once the CDC worked out the problems with its contaminated tests, and the kits were being shipped by its manufacturing partner IDT, there was no process for determining where to send the kits. The career staff in the FDA's device center again stepped in to play the role of traffic cops, helping IDT figure out where to ship kits based on calls from labs that were in need of the supplies. Otherwise, HHS had no process to manage the distribution. The improvised role being played by the FDA's professional staff showed the benefits of having some central coordination over these efforts in a crisis. Or else, everyone was on their own.

There was no national lab policy that envelops the entire continuum of all of the kinds of labs and tests that are needed in a crisis. The closest thing we had was the Laboratory Response Network, a collaboration between the CDC and the public health labs established in 1999 that serves as hundreds of sentinel sites for the detection and identification of a narrow set of pathogens that may be used in a bioterrorist attack or other biological incident.[42] It was another measure of how much of our limited infrastructure was focused on the risk from special pathogens that would be used in a deliberate attack, and not the natural evolution of a pandemic virus. Moreover, the network was not interconnected with the commercial labs like Quest and LabCorp or the clinical labs inside academic medical centers. So, it wasn't prepared to provide the kind of coordination that was needed across the existing layers of the nation's system for diagnostic testing.

In a fast-moving epidemic, there needs to be synchronization across labs for training, sharing samples, and distributing supplies. In some states, this sort of alignment exists. It's typically organized by the public health labs. They could play this coordinating role on a more national basis, serving as a central hub for the various commercial, clinical, and small independent labs doing the testing within each state. They would need to have proper support to pursue this mission.

If we had a system for coordinating lab activity, we could share testing supplies more efficiently. During COVID, many labs hoarded material, ordering way more than they needed in hopes of getting a little, and sometimes ended up with a lot. Suppliers were developing "priority" customer lists for labs, and the commercial labs were leveraging their existing relationships to move up on these supply lists. It's likely that some of that material they amassed expired before it could be used. It's another example where more coordination was needed.

To give a measure of how this kind of organization can be helpful, consider an episode that occurred in the spring of 2020. The CDC was going to advance an initial COVID test that would have forced labs to rely on a special system for extracting the coronavirus RNA from patient samples. The problem was that most of the public health labs didn't have access to the tools that the CDC's approach required. So they wouldn't have been able to follow the agency's prescribed steps for how the test could be reliably conducted. But there was an established channel of communication between the CDC and the public health labs, and through that channel the agency was able to get feedback that its protocol would create obstacles for the labs. The CDC changed its approach. This kind of know-how, on the testing methods we should advance in order to maximize the capacity of the existing labs and make fullest use of the installed base of testing instruments, needs to be institutionalized into our pandemic response. It came into play here because the public health labs are the one segment of the lab community with whom the CDC already coordinates. So the agency was able to get that feedback from the public health labs. But the public health labs represent just a small fraction of America's overall testing capacity, and the CDC lacks the same visibility into and interconnectedness with the other segments of the US laboratory system.

A networked system would have other benefits. For example, another challenge we faced during the pandemic was that many labs were overwhelmed, while other labs had capacity that wasn't being fully used. Many of the commercial labs were slammed with samples from physicians who were accustomed to sending their order sets to these companies while some academic labs had available capacity. A better-networked system could have helped make sure that samples got distributed more evenly, so we made the best use of whatever testing capacity exists.

In contrast to its work with the public health labs, the CDC hasn't tried to leverage its relationships with commercial, academic, and smaller independent labs, even though it has an established memorandum of understanding with the commercial lab association. The fragmented US testing system was not tethered together in a way that the resources could be put to maximal use in a moment of crisis. This is why it often felt like there was a shortage of testing even when there were a lot of test kits available and a lot of testing platforms sitting idle. Some labs were overwhelmed with samples. Others were operating at a fraction of their total capacity. Another problem was the fragmentation in the installed base of testing equipment across different labs. The big labs had stockpiled the supplies for the capital equipment they already owned, and supplies for the lower-throughput, manual testing equipment were generally more available in the crisis while supplies for the high-throughput machines that we needed more of had been hoarded and were now in short supply.

No single agency had visibility across the entire system. The FDA, which had perhaps the most insight into the entire market, regulated the medical devices used in testing. It wasn't in a position to play an operational role by helping to coordinate activity across the market, even though, in some narrow circumstances, that's precisely what it started to do. The White House tried to launch a new office to coordinate the sourcing and distribution of testing supplies, but it didn't have visibility into the testing landscape, either. (This office was going to be directly supervised by Jared Kushner and was staffed by some business executives he brought in from outside the government.) At times it felt like there was a testing famine amidst a bounty of plenty of testing equipment sitting idle in labs. We had a lot of labs. We had a lot of testing equipment and even residual capacity. But all the installed capacity wasn't lined up to support the surge in demand.

A national lab strategy would create an organized system to apportion testing supplies and testing capacity across the entire market. It would assist labs in developing their own laboratory developed tests, and direct patient samples to labs with available capacity. In a crisis, we needed to squeeze all of the testing supply out of the system. We fell short.

Following the Zika crisis in 2016, there was a nascent effort to create a more organized system, and a tri-agency coordinating group was formed to improve crisis response, linking the CDC, the FDA, and the Centers for Medicare & Medicaid Services.[43] The new effort was announced when I was FDA commissioner and it was a step in the right direction, and a model that could have been expanded. However, at its inception, the coordinating committee was a low-level work group and didn't have sufficient reach within its parent agencies to be the vehicle for the kind of cross-agency collaboration that was needed. It wasn't intended to play that sort of role.

Instead, we need an agency that's empowered to organize a testing response. It has to have a more operational function able to synchronize across labs in a time of crisis. It must be able to work with manufacturers to advance the development of test kits, sharing samples, supplies, and information to expand capacities. It needs also to serve as a clearinghouse to maximize the use of the available testing platforms and to make sure samples are routed to the labs that have available capacity so that more of the nation's testing capacity is brought into the fold. Writing in *The New Yorker* in the fall of 2020, the physician and author Dr. Atul Gawande compared the general concept to load balancing among power utilities: "Decades ago, electric companies were organized in the same way that laboratory testing is organized today. They were vertical monopolies that ran their own power plants, transmission lines, and customer operations. That arrangement got the job done, but it meant that many communities endured brownouts and blackouts from a shortage of capacity, while others had an oversupply. . . . The creation of a national electric grid that physically connected the electricity supply . . . opened the door for load balancing, increased supply, lower costs, and alternative energy production. We have no national grid for the generation, transmission, or distribution of our testing supply. . . . Now we're paying the price. In power generation, the worry is that our national grid is aging; in health care, the worry is that we have no grid at all."[44]

Inside HHS, the Office of the Assistant Secretary for Preparedness and Response might have performed this function. This office was created after 9/11, but it had evolved into more of a policy function, with its focus oscillating over time between prescribing a policy for countering bioterrorist threats, to a focus on pandemic preparedness, and then back again to a preoccupation with the risk from bioweapons that might be used in a hypothetical war with North Korea.[45] The office was originally called the Office of Public Health Emergency Preparedness, with no response in its title, because it wasn't envisioned to have an operational role. Later on, it was renamed with a view that it would develop an operational capability, but that mission was never fully realized. The office can be refocused to take on the operational role envisioned for it, working to coordinate response across the different health silos that are central to coping with a biological threat.

Another option that some might consider is to give this mission to a newly constituted unit inside the CDC. To make that work, however, the role and operational ethos of this new organization will have to be made deliberately distinct from the existing CDC culture, where functions are bunkered inside rigid silos in a way that can impede multidisciplinary collaboration, and where the agency's academically oriented customs supersede the more operationally focused, national security mind-set that will be key to crisis response in a biological disaster. South Korea found that its own CDC fell short during the MERS epidemic in 2015 and revamped it to better respond to crises. That's why the South Korean CDC was able to play a more hands-on role in advancing diagnostics to respond to SARS-CoV-2. Replicating that experience in the US would require a change in the mission and mind-set of our CDC. One consideration would be to split the health prevention role out from the CDC and embed it within a new institute in the NIH, or a new division within the Office of the Assistant Secretary for Health inside HHS (where similar policy issues are already housed), or else entrust the mission to a new agency dedicated exclusively to health prevention. Then, more of the CDC's mission and mind-set can be focused on its functions aimed at identifying and responding to the spread of communicable disease. It could be resourced to focus on its response mission related to emerging diseases and its national security undertaking. Congress would have to reprogram the agency.

For now, the CDC lacks that singular focus. When COVID struck, someone needed to coordinate the industrialization of a brand-new, national testing enterprise. "Nobody said, it's January 30, where are we on testing?" one person involved in these efforts at the CDC told me. "There was an assumption that CDC was going to develop a test and it was going to go to every hospital in the U.S. and that was never going to be the case. HHS didn't seem to understand the CDC's role."

The need to develop such a test was simultaneously everyone's problem, and nobody's problem. In many respects, the later creation of Operation Warp Speed was a recognition of the challenge of securing commercial-scale production and distribution of a brand-new product that would be needed to respond to the crisis. However, Operation Warp Speed would focus almost exclusively on drugs and vaccines, and put scant money behind the development of new diagnostics.[46]

Once again, our focus on preparing for a pandemic flu may have caused us to overlook a key piece of our response. With a flu, establishing a laboratory-confirmed diagnosis is not as critical as was the case with COVID, because the incubation period for flu is short, so contact tracing isn't as valuable. By the time you test and trace someone's contacts, they have already contracted influenza—and, if they were going to be the next link in a chain of transmission, propagated the infection. But SARS-CoV-2 had a wider range of incubation periods. The length of its typical incubation period was longer than for flu, and the coronavirus had a range of presentations that did not always lead to easy diagnosis. Accurate, fast diagnostic testing could be important to identifying chains of transmission and intervening in time to break off further spread.

Instead, whenever we considered diagnostics as a potential countermeasure, it was typically in the context of a bioterrorist attack. In an attack, we'd need the capability to blanket a city with testing in the event of an anthrax attack or some similar contingency. The attack would be over an identifiable geographic area. It would have well-defined boundaries. The structural issues of needing enough testing capability to deploy a novel diagnostic on a national scale and having the capacity to conduct millions of tests a day were never fully considered. For a pandemic flu, which is the only context in which pandemic contingency planning was seriously engaged before COVID, we already had an installed base of flu testing that would be sufficient.

Going forward, we need to provide for a national-level capability to field and deploy clinical tests for novel pathogens. Wherever this function is housed, it needs greater resiliency in a crisis.

Kushner's team would work on a plan later that spring to establish a more coordinated approach to the US testing system.[47] They were trying to write the blueprint for a national response while we were already in the middle of the crisis. That playbook should have been sitting on a shelf somewhere, with agencies already staffed and resourced to implement its instructions. On April 27 President Trump stepped to the Rose Garden podium to announce a national testing plan. But it wasn't the one that Kushner and his team had spent weeks working on.[48]

I had been an informal adviser to that process. The initial plan would have created a coordinated, national system for securing and distributing the different testing components that were in short supply. The federal government would have assumed a lead role in helping increase the supply of these parts. But this plan was scrapped.[49] Its architects no longer thought the federal government would be able to do much more than was already under way to secure these things in the tight timeframe needed. Instead, the plan put forward by President Trump envisioned a much smaller role for the federal government, allocating testing supplies among the states and pressing harder on an FDA-led effort to authorize access to more COVID tests. The main elements were a plan to help ramp up the manufacturing of nasal swabs and broker access to testing kits and some of the reagents that were in short supply. By June, a new wave of infection would sweep across the Sunbelt. Once again, the states would run out of testing supplies.[50]

CHAPTER 10

PREPARING FOR THE WRONG PATHOGEN

Great Understatements in History:

> Napoleon's retreat from Moscow—"just a little stroll gone bad"
> Pompeii—"a bit of a dust storm"
> Hiroshima—"a bad summer heat wave" and
> Wuhan—"just a bad flu season."

So read an email from Dr. James Lawler, an infectious disease expert at the University of Nebraska who served in the White House under Presidents George W. Bush and Barack Obama.

The note was sent on January 28, 2020, to a prominent group of current and former health officials. All were experts in pandemics and would become key figures in the COVID response. They had begun exchanging private notes about the unfolding danger on a secretive email chain that they dubbed "Red Dawn," named after the 1984 film that depicts World War III, where the US is invaded by the Soviet Union and a group of American high school students resists the occupation with guerrilla warfare. I was on some of these emails.[1]

Writing earlier that day, in the email that would elicit Lawler's memorable reply, one of the most active members on the email chain, Dr. Carter Mecher, a physician and senior medical adviser at the Department of Veterans Affairs, summed up the impending danger this way: "No matter how I look at this," he wrote, "it looks [to] be bad. The projected size of the outbreak already seems hard to believe, but when I think of the actions being taken across China that are reminiscent of 1918 Philadelphia, perhaps those numbers are correct. And if we accept that level of transmissibility, the [case fatality rate] is approaching the range of a severe flu pandemic. . . . Any way you cut it, this is going to be bad."

Mecher's rough analysis turned out to be prescient.[2] SARS-CoV-2 looked more transmissible than a pandemic flu and a lot more deadly than seasonal flu.[3] It was the end of January. The nation was just becoming more aware of the events unfolding in China. The White House was starting to hold daily meetings on the crisis. But many of the experts on the Red Dawn emails were already convinced that the coronavirus would soon overwhelm America. And they already harbored apprehensions that the pandemic plans we had prepped might be no match for it.

Some of the most prescient notes concerned the use of community mitigation measures like school closures, also known as nonpharmaceutical interventions, or NPIs for short, as a way to confront COVID. "NPIs are going to be central to our response to this outbreak (assuming our estimates of severity prove accurate)," Mecher wrote to the group on February 17. We might need to "close the colleges and universities," he wrote.[4] It was a full month before states began to implement stay-at-home orders, and a week before we identified the first community spread.

"Nonpharmaceutical interventions" is a catch-all phrase for measures aimed at controlling an epidemic that didn't involve the use of drugs or vaccines. "I anticipate we might encounter pushback over the implementation of NPIs and would expect similar concerns/arguments as were raised back in 2006 when this strategy first emerged," Mecher also wrote on February 17. He had served as the director of Medical Preparedness Policy on the National Security Council in 2005. The modern application of these mitigation approaches was first envisioned in 2005 by a team that Mecher helped lead. The aim of that 2005 effort was to develop a plan that would slow the spread of a pandemic, save lives, and buy time for the US government to come up with medical countermeasures to fight a new and deadly strain of influenza.[5]

In 2005, the fear was that a novel strain of bird flu, known as H5N1, would soon cause a deadly pandemic. The virus had begun to circulate in other parts of the world, claiming the lives of more than 50 percent of its victims. Adding to the concerns was mounting evidence that these dangerous bird flus had found a home in wild waterfowl, especially ducks and geese, many of which migrated over long distances.[6] So the global reach of a new flu wouldn't have to rely on airplanes and traveling human hosts; migratory birds could do that work just as efficiently.[7] The 2005 plan was drafted to counter this threat. The NPIs were viewed as a holding step in the event that a pandemic emerged and we didn't have a vaccine or sufficient amounts of antiviral drugs. The tactics were seen as a way to slow the spread until we could develop those countermeasures. The threat was judged to be so immediate that the team who drafted the plan didn't think of it as a "general" pandemic plan. Theirs was a plan to counter a specific pandemic involving H5N1. It was only "general" in the sense that some of the measures, particularly the NPIs, could be dialed up or down depending on how severe it was.

While prior variants of H5N1 were seen before, the old strains were markedly different from the version of H5N1 that emerged in 2005. There are four main types of influenza: A, B, C, and D. The two strains that infect humans and trigger seasonal epidemics are influenza A and B. Influenza C generally causes mild illness and is not thought to be responsible for human epidemics, and influenza D primarily affects cattle. It's influenza A that gets most of the attention—the most common strain of influenza and the only one that's known to have triggered past pandemics.

Flus are named for the properties of two key proteins that appear on their surfaces, hemagglutinin (H) and neuraminidase (N). The two numbers used to describe a strain of flu correspond to which variant of these two proteins a virus harbors. There are eighteen different subtypes of hemagglutinin and eleven different subtypes of neuraminidase. So, H5N1 is named for the fact that it combines the fifth subtype of hemagglutinin with the first subtype of neuraminidase.

Each surface protein performs a different function. Hemagglutinin helps the virus attach to the surface of our cells, while neuraminidase helps it replicate once it's settled inside our bodies. These two proteins sit visibly on the virus's surface and make conspicuous targets. So, to develop immunity against a particular strain of flu, our bodies generate antibodies that target each of the proteins. Most of our immunity to flu is derived from the antibodies that we generate against hemagglutinin, which ends up being a more prominent target. But the genomes of flu viruses are rather unstable, especially compared to coronaviruses. That's because the flu genomes are comprised of eight separate segments, and this segmentation adds to their tendency to mutate.[8] Like other RNA viruses, influenza mutates inside each of these segments. But in addition to mutations within each segment, the different segments can also swap genetic material with one another. This stands in contrast to a coronavirus, which is not segmented. Since a coronavirus cannot as easily reassemble the genes inside its own genome, it doesn't tend to mutate as readily as influenza. And because the genetic code of a flu virus is prone to constant changes, its surface proteins also more easily evolve, often in ways that can evade the antibodies we may have developed after infection with an earlier strain or through vaccination.[9]

Data on the genetic sequences of the H5N1 infections that had been identified in 2005 showed that this particular strain was mutating in ways that could make it more menacing to people.[10] It was believed that the H5N1 virus had first jumped from birds to humans in Hong Kong in 1997, in an outbreak that infected eighteen people, six of whom would die.[11] The infection was fearsome, causing severe inflammation of the lungs, a condition known as pneumonitis.[12] When it first emerged, people were stricken only after close contact with infected chickens; there was no evidence of person-to-person spread. The outbreak was ultimately stopped. But at a big cost. Among other measures, Hong Kong slaughtered more than 1.5 million chickens.[13]

Still, it wasn't the last word on the lethal virus. H5N1 reemerged in January 2004 in a big way. An outbreak struck poultry farms in Vietnam and Thailand, and within weeks, it spread to farms in ten countries and regions in Asia, including Indonesia, South Korea, Japan, and China. Twenty-three initial infections resulted in eighteen deaths.[14] Domestic ducks in southern China had become key players in sustaining the virus and helping it evolve. China also contributed to its spread with poor animal husbandry. I was working at the FDA over this time period, and my colleagues and I were particularly concerned by China's use of antiviral drugs in the feed of chickens, in an effort to prevent flu viruses from decimating their flocks. By exposing tens of millions of chickens to the only antiviral drug known to potentially work against H5N1, China was conducting a massive scientific exercise to select for mutant strains that would evade perhaps our only available countermeasure.[15] Vietnam and Thailand had reported eleven confirmed human cases, but China, with larger outbreaks in its poultry population, said it didn't have any spread to humans.[16] "The epidemic situation . . . is under control and no infections in human beings have been found," China's state news agency announced.[17] Yet, a year later, it would finally be revealed that indeed, H5N1 had also spread to people in China. An entry in the *New England Journal of Medicine* also detailed the death of a man in Beijing in 2003 that would initially be misattributed to SARS-1, but was later revised and ascribed to H5N1.[18] The entire chain of events would raise questions about how many other cases of human H5N1 infection might have gone unreported in China, or been wrongly linked to other causes.[19] Following a trail of genetic changes that the virus had undergone, investigators traced the new strain back to the original outbreak in Hong Kong in 1997.[20] The virus had continued to lurk furtively, probably in its animal hosts, and had evolved to become even more lethal, claiming the lives of nearly 60 percent of its victims.[21] It still wasn't able to pass easily between people, but the risk was clearly growing that it would become better adapted to its human hosts.[22] Fearing its continued spread, the WHO warned that the next pandemic could cause the deaths of up to 7 million people globally. From that point on, experts believed that containment could delay H5N1, but many were convinced that a pandemic was inevitable.[23] As a result, everyone's focus turned to countering bird flu.

In January 2006, the H5N1 strain reemerged, and once again it had become more menacing. Migrating birds spread the virus from Asia to Europe and Africa, killing domestic stocks of chickens and permanently changing the way poultry was farmed. When human cases started to appear in Turkey without a known link to birds, American officials worried that chains of human-to-human transmission had already begun. The cases stopped, and ultimately they were traced to sustained contact that these patients had with birds.[24] But by April, scientists concluded that efforts to contain the spread of H5N1 had failed owing to the role being played by migratory birds.[25] By June, the virus had infected 228 people worldwide and had killed 130, most of whom had close contact with infected birds. However, now other mammals such as cats and pigs had also become infected. That month, the WHO once again predicted that H5N1 would start to infect humans. Many experts believed it would soon appear in North America.[26]

In the US, attention turned to these threats in earnest during the summer of 2005. President Bush read a new book, John M. Barry's *The Great Influenza,* while on vacation at his ranch in Crawford, Texas. The book recounted the emergence of the 1918 Spanish flu, and the country's failed efforts to contain it. Its publication coincided with the outbreak of H5N1 that year in Vietnam and, taken together, the book and the outbreak stoked concerns that a deadly pandemic was inevitable. If one occurred, our vulnerability was crystallized by events that happened the previous year, when nearly half of the nation's flu vaccine supply had to be discarded because of systemic problems with quality and sterility at a manufacturing plant in Liverpool, England, run by Chiron, a drug maker that produced half of America's flu vaccine each year. Chiron was one of only two large manufacturers supplying the entire US market with injectable flu vaccine.[27] I was working at the Centers for Medicare & Medicaid Services at the time, and I remember the long lines that started to form in the US as people rushed to clinics to get a flu vaccine before the limited supply that season ran out. States threatened to levy fines on doctors and nurses who gave shots to anyone not in a high-risk group. The problems underscored our vulnerability.[28] We weren't prepared.

When President Bush returned to Washington, he called his top homeland security adviser, Fran Townsend, into the Oval Office and gave her his copy of Barry's book. "You've got to read this," he told her. "Look, this happens every 100 years. We need a national strategy."[29]

The world had seen three pandemics sweep the globe over the last century. Bush told his team that it was just a matter of time before it happened again. His health advisers agreed, and the H5N1 avian flu was judged to be the most proximate threat.[30]

At an Oval Office meeting on the morning of October 14, 2005, Dr. Rajeev Venkayya, a physician trained in pulmonary and critical care medicine and Bush's special assistant for biodefense, summarized the work that was already under way to prepare for a pandemic. Venkayya was the head of the biodefense directorate on the Homeland Security Council. The group was mainly focused on agents that terrorists might employ as biological weapons, principally ricin and anthrax. Venkayya briefed Bush on the limited planning that had already been undertaken to prepare for the possibility that a pandemic flu would emerge. The arrangements had focused mostly on how the country would rush the development of drugs and vaccines. The president wasn't satisfied.[31] Bush leaned toward the group and said, "I want to see a plan."[32] Venkayya traveled to his parents' house in Xenia, Ohio, and wrote an initial blueprint. When he returned to Washington, he had a draft of a twelve-page "National Strategy for Pandemic Influenza."

The new document encompassed three pillars.

First was preparedness, which would include the production and stockpiling of vaccines and medical countermeasures, and the creation of diagnostic tests.

Second was surveillance and detection, which relied on our ability to detect and report outbreaks, and to use testing and tracing to limit spread.

The final piece was response and containment. This spoke to the need for a surge capacity in our healthcare system that could care for waves of sick people, and a plan for mounting a coordinated response among the states to limit the national spread. President Bush unveiled the outlines of that new blueprint in a thirty-minute speech on November 1, 2005.[33]

As part of these efforts, President Bush already secured a new International Partnership on Avian and Pandemic Influenza at the opening of the United Nations General Assembly in September.[34] The aim was to improve international readiness for a flu pandemic by promoting closer collaboration among nations. It built on years of efforts to get countries to work together to identify and contain these threats. But all of these commitments depended on transparency. A nation that was host to new viruses had to be willing to share details of this outbreak, so that preparations could be made to prevent global spread. However, the need for early awareness is usually where these agreements broke down, because too many nations were slow to share the news when they identified an outbreak with a novel or dangerous pathogen.

We had learned that most recently after the troubling events surrounding SARS-1, and China's effort to conceal that contagion. To address these earlier breakdowns, and put nations on a more responsible footing, in June 2005, the WHO had already revised the International Health Regulations that governed global cooperation around pandemics.[35] The idea was to provide more specificity about what countries were obligated to do when they identified a new virus. The international rules needed more teeth. The language was toughened to require mandatory reporting in circumstances when a country identified a completely novel pathogen that posed a risk to humans. The aim of the agreement that Bush was announcing four months later was to build on these new commitments, by crafting a specific set of obligations when it came to influenza.[36]

Even though the government had been worried about the risk of a pandemic flu for a number of years, up until that point relatively few resources were focused on developing reliable countermeasures. The NIH's entire flu research budget for 2005 was about $119 million. HHS had created an initial stockpile of just 2.3 million treatment courses of oseltamivir, the only approved drug that had shown activity against a range of potentially dangerous influenzas.[37] It was a paltry amount for a country of nearly 300 million people. In a real pandemic, it was estimated that at least 100 million doses would be needed, and perhaps more if it took a long time to develop a vaccine.

Making a vaccine can be hard. The development work might take years. Part of the challenge is the way that flu vaccines are manufactured, using chicken eggs to culture the virus and grow the critical proteins that are used as the stock for the vaccines. Under this process, a fertilized chicken egg is inoculated with a strain of influenza. The virus is allowed to replicate in the yolk, before it's harvested, inactivated, and incorporated into vaccines. In this way, the eggs serve as incubators for the development of the viral proteins that are used for the vaccine.

This outdated approach is the backbone for producing the seasonal flu vaccine. But there were serious concerns in 2005 that it wouldn't be adequate—or even feasible—in the event of a pandemic bird flu. For one thing, there was fear that there wouldn't be enough chicken eggs if there was a surge in demand for the vaccine. This prompted federal officials to finance the formation of secretive flocks of chickens located in sprawling, indoor facilities that were spread across thirty-five sites in the eastern United States.[38] About $44 million was invested in a five-year program to harden the sites, to make sure there would be a sufficient supply of eggs.[39] The US government even classified these chicken farms as part of the nation's "critical infrastructure," shielding them from public view. One person who worked for me at the FDA, and toured the locations, described walking into one cavernous installation and seeing an endless sea of white chickens. The hens would all run away from you, she said, and the roosters would all charge toward you.

The next concern was whether the chickens—and their eggs—could survive a pandemic bird flu long enough to support the production of a vaccine. A bird flu, by its very nature, kills poultry. It could wipe out the chickens. The same strain could prove deadly to the eggs and to the developing embryos that are needed to grow the viral stock that's used to produce the vaccines. Getting a developing chicken egg to survive inoculation with a bird flu may not be feasible. So, as a fail-safe, the team established a backup plan to build manufacturing sites that wouldn't be dependent on chickens or their eggs, relying instead on cells derived from mammals rather than chicken embryos as the medium on which to grow the influenza virus and harvest its proteins.

The presumption is that mammalian cells would be less likely to be killed outright by a strain of bird flu. Moreover, because the cell cultures could be incubated in large bioreactors, it made it potentially faster, and easier, to produce the vaccine stock in large quantities. These processes were also easier to expand if we needed to surge the production of vaccines. In the manufacturing of pharmaceutical products that are derived from biological sources, it's often said that the process is the product. The characteristics of a vaccine or other biological are closely tied to how it's manufactured. There are a lot of things that can go wrong, and many of them are hard to predict—and detect. It's estimated that 70 percent of the time and resources related to the manufacture of a vaccine is focused on quality control.[40] The biggest risk often comes from "scaling up" production—expanding an existing manufacturing line to boost its output—and "scaling out" a process—partnering with other manufacturers to broaden your production capabilities and teaching other plants how to reliably make the complex product. The cell-based process was easier to expand; lots of manufacturing plants could accept cell lines and grow them in their existing vats, often with minor modifications. You didn't need to wait for eggs to be laid.

There were other advantages to the cell-based process. The virus obtained in cell cultures can be made to more carefully match the circulating strain of flu. The mammalian cell cultures can more closely mimic the human cells that the virus is ultimately programmed to target. In this way the virus that gets churned out from these cultures can also be made to look just like the viral copies that will be produced when the virus naturally infects its human host. A virus manufactured in a mammalian cell culture doesn't need to adapt as much to its new growing environment as a virus grown in eggs. If the virus was already trained to infect human cells, then replicating it in a mammalian cell culture doesn't require it to change its ways.

By contrast, virus manufactured in eggs often undergoes adaptation to the chickens, a process called "antigenic modification." These modifications are small changes that make the virus more closely suited to growing in a chicken egg. But the adaptation can also change the conformation of the virus's proteins, making them less likely to be recognized by a human immune system, and more likely to be recognized by a chicken's. That can make the vaccine less potent in stimulating our immune system. If these viral proteins undergo too many changes, then once they are injected into our arms, the proteins could end up stimulating the production of human antibodies that aren't as well suited to attacking the real virus.[41] They become adapted to attacking a version of the virus that had itself been modified to replicate in chickens. This is what happened in 2017, when I was FDA commissioner. The flu vaccine didn't work well that year, probably because it had undergone too much "egg adaptation." The vaccine was only about 25 percent effective against the principal strain of influenza that circulated that winter.[42]

Recognizing these vulnerabilities, and the unique challenges they'd pose in a pandemic, there were earlier efforts to expand our domestic capacity to make flu vaccines in cell cultures and not chicken eggs. In the early 2000s, as part of this effort, the government refurbished and enlarged some domestic manufacturing sites.[43] Building these plants would take many years and require an enormous investment of capital. Yet, after some initial efforts, attention would turn away from these priorities. The focus on building out our domestic capability to make flu vaccine, in facilities that could be impervious to a pandemic strain of bird flu, would ultimately give way to other priorities. We didn't stick with it. Even with a lot of planning and a huge sum of money, only one significant facility would result from this effort; a single cell-based manufacturing site located in North Carolina.[44] The plant was built for the production of a seasonal flu vaccine (and was eventually sold off to a foreign owner and wasn't leveraged by the US government to boost vaccine manufacturing during COVID). Even after the plant was built, developing flu vaccines from this cell culturing process would run into regulatory hurdles. The first seasonal flu vaccine based on these new cell-culturing techniques wouldn't be licensed by the FDA until 2012, almost a decade after these efforts first got under way.[45]

The path to building better resiliency would be protracted and uncertain, and ultimately many aspects of our plans would be set aside. The dangers persisted longer than our attention to the risks.

CHAPTER 11

STAY-AT-HOME ORDERS

Back in 2005, after President Bush had focused his health advisers on the threat posed by the H5N1 bird flu, they turned their attention to a thorny question of logistics. If the US suffered a pandemic with a novel flu, even in a best-case scenario, everyone figured it would take at least six months to develop a vaccine, and that depended on everything going right. It depended on being able to culture the new virus and grow it inside the chicken eggs without killing the eggs. It depended on generating sufficient yields to make enough shots, and on the vaccine being protective enough to arrest the spread. And then we had to distribute it to patients, which could take months.

Even today, we don't have enough cell-based manufacturing as a hedge against a pandemic strain that might be hard to grow inside the eggs, and certainly not enough cell-based manufacturing to cover the American population. We'd still be dependent on growing the new vaccine in the chicken eggs. In 2009, when the H1N1 swine flu struck, almost everything went largely right, and it still took about nine months to develop and manufacture a vaccine. For a new flu vaccine, a six-month development process is a normal timeline—and in the vaccine world, that's considered fast. Since the procedure for making flu vaccines is so well established, tweaking the vaccine each year, to insert a new strain of influenza, is a relatively straightforward process. But six months is still a long time in a deadly pandemic, and with a novel strain that might be hard to grow in chicken eggs or cell cultures, it could take much longer. So, if a pandemic struck, we needed a way to slow the spread while we raced to develop a vaccine. That's where the idea of using nonpharmaceutical interventions came in.[1] What we needed was a holding plan.

The task of coming up with one was given to Carter Mecher and Dr. Richard Hatchett, who was serving as a White House biodefense policy adviser. Mecher was recruited to the effort because of his experience in healthcare delivery and his knowledge of how hospital systems functioned.[2] Hatchett brought deep expertise in public health preparedness and had worked with Rajeev Venkayya on other projects, including the effort to establish the Cities Readiness Initiative, a federally funded program that would provide the logistics needed to deploy medical countermeasures from the Strategic National Stockpile to local communities in the event of an emergency.[3] With Venkayya's support, Hatchett and Mecher formed a new work group to devise a plan for a worst-case scenario, a situation where a deadly flu pandemic started to sweep across America, and our drugs and vaccines couldn't interrupt its spread.

The team started with a large modeling exercise to gauge the impact of mitigation and the risks that these measures could pose. The researchers wanted to estimate how nonpharmaceutical interventions (NPIs) would work in the real world. They had modeled three different scenarios based on achieving different levels of community-wide social distancing. Taking into consideration the transmissibility of a pandemic strain, they found that timely adoption of these measures could substantially limit the spread of disease. The models also found that speed mattered. Implementing measures earlier would have the biggest impact on reducing transmission.[4] Another key was coordination. The virus crossed state borders, so mitigation measures needed to be adopted wherever the virus was spreading.

The modeling and especially what it revealed about the correlation between the timing of interventions and the different outcomes gave the team the insight to go back and reexamine how the approaches were used in the past. The team now knew what to look for. They wanted to see if they could discern patterns in the real-world application of these mitigation tactics that had escaped previous notice. Some of the elements of the overall approach had been used before, most prominently during the 1918 Spanish flu. The prevailing accounts about why different cities experienced different outcomes was that the virus had mutated as it moved from city to city. The modeling revealed something different. It showed that perhaps the mitigation tactics worked; however, timing mattered. The same sets of interventions, implemented at different times relative to the different epidemics in each of the cities, could result in vastly different outcomes.

Hatchett and Mecher teamed up with three other scientists with diverse expertise in pandemic response: Dr. Marc Lipsitch, Dr. Martin Cetron, and Dr. Howard Markel. Lipsitch is an epidemiologist focused on understanding mechanisms of immunity and would become one of the key experts to model the spread of COVID. Cetron is the longtime director of the CDC's Division of Global Migration and Quarantine, which oversees the agency's response to outbreaks. At the outset of the COVID pandemic, Cetron would take a lead role in repatriating Americans from China. Markel, a medical historian at the University of Michigan, focuses on the social impacts of epidemics.[5] Together, the group set out to see what lessons history offered.

Mecher and Hatchett formed one work group with Lipsitch, while Cetron and Markel worked independently on a similar effort. Each group aimed to dissect the steps that had been taken to contain the 1918 flu and the results of those efforts. A CDC team led by Cetron did an extensive analysis of data from about forty cities.[6] Hatchett's group examined newspaper clippings and other archives to see what was happening inside American cities as the virus spread across the country.[7] Hatchett's team initially looked at seventeen different cities but chose to focus on two of them, Philadelphia and St. Louis. The contrast between these two cities, both in terms of the timing of their interventions, and the scope of the epidemics that they'd endure, yielded critical insights. The goal was to develop an analysis on what worked and then channel the learnings into an actionable plan. Two independent researchers, Dr. Neil Ferguson and Dr. Martin Bootsma, conducted a third analysis (their paper would end up being published back-to-back with the analysis done by Hatchett's group).[8] The three groups would reach identical conclusions: there was a tight correlation between the number of interventions that cities would adopt, the speed by which these measures were implemented, and the outcomes that different cities experienced.

The Spanish flu was the deadliest pandemic in modern times, striking the world in three waves. In Philadelphia, the first cases of flu were reported on September 17, 1918, but local authorities were not overly alarmed. The city's director of public health and charities, Dr. Wilmer Krusen, was initially worried about an outbreak at the Navy's Fourth Naval District, which comprised the League Island Navy Yard in Philadelphia. On September 18, he set aside 150 beds at the city's Municipal Hospital for Contagious Diseases to care for navy personnel who had been stricken by the virus. That same week, the city had already recorded its first influenza death—an eighty-nine-year-old banker. But Krusen continued to believe the flu posed little threat to Philadelphia so long as sick patients were isolated.[9]

As cases grew, so did concern. The city's board of health made flu a mandatory reportable disease on September 21, reflecting these growing worries. They wanted to be able to track the epidemic.[10] Yet Krusen overruled requests for more forceful action. Schools held classes, church services continued, shops were open for business. The city also continued to allow large public gatherings to take place. On September 28, the city allowed the Fourth Liberty

Loan Parade to continue as planned. The parade was two miles long, guided by a marching band led by John Philip Sousa, and featuring biplanes, a new wartime innovation, that had been built in Philadelphia's Navy Yard.[11] The parade and accompanying festivities raised money through the sale of Liberty Loans—government bonds issued to help pay for World War I.[12] No measures were taken that day to limit contact among the 200,000 spectators who massed along Broad Street to cheer on the marchers. The flu, for a fleeting moment, seemed like an afterthought.

Like the Sturgis Bike Rally, which in the summer of 2020 brought together an estimated 460,000 motorcycle vehicles for a ten-day annual celebration held in Sturgis, South Dakota, just as COVID was beginning to surge across the Midwest, the Liberty Loan Parade was a fateful gathering, timed at precisely the moment that the flu was about to explode.[13] The parade was the tinder that ignited an epidemic in Philadelphia that would rage out of control. It would become the largest known "superspreading" event of the 1918 pandemic.

In Philadelphia, even after every bed in the city's thirty-one hospitals became filled, local officials were slow to act. On September 30, local health leaders met to discuss the crisis. They argued over whether the flu posed a significant threat to the population. They argued whether it merited stronger actions that would disrupt social and economic life. They couldn't agree, and little else was done. A circular was sent around to state health agencies charting how providers should care for patients. Some funds were allocated to the effort. That was it.[14]

After the parade, the situation deteriorated. By the first week in October, about 2,600 Philadelphia residents had died. A week later, the number grew to 4,500. Many of Philadelphia's health professionals were pressed into military service, and the city's remaining doctors couldn't handle the waves of illness.[15] With hospitals stretched and casualties mounting, city leaders finally started to take more decisive measures. The virus was out of control.

On October 3, the Pennsylvania health commissioner, Dr. Franklin B. Royer, issued a statewide order mandating that all places of public amusement, including theaters and saloons, suspend operations. "Never since the city was scourged with infantile paralysis," Krusen said in response to the order, "has it been necessary to take such stringent measures for the public good." The closure of schools and churches was initially left up to the discretion of local authorities. Many houses of worship decided to shut down, but some schools remained open. As the epidemic continued to worsen, and after four thousand pupils had become ill with the virus, the city's board of health intervened and ordered all schools closed, along with all places of worship.[16]

"The disease has been raging more freely in Philadelphia than in any other city in the state," reported the *Philadelphia Inquirer.* The actions were "the best means of safeguarding the people of the state," said Dr. Royer.[17]

But even against the backdrop of rising death and disease, and an especially cruel impact that the virus was having on Philadelphia's poorest residents, the steps stirred anger and opposition.

On one side of the debate, the *Philadelphia Inquirer* said the orders went too far, calling them a product of, and a currency for, the unfounded panic that accompanied the flu. On the opposite side was the *Philadelphia Evening Bulletin*, which wrote that the closures were crucial because group gatherings were driving transmission.[18] The epidemic, the newspaper said, would get worse before it got better. The city needed to control the spread.

Cases continued to surge. And yet many civic leaders would argue that the Spanish flu was driven by misguided fears and wasn't that dangerous. In 1918, one of the most prominent of these individuals was Krusen, who would declare that the end of the pandemic was near and that the cases had reached a "crest." Dr. John W. Croskey, president of the West Philadelphia Medical Association, similarly said that "the public should be educated to the fact that the disease is not as deadly as many believe it to be." However, Croskey had grossly underestimated the severity of the flu, putting the case fatality rate—the percentage of people who developed symptoms of flu and would die from the disease—at about 0.5 percent, which was far less than its real fatality rate.[19]

The controversy and the arguments for prudence had eerie echoes to the debate that would unfold in the face of COVID, with a vocal minority arguing that the danger posed by the flu had been exaggerated, and that mitigation was imprudent, unnecessary, and far too costly for what was presented as a limited threat. However, flu cases, and deaths, continued to rise, overwhelming the local hospitals. With doctors in short supply, medical students were pressed into service. Finally, by the end of October, new cases started to slow. On October 30, theaters and saloons were reopened. Church services and schools were restarted. The epidemic, for now, was receding.[20]

Philadelphia was slow to act and paid a heavy toll. But Mecher and Hatchett found a different experience in St. Louis, where the city's health commissioner, Dr. Max C. Starkloff, watched the virus sweep across the East Coast and knew it was just a matter of time before it spread west. Early on, Starkloff ordered local doctors to report cases of flu to his office and published an article in the *St. Louis Post-Dispatch* warning residents to avoid crowds and stay away from those who were ill.[21] The first seven cases of Spanish flu emerged on October 5, all in the same family.[22] The next day, another fifty cases were reported to city health officials. Starkloff knew the city was at a tipping point and that he needed to act aggressively. He asked that an upcoming Liberty Loan drive be canceled—in sharp contrast to the procrastination in Philadelphia. In the next two days, as cases grew (reaching one hundred civilian cases and another nine hundred at a local barracks), Starkloff asked the city's mayor and other leaders for legal authority to issue public health edicts. His request was granted. Starkloff's actions were swift and forceful. Starting on October 8, theaters, pool halls, and other public amusement venues were ordered shut. All public gatherings were banned. Churches were also shut. Schools were ordered closed the next day.[23]

The difference in the response times between Philadelphia and

St. Louis amounted to fourteen days when measured from the first reported cases—but those two weeks represented about three to five doubling times for a flu epidemic. For Hatchett and Mecher, the lessons were clear. The cost of even that brief delay was severe.[24] St. Louis got control of its epidemic much earlier, and with far fewer deaths. They found that cities like St. Louis, where multiple mitigation steps were taken early, had peak death rates that were about 50 percent lower than cities that acted more slowly and sparingly. They also had lower cumulative deaths. Early action mattered.[25]

Few cities maintained the interventions for longer than six weeks. So, it would be hard to know what the impact would have been if local leaders had stuck with the measures. But the findings showed that if social distancing had been implemented immediately after the flu first emerged, it could have significantly reduced death and disease. Hatchett and Mecher recognized that the virus would reemerge once the measures were lifted.[26] All the more reason it was just a bridge to a vaccine.

Hatchett and Mecher drew on their observations to formulate a plan, encompassing a combination of tactics. First were case-focused interventions such as isolation, household quarantine, and treatment with antivirals. Still, they knew that these measures would be insufficient because of the asymptomatic or minimally symptomatic cases that would escape detection. The next step was to lean on different forms of social distancing interventions. These measures would include closing schools, keeping kids at home, social distancing in the workplace, canceling mass gatherings, and similar approaches. A communicable disease spread through social networks, and key to its control were measures that would disrupt those social connections.[27] Additional mitigation tactics could include everything from altering work schedules, closing certain businesses or regulating the hours they could be opened, restricting public transportation, and mandating the use of face masks.[28] The measures could be imposed with varying degrees of intensity.

For example, social distancing could include an order to work from home, or a ban on gatherings. It could include strategies to create physical separation when people were in close settings—spacing out tables six feet or erecting physical barriers when separation wasn't possible.[29]

The analysis by Cetron and Markel, completed later, supported the work by Hatchett and Mecher. It found "a strong association between early, sustained, and layered application of nonpharmaceutical interventions and mitigating the consequences of the 1918–1919 influenza pandemic in the United States." The two authors had looked at forty-three US cities from September 8, 1918, through February 22, 1919, to see whether city-to-city variation in death and disease was associated with the timing, duration, and combination of mitigation tactics that were adopted. They found that "combinations of nonpharmaceutical interventions including school closure and public gathering bans appeared to have the most significant association with weekly excess death rates." The earlier the schools were closed, the stronger the association.[30]

Until Hatchett and Mecher undertook their efforts, the received wisdom was that nonpharmaceutical interventions didn't work and that the Spanish flu was "unstoppable." Hatchett and Mecher didn't see it that way and thought the tactics were underestimated. They wanted a "break the glass" plan, Hatchett said—a way to slow the spread of a novel flu while researchers were chasing medical countermeasures.[31] Hatchett's team dubbed the approach they had come up with as "targeted layered containment." However, they knew that the mitigation would only reduce the pace of transmission. In a pandemic, it wouldn't prevent most people from eventually getting infected. What mitigation could do is cut the peak number of new infections occurring at any one time, to keep the waves of infection manageable. The aim was to use these approaches to keep the healthcare system from becoming overwhelmed by a sudden rush of sick patients. So long as hospitals could keep up with the spread and maintain enough intensive-care beds and ventilators to care for the critically ill, the belief was that you could keep the death rate down.

In modern times, when seasonal flu would become epidemic in a specific region, elements of these mitigation tactics were sometimes used on a small and mostly uncoordinated scale. Schools would be closed temporarily in the event of a major flu outbreak. Public events with large crowds would be canceled or postponed when communities experienced a lot of spread.[32]

In more recent years, while the evidence of the effectiveness of these tactics was mixed, there was reasonable support for their use. The merits of closing schools as part of the pandemic plan was one of the biggest sources of controversy, and would again emerge as a key uncertainty when it came to our response to COVID. It has been accepted that closing schools in the setting of a flu outbreak could reduce local spread since schools can accelerate transmission in a community.[33] The fact that the incidence of infection increased when schools reopened lent further support to the belief that school closings could reduce flu outbreaks. However, in most cases, schools weren't closed in advance to prevent amplification of flu, but shut reactively, only after an epidemic was well under way and it became hard to keep them open.

School closures were also adopted on a limited basis in 2009 to confront the swine flu, with mixed results, in part because the closures were enacted reactively, after epidemic spread had already ensued.[34] The 2009 experience informed a lot of our thinking on schools and epidemics. The tactics were usually proportionate to the perceived risk. For example, classes would be canceled if more than a certain number of students were absent, or the entire school would be closed if the rate of absentees rose above some threshold. When it came to flu—and especially a pandemic strain—children could be severely affected, and schools were also perceived as being sources of community spread. The behavior of children and the density of school settings meant that schools were seen as the social compartments with the highest attack rates.

Following the work done by Hatchett's team in 2005, there emerged a belief that closing schools could reduce death and disease among children and reduce the scope of the epidemic. But the evidence for these measures was mixed, and in almost all cases, the data were specific to a pandemic flu. For epidemics in which children are less severely affected, and less prone to catching and spreading the infection, it's likely that closing schools would have less impact on spread.[35]

In reality, the full effectiveness of isolated school closures in an epidemic was never fully measured in the real world, even in flu, because it was seldom used on a large scale. During a 2008 outbreak of flu in Hong Kong, kindergartens and primary schools were closed for two weeks after two children died from influenza. Researchers concluded that the closures came just as the epidemic was already peaking and had no impact on spread.[36] Other experiences would point in the opposite direction. In Israel, a teacher strike began in the last week of December 1999, causing elementary schools to close nationwide. The strike came right in the middle of a furious flu epidemic. Flu cases fell sharply when the strike forced the schools to close. And when the strike ended and kids returned to classes, flu cases rebounded sharply. The experience suggested that the strike, and the forced school closures, likely had a suppressive effect on the rate of flu transmission.[37] The experience in Russia, where local officials would close the country's schools during especially intense flu outbreaks, also indicated that the measures could slow community spread.[38]

Modeling suggested that timing matters: the interventions had their greatest impact if schools were closed before 1 percent of a local population was infected.[39] More systematic studies had found that in the setting of flu epidemics, closing schools for long stretches of time reduced the total number of community cases.[40] These steps could also reduce peak attack rates by up to 45 percent among a community, according to the research (and by as much as about 50 percent among children).[41] Real-world studies, including surveys done after the 2009 pandemic that analyzed the influence of school closure on transmission, supported these conclusions.[42]

These findings were not without debate and some controversy, but there was general agreement that schools could be amplifiers of spread in the setting of a flu epidemic. Adding to the importance of focusing mitigation steps on kids was the fact that children were seen as being uniquely vulnerable to a pandemic strain of flu. In all of the past pandemics, the flu had a disproportionately heavy toll on kids. American children under the age of nineteen had a higher risk of being infected with flu, and a higher risk of bad outcomes, than adults of any age during prior epidemics that had involved influenza A in 1977–78 and 1980–81, as well as the epidemics that had involved influenza B in 1976–77 and 1979–80. In the first ten months of the 2009–10 H1N1 swine flu pandemic, the hospitalization rate among stricken kids under the age of five was more than double that of any other age group.[43] The younger the kids, the more likely they were to be hospitalized from flu and its complications.[44] Similar findings were seen in other nations.[45] In Germany, death rates during the 2009 pandemic were highest among kids under age one and adults aged thirty-five to fifty-nine.[46]

Yet the recommendation to close schools was the piece that would galvanize the most criticism of the strategy of mitigation that Hatchett and Mecher had fashioned. One of the most outspoken critics of the overall plan was Dr. D. A. Henderson, a famed American physician who led the global effort to eradicate smallpox when he served as chief of the CDC's virus disease surveillance programs in the 1960s. More recently, after the 9/11 attacks, he had advised the White House on efforts to improve the nation's defense against bioweapons.[47]

Henderson argued that mitigation tactics, used to confront a pandemic flu, would "result in significant disruption of the social functioning of communities and result in possibly serious economic problems." He said the NPIs could cause more problems than they solved. This was before Zoom, before technology could be used to help substitute for onsite work, and before states had thought about the implications of closing schools and shutting children off from subsidized lunches that many poor kids depend on.

Writing in opposition to the plan that Hatchett and Mecher were advocating, Henderson said that, except in the most extreme circumstances, community mitigation never before had a significant impact on altering the course of a pandemic. He said that we would have to let a virus run its course if we didn't have the medical countermeasures to confront it directly. "Few analyses have been produced that weigh the hoped-for efficacy of such measures against the potential impacts of large-scale or long-term implementation of these measures," he wrote.[48] "A fundamental premise of disease mitigation that has been advanced by some in the policymaking community is that a less intense but more prolonged pandemic may be easier for society to bear," he said, but this too was "speculative." Others shared Henderson's views. Leaders from the Department of Homeland Security and the Department of Commerce were concerned about the negative impact that the mitigation tactics could have on the economy.

However, Hatchett and Mecher's modeling, and the deep dive into the history of the Spanish flu, started to change minds inside the public health establishment, and especially the CDC. The findings on the NPIs from 1918 were so striking that they surprised the team. The nonpharmaceutical interventions had a profound effect on slowing spread, but they needed to be adopted early in the course of a pandemic. The best way to contain a pandemic would remain through vaccination. But it might be months, or longer, before a vaccine could be made available. In the meantime, authorities would have to lean heavily on detection, testing, tracing, and isolating the infected. But they could also use nonpharmaceutical interventions as a way to reduce spread.

The pandemic plan that ultimately emerged had an entire chapter focused on mitigation. Included were measures such as advising people to stay at least three feet apart, relying on telecommuting for work, closing schools, canceling nonessential public gatherings, and restricting long-distance travel. The plan also advised that local authorities consider "snow day" restrictions—mandates by community authorities that everyone stay home for two incubation cycles (about four days for typical flu viruses).

The history of the "snow day" concept is interesting. It was imported into the COVID response, and the Coronavirus Task Force was poised to propose federally directed snow day restrictions in the early-outbreak cities like New York before these cities implemented their own stay-at-home orders. The snow day concept was first advanced in the 2003 time frame by Stewart Simonson, who served as the assistant secretary of Health and Human Services for Public Health Emergency Preparedness (now known as the Office of the Assistant Secretary for Preparedness and Response), as part of his thinking about anthrax response, on the grounds that everyone can shelter in place for three to five days, more or less at a moment's notice (as people do when a "snow day" is declared). The approach grew out of a tabletop exercise called Scarlet Cloud, where officials simulated an anthrax attack on a major city. The benefits of sheltering in place after an aerosol attack with anthrax are multiple. Doing so will keep people off the streets and in place while vaccine and drug supplies are mobilized and a treatment campaign is implemented. If people don't move around a lot, it will be much easier to deliver antibiotics as post-exposure prophylaxis to those who need it within an interval when it can work. So, it was a very specific and germane response to a very explicit biodefense problem—before it was imported as a more general concept for dealing with a crisis involving a communicable disease. But it was never envisioned as lasting more than a week.

The recommendations that Hatchett and Mecher helped fashion were ultimately incorporated into two lengthy reports, published in 2006: the Department of Health and Human Service's Pandemic Influenza Plan and the Homeland Security Council's National Strategy for Pandemic Influenza: Implementation Plan.[49] These documents formed the crux of the nation's blueprint for dealing with a pandemic. The secretary of Health and Human Services at the time, Michael O. Leavitt, was a key architect of the plan and helped lead the effort that brought it together.

In the final documents, the inclusion of nonpharmaceutical interventions got little public attention. People focused on other proposals, especially the enormous sums that would be spent to build a national stockpile of countermeasures and the proposed investments in advanced manufacturing for vaccines. During the COVID pandemic, Secretary Azar would tell the Coronavirus Task Force that both documents helped inform the plan that was ultimately employed to confront COVID.[50] But back in 2006, the CDC still hadn't fully bought in to the concept of using non-pharmaceutical interventions, even though the framework was included in the final plan.[51] Hatchett and Mecher had nailed down the key findings showing the benefits of the approach, but they hadn't persuaded everyone. In the end, the CDC would come to champion the proposals. However, broader acceptance didn't come about until a year later, in 2007, with the publication of the CDC's Community Mitigation Guidance, in which NPIs held a prominent role.[52]

The approach outlined, however, differed from the strategies that would ultimately be imported into the COVID response. In workplaces, for example, Hatchett and Mecher didn't envision that businesses would be closed entirely. Instead, businesses would follow plans to limit spread through social distancing. The plan Hatchett and Mecher crafted had discussed recommendations to close certain hospitality venues like theaters or bars, but broad stay-at-home orders, or "lockdowns," were never considered as an option. The 2006 plan was thorough, but also limited in some important ways. It focused entirely on the risk from a pandemic flu, and didn't contemplate a coronavirus, or even contain the word anywhere in its 233 pages. Since the 1918 pandemic, there had been two major global pandemics. Both involved influenza A.[53] The next pandemic, it was assumed, would also involve a novel flu. So the plan was written with that fear in sight. But it would be adopted to counter COVID nonetheless.

On February 20, 2020, inside the White House Situation Room, Secretary Azar delivered to the White House leadership what he'd frame as the "doctrine" that we'd follow for addressing COVID. Also baked into his presentation were findings HHS had generated from a "tabletop exercise" run the previous year, as part of the routine planning that has gone on since 2005 to help prep the nation for a pandemic. The simulation, dubbed Crimson Contagion, was a joint exercise conducted from January to August 2019 that aimed to test the capacity of the federal government and twelve states to respond to a severe pandemic flu originating in China. In the scenario they drilled, tourists returning from China spread a respiratory virus in the US, beginning in Chicago. In less than two months, the virus had infected 110 million Americans, killing more than half a million. The report issued at the conclusion of the exercise was ominous.[54] Federal agencies fought over who was in charge. There were shortages of protective gear like N95 respirators and ventilators. States went their own way on mitigation, with some states refusing a CDC directive to close schools as a way to limit spread.[55]

The plan that Secretary Azar briefed that day would shift a lot of the responsibility to the states for taking steps to respond to COVID. It gave the states a lot of leeway to fashion different strategies. However, the idea of using mitigation as a way to slow the spread remained a core tenet.

By the beginning of March, as community spread was under way, and without a single national strategy, cities and states began to go their own way. Governors were implementing local policies that were beginning to produce regional outcomes on what would be a national epidemic.

CHAPTER 12

A PLAN GONE AWRY

On March 13, walking around the West Wing of the White House, you wouldn't know we were about two weeks away from a synchronized, national shutdown of nonessential businesses.

I was at the White House that day for a scheduled meeting to see President Trump to share my views on the unfolding crisis. We were three days away from the lockdown of San Francisco, seven days away from the stay-at-home order in New York. Our models would later show that the epidemic was already spreading uncontrollably across the nation.[1]

And yet a lenience about the personal dangers pervaded the building among the White House staff that I met with that morning. Vice President Pence insisted on shaking my hand and seemed nonplussed when I immediately dabbed my palms with Purell. Young staffers were crammed into small offices. More than a dozen executives from health-related companies like CVS and Walmart were crowded into the small West Wing lobby, waiting to go into a meeting with the president in the Roosevelt Room. Later, they'd join the president in the Rose Garden for an announcement on new efforts to create drive-through testing sites. Roughly eighty would be established, although thousands would be promised. Supporting these eighty sites would consume about 30 percent of all of the scarce medical supplies—including N95 respirator masks, gowns, and gloves—that remained in the national strategic stockpile of medical equipment.[2]

It wasn't just inside the White House—across the nation, we were just coming to terms with the extent of transmission that was already well under way, and the imminent danger. The president was set to declare the pandemic a national emergency that day. It was the first time such a declaration had been issued for an infectious disease outbreak since the H1N1 pandemic in 2009.[3] Even so, no one was tested for the virus before they entered the West Wing. There were no temperature checks. No health questionnaires. No masks. And no social distancing. This was before the epidemic spilled into open view, but while it was spreading quietly and widely. In fact, the only indication of the close reach of the virus inside the White House came from a FedEx package that I was handed early that morning, in a pre-meeting before my Oval Office visit. I was asked, somewhat mischievously, to open the box. I did. The package contained two plastic containers, each filled with the liquid medium that's used to carry respiratory samples gathered from patients for testing. These are the reagent fluids that are used to preserve viral samples until they can get to a lab for processing. These two test kits, I was told, were for the president.

President Trump had called me the previous day and asked me to come to Washington to meet with him. He wanted to discuss the unfolding crisis and get my take on the outlook. Trump had taken an interest in my work at the FDA and had supported my efforts to rein in teen use of tobacco products, to advance policies that promoted generic drug competition, and other public health initiatives. I had stayed in touch with senior White House staff and was grateful to the president for the active support he showed me when I served as his FDA commissioner.

When I met with senior White House staff that morning, they wanted something different: my help in trying to convince the president that we needed to adopt some form of mitigation to slow the spread. In the Oval Office that day, I made a brief pitch for mitigation steps—to restrict large gatherings and urge social distancing measures. But the president seemed already convinced that we needed to urge the public to adopt the tougher and more cautious measures. If he needed any further persuasion, it felt like someone else had already done the work before me. The COVID test was mostly what he wanted to talk about.

The president was being tested because earlier that week he had been in close contact with Fabio Wajngarten, the press secretary for President Jair Bolsonaro of Brazil, and Wajngarten had developed COVID. A photograph was circulating in the media showing Trump standing next to Wajngarten at a soiree at Mar-a-Lago the previous weekend. No precautions had been taken at the event, or the following day, when Trump hosted a fund-raising brunch for about nine hundred attendees.[4] The two events were held just as America was beginning to become aware of the risks, and that SARS-CoV-2 could already be spreading in parts of the US. Wajngarten began showing symptoms three days after meeting with Trump and tested positive for the virus shortly afterward.[5] It would emerge a few days later that, in total, twenty-two members of the Brazilian delegation were diagnosed with the coronavirus when they returned home from the US trip.[6] Based on what we were learning about the virus's incubation period and the fact that people could be most contagious just before or right after they first developed symptoms, Wajngarten could have passed it on to Trump.

In the Oval Office, I still made my case for some of the steps that could be adopted to slow the spread of coronavirus, to prevent the healthcare system from becoming overwhelmed by patients rushing to hospitals all at the same time. The problem was that without a diagnostic test, we had no way of knowing how widely the virus was spreading and which cities were already heavily seeded. I told him that because the US didn't have a reliable test for SARS-CoV-2, and we were not screening patients, the virus could be spreading widely in our communities and America might be unaware.

It was clear that New York was on the cusp of a major crisis. A handful of other cities looked like they were staring over the same precipice. Health officials were now working on a belief that the country was being heavily seeded. In retrospect, only certain regions, and cities, were in imminent danger. In New York City, perhaps the largest and best-equipped healthcare system in the world would be brought to the edge of collapse in the coming weeks.

The president seemed to agree on the need for stronger action. He was well briefed before my meeting. Even though some of his public statements over that time period would have suggested he didn't see the looming calamity, or the need for strong action, his private demeanor that day left me with the clear impression that he recognized the grave risks, he was more solemn than previous meetings, and when it came to the question of mitigation, he was mostly sold on the ideas before I had arrived.

It would be revealed seven months later that the president was sitting down for a series of taped interviews with *Washington Post* journalist Bob Woodward around the same time. Trump revealed his fears to Woodward and would face understandable criticism for, at times, publicly downplaying the risks while privately he was worrying about the dangers.

Six days later, Trump told Woodward he was expressly downplaying the dangers to avoid creating a panic.[7] The president, Woodward would write, understood the gravity of the threat we faced. When I had seen Trump on March 13, I told him that New York was on the brink of a major epidemic that could dangerously strain its hospitals. Other cities were not far behind. But without broader testing, we didn't know how much virus was already circulating.

Five days earlier, I had aired the same message on the Sunday morning CBS news show *Face the Nation*. Referring to the mitigation steps that I believed we would soon be forced to adopt, I told host Margaret Brennan that "we're past the point of containment. We have to implement broad mitigation strategies. The next two weeks are really going to change the complexion in this country. We'll get through this, but it's going to be a hard period."[8]

The fundamental challenge remained that we were implementing a plan that was crafted with flu in mind, not a coronavirus. All the work that Hatchett and Mecher had done in 2005 and 2006 would be put to the test for the first time, and on a wide scale, against a pathogen for which it wasn't fashioned. But more critically, the historic failure to deploy a simple screening test, and thus our inability to identify the virus and slow its advance through testing and tracing, meant that mitigation would be used far more strictly, and indiscriminately. The nation would go well beyond that 2005 plan and implement a national stay-at-home recommendation that was never envisioned in any of our pandemic playbooks.

Back in 2005, a national stay-at-home mandate had not even been considered as part of the plan, in part because the plan anticipated that we could test for the virus and rely on more targeted interventions. After my meeting with President Trump, he was scheduled to deliver remarks in the Rose Garden. He would make the first nod toward the measures that would soon constitute the administration's plan to "Slow the Spread": a mix of actions to keep people apart until chains of viral transmission could be broken and the epidemic curve would start to flatten. The White House would soon issue guidance, he said, directing nursing homes to prohibit nearly all visitors that were not deemed medically essential.[9] Trump was being briefed on new projections from Imperial College London that showed millions of people dying if more extensive measures were not taken. Chastened by the new data, the president showed a changed demeanor.[10]

Health and human rights lawyers often talk about using the least restrictive measure necessary to achieve the public health outcome in the setting of a crisis.[11] Blind to COVID's spread, this was not what the US government adopted three days later, on March 16, when President Trump announced his policy of fifteen days to "slow the spread."[12] It was the beginning of our counterattack against the virus.

The measures that we'd be forced to embrace would be far more sweeping, more indiscriminate, and more burdensome, because we had no idea where SARS-CoV-2 was already spreading widely, and where we could still try to use testing and tracing to contain it.

We had no way of detecting the virus.

So, instead of relying on testing, tracing, and isolation, the plan urged vulnerable Americans to stay home, and everyone else to essentially self-quarantine. "If someone in your household has tested positive for the Coronavirus, keep the entire household at home," the new federal guidelines urged. Americans shouldn't gather in groups larger than ten, and "if you are an older American, stay home and away from other people," the guidance said, adding, "If you are a person with a serious underlying health condition—such as a significant heart or lung problem—stay home and away from other people."[13]

"With several weeks of focused action, we can turn the corner and turn it quickly," the president said in announcing the fifteen-day plan.[14] At the same time, the White House was working on the largest-ever economic recovery bill to offset the hardships that people would bear by adhering to the new guidelines. The mitigation plan slowed the spread. But at a huge cost. It remained contentious.

I talked to Vice President Pence by phone a few weeks later, and he told me that he had just finished another telephone call, where he had been briefed on a plan where doctors could convert machines used to deliver anesthetics during surgery into ventilators.[15] These machines are constructed to deliver a mixture of oxygen and gas to place patients under general anesthesia, but Pence said that they could be modified with a simple instrument to provide the basic components of ventilation.

At that time, there was palpable fear that the United States would run out of ventilators to care for a rush of desperately ill COVID patients. As the virus surged across major American cities, nobody knew how bad it would get. The plan to retrofit the anesthesia machines struck me as a jarring reminder of the precariousness of the situation we were in, and our limited options. The vice president derived an understandable measure of relief from this plan. It could be a critical stopgap in extending the available pool of ventilators. However, the makeshift effort was hardly ideal, and it was a testament to the challenging circumstances that we faced with our own supply chain. This is what it had come down to: converting anesthesia machines to make ventilators.

When the epidemic was already spreading widely in New York, several senior administration officials told me at the time that they shared a deep concern that other American cities would soon experience spread on the same scale. A number of cities, including Chicago, Detroit, and New Orleans, were all seeing accelerating transmission. One senior federal health official fearfully said to me that if any one of these other cities "fell," then the federal government wouldn't be able to give it "the New York treatment." The comment stuck with me. The federal government, he said, had forward-deployed so many resources to New York—masks, ventilators, drugs, and other supplies—that, by most measures, it was tapped out. In fact, by March 24, Washington had shipped four thousand ventilators to New York. It was a meaningful portion of the total supply of working ventilators that the government had in its national stockpile.[16]

The US never did run out of ventilators, in part because of the extraordinary action we took to require the vast majority of Americans to largely remain at home, steps that slowed the spread of the virus long enough to give our healthcare system—especially in the hard-hit Northeast—some breathing room. The federal government also undertook a major effort to secure more ventilators. Large industrial manufacturers, including the automakers, got into the market, repurposing components similar to those found in ventilators.[17] By the fall, General Motors and Ford had delivered more than eighty thousand ventilators to the US government.[18]

Even so, New York nearly collapsed. Other cities were on the brink. I was told by members of the White House staff that the fear that another city could follow a similar fate as New York was one factor that persuaded President Trump to reluctantly support a national plan that called on nonessential workers to stay at home for forty-five days.[19] Without any reliable way to uncover the true number of infections, the Trump administration had to plan for the worst. They knew that if multiple cities faced epidemics on the scale of New York, it would overwhelm the nation.

In these settings, you're trying to make appropriate decisions in the absence of adequate information. These were the words of Dr. William Foege, an American physician credited with devising the global strategy that eradicated smallpox in the late 1970s.[20] On March 16, the executive director of the WHO emergencies program, Dr. Michael Ryan, put his own spin on this observation. "Perfection is the enemy of the good when it comes to emergency management," he said. "Speed trumps perfection, and the problem in society we have at the moment is everyone is afraid of making a mistake, everyone is afraid of the consequence of error. But the greatest error is not to move. The greatest error is to be paralyzed by the fear of failure."[21]

The stay-at-home order was critical to stem the spread in New York, but investigations would later show that the measures were applied too slowly, contributing to the city's devastating death toll.[22] For a period of time, the New York hospitals became a COVID-only healthcare system, barely able to handle medical emergencies like heart attacks and strokes. New York's total hospital capacity was breached, forcing patients to be moved into makeshift hospitals at the Javits Center and in tents pitched in Central Park. By June, almost nineteen thousand New Yorkers would be dead from COVID.[23]

The impact on New York's communities of color was especially brutal. Many of the Americans who were hardest hit by the pandemic were those who already faced racial disparity in healthcare access and delivery. Many lived in crowded multigenerational homes where one infection could put an entire family at risk, and they didn't have the social capital at work to insist on protective personal equipment and other measures that could help keep a person safe at their place of employment. An analysis conducted by the office of Scott M. Stringer, the New York City comptroller, found that 75 percent of front-line workers in the city are minorities. They were on the front lines of keeping the country functioning through COVID. And in the process, they were made excessively vulnerable to the virus.[24] During the epidemic's initial wave in the spring, Black New Yorkers were twice as likely as white New Yorkers to die from COVID and more than twice as likely to suffer a nonfatal hospitalization.[25]

Other cities faced similar threats. New Orleans, Detroit, Chicago, Boston, and Philadelphia also had to take strong mitigation steps to limit the scope of dangerous epidemics. However, not every part of the country faced equal risk. The problem was, we didn't have testing in place that would let us know everywhere the virus was spreading, so we couldn't target our mitigation.

Policymakers were forced to rely on incomplete data and a fair degree of scientific conjecture. For example, New York never shut its subway system, and the Coronavirus Task Force believed that contaminated surfaces (fomites) on mass transit systems may have been contributing to spread. It was a fact that was initially implicated in the city's devastating epidemic. The view was that shared surfaces in buses and subways were becoming contaminated with viral particles left behind by infected patients. These respiratory secretions were then picked up by other passengers who would touch the dirty surface. (The New York City Metro Transit Authority closed its subway system every night to treat passenger surfaces with antiseptics.)[26]

The theory sprang from an observation inside the CDC that there were a lot of people falling ill with COVID, where public health workers couldn't trace their illness back to some symptomatic patient they'd been in contact with. Underlying all this, the

CDC believed that the coronavirus was behaving like influenza, and so the most plausible explanation for these mysterious chains of transmission must be some contaminated surface that patients had touched, where they picked up the virus through respiratory droplets left behind on a subway pole or a door handle.

In reality, a lot of those puzzling chains of transmission weren't the result of fomites, but rather, asymptomatic carriers who had gone on to unknowingly infect those around them. Since the CDC believed that SARS-CoV-2 spread like flu, however, they discounted the role of asymptomatic transmission. And since doctors couldn't test people for the virus, nobody could firmly uncover those asymptomatic cases. Remember, New York never shut its subway system, and early on kept it running with normal operations, without the deep cleanings. It was a fact that the CDC wrongly implicated in the city's devastating epidemic. Showing how weak data could, in turn, lead to misguided policy decisions, this preoccupation with fomites led the task force to be especially worried about cities that were operating large mass transit systems like Chicago and Boston. In retrospect, mass transit may have fostered spread. However, data would show that the spread of the virus probably wasn't from shared surfaces but from the aerosolization of respiratory droplets in confined space.

Because the US wasn't doing enough testing and contact tracing, however, public health officials and the CDC weren't able to identify the real patterns of spread. It caused us to overestimate the impact of contaminated surfaces as a source of transmission because we were missing all of the asymptomatic cases. That caused the task force to focus its concern on some of the wrong cities. I was told that part of the impetus to implement the national stay-at-home recommendation was tied to the concern around the shared surfaces being a major source of early spread. The task force feared that many cities had continued to operate large mass transit systems at the outset of the pandemic and, like New York, these cities too would soon experience uncontrolled spread.

Health officials may have made the right policy decision to pause activity, but they did it for some of the wrong reasons. The cities that were most at risk wouldn't be measured by the size of their mass transit systems, because the principal driver of transmission wasn't contaminated shared surfaces. The CDC's revised guidance would later declare that most of the transmission was through respiratory spread. However, it would take the CDC almost a year to fully revisit the risk from fomites.[27] In the meantime, businesses and municipalities overinvested in cleaning programs. Americans wrongly shunned food deliveries, wiped down their groceries unnecessarily, and focused limited effort and resources on mitigating the wrong drivers of spread.[28]

The inability to uncover the real patterns of spread, and a preoccupation inside the CDC that the coronavirus would behave like influenza, caused the country to underestimate and overestimate the virus in ways that had tragic consequences. The CDC's preoccupation with fomites probably contributed to the agency misjudging the contribution of aerosol transmission and discounting the use of masks early in the pandemic. It became increasingly clear that a lot of spread wasn't merely through droplets but also through aerosol transmission.[29] Droplets are larger particles we secrete when we talk, cough, or sneeze.[30] They are heavier and can't travel long distances. Aerosols are smaller particles that emerge after droplets start to evaporate. They can spread through the air over longer distances and stay suspended for prolonged periods of time. This becomes especially true when the aerosols get into sealed, indoor spaces. If the right patient was in the right confined space at the right time in their illness, typically a day or two before symptoms set in and when they were maximally contagious, then the pattern of spread they could spark could look more like measles (which spreads through aerosols) than flu (which spreads primarily through droplets). Yet, in considering COVID, the CDC had flu in mind.[31]

The early fixation on droplets and fomites as the primary modes of spread would persist, even as new evidence began to challenge these assumptions. The CDC gradually reconsidered its perspective on the relative risk of fomites as a source of spread. However, the first update came in May, far too late in the pandemic, and long after a lot of energy had been misdirected on scrubbing down surfaces, efforts and resources that could have been focused on activities more likely to reduce risks (like wearing masks, or improving airflow and filtration in confined spaces).[32] Eighteen months after the pandemic began, on May 7, 2021, the CDC finally updated its explanations on how SARS-CoV-2 is spread, citing inhalation as the main mode of transmission and emphasizing the role of aerosols, and placing less emphasis on fomites.[33]

The CDC's revision of its recommendations for people to remain six feet apart, in March 2021, also came a full year after the pandemic began.[34] The six-foot distancing requirement was a primary reason many schools cited for why they couldn't open for full-time classroom instruction in the winter and spring. They didn't have enough space to create that much distance between students. It was probably the single most costly intervention the CDC recommended that was consistently applied throughout the pandemic, causing schools and businesses to have to forgo a lot of activity to create that much distance between students and patrons.[35] It was another reflection of how the CDC's focus on flu as a model for COVID spread ended up being a costly misjudgment.

Some of the small number of large school districts that were able to remain open full-time since the fall achieved the goal by rejecting the six-foot requirement, including schools in Florida, where the Florida Department of Education allowed students to be closer than six feet.[36] And nobody knew exactly where the guidance on six feet of distance came from.

"The origin of the six-foot distancing recommendation is something of a mystery," wrote the *New York Times* in March 2021, a year after the pandemic started. "When the virus first emerged, many experts believed that it was transmitted primarily through large respiratory droplets, which are relatively heavy. Old scientific studies, some dating back more than a century, suggested that these droplets tend not to travel more than three to six feet."[37] One senior HHS official told me that the six-foot recommendation was a compromise reached between the CDC and officials in the Office of Management and Budget (OMB). Initially, in February 2020, the CDC had recommended eight to ten feet of distance. However, the acting head of OMB, Russ Vought, said that such a requirement would be inoperable, hard to follow, and as a practical matter, couldn't be uniformly implemented. So, the CDC compromised with the White House and settled on six feet.

But the CDC's recommendation for six feet of distance would persist well past the scientific realization that aerosols might be responsible for a lot of the transmission, and that the requirement for six feet of distance would be less relevant for this mode of spread.

That requirement for six feet of distance was particular to the US. The WHO recommended one meter, or 3.3 feet.[38] China, France, Denmark, and Hong Kong went with one meter. South Korea opted for 1.4 meters; Germany, Italy, and Australia for 1.5 meters.[39] (The European CDC continued to recommend maintaining physical distance of ideally two meters.)[40] Even the CDC's decision in the spring to revise the recommendation from six to three feet seemed arbitrary. While the CDC published some accompanying studies to support the move, it wasn't any better established that the science had firmed around the new requirement. I was told by one HHS official that the revised three-foot number was based on an experiment that the CDC had shared with the senior leadership of HHS in the fall, six months earlier, showing that two people who were masked and standing three feet apart reduced spread by more than 70 percent. If that experiment formed some of the basis for the change in the recommendation, it begs the question, why did it take the CDC six months to issue the revised guidelines?

If we had more insight into where and how the virus was spreading, we would have been able to reserve the most stringent measures, like stay-at-home orders, only for cities where the virus was already epidemic. We would have been able to adopt mitigation tools that had the best opportunity to interrupt spread based on how the virus was being transmitted. That's what happened with some success in 1918. That's what the 2005 pandemic plan had prepped for. It envisioned that mitigation would be directed to places where there was active spread, with the strictest measures reserved for cities where the virus was already out of control.

Such targeting would have spared areas of the nation that were not yet at significant risk, where containment, by testing and tracing sick people and then isolating infected patients, was still possible. That would have reduced the national burden we incurred. It also would have preserved more credibility for public health officials to adopt these measures in places where stronger action was needed later, when the virus finally became epidemic in the South and Midwest.

In July and August, when a heat wave struck the Sunbelt and people were driven inside for air-conditioning, the crowding fueled a dense epidemic.[41] However, at that point, people in states like Arizona, Florida, and Texas were psychologically done with "lockdowns," having shut down during the spring, when the virus wasn't yet spreading widely in those regions. They resented having to close their businesses in the spring when their cities didn't yet face significant risk from the virus. So, when that risk finally arrived in the late summer, and the epidemic spread to Houston, Phoenix, and Miami, governors were hesitant to reimpose tough measures. There was no popular support for more shutdowns. The people said: We already did this; we're not going to do it again.

By the summer, Americans were fed up with the isolation. People started to go out again, creating informal new norms, and governors sought policies to conform safely to what people were already doing. Few states met the criteria that had been established for reopening. It didn't matter. People were done with the closures of businesses and limits on large gatherings. As befits a free society, governors mostly followed the public's inclinations.[42] States such as Florida, Texas, and Arizona would have benefited from tougher mitigation once they faced epidemics in the summer, but there was no backing for the measures. These southern states would have been better served if the initial round of shutdowns had been more targeted, without their major cities incurring stay-at-home orders in April, when they faced less risk. As it was, the governors from these states had already spent the political capital that would be needed to gain support for the aggressive measures in July, when they finally suffered pervasive spread.

Once the initial policy of "fifteen days to slow the spread" was implemented, and lapsed, the White House began discussing a more targeted approach to the mitigation. The idea was to maintain social distancing, work from home, and other measures in places where the virus was out of control, but relax the measures in places where there was less spread.[43]

The president's advisers were deeply divided on the national stay-at-home order and badly wanted a less taxing alternative. The president, too, wanted a middle ground. Trump told the nation's governors on a private phone call that his administration would propose three categories for ranking counties, a color-coded scheme that would rate different regions red, yellow, and green based on their burden of infection. The idea was to adjust the mitigation locally, based on measurable risk on the ground. I had discussed this plan with the White House staff at the time. The idea had a lot of merit, but it could never be put into practice, because it relied on data that the president's own experts said wasn't yet available.[44] The CDC and task force didn't know which local regions had runaway transmission, or where SARS-CoV-2 spread was still quiescent. The president's ambition was ahead of what his public health authorities could deliver.

So, the White House tried to develop the data to support the new framework, but it couldn't be built on the fly. To advance the color-coded scheme, the White House had invested in a major collaboration with the nation's large commercial labs, including Quest and LabCorp. Since the federal government was still struggling to get more testing in place to diagnose active infection, the idea was to use a different test, one that screened routine blood draws for the presence of antibodies to the virus, to build a map that identified areas where a lot of people had already been exposed.[45]

The effort layered the antibody tests onto normal blood draws that were already being processed by the country's two major commercial labs, Quest and LabCorp. The tests for COVID antibodies were added to blood samples drawn on people who had come into these labs to be tested for some other reason. The aim was to use this information to develop the color-coded map that the president had pitched to the nation's governors.

The data were collected, and even made available on the CDC's website.[46] However, it was never able to support what the White House envisioned. Screening for the prevalence of COVID antibodies was a lagging not a leading indicator of epidemic spread. It could tell where the virus had been, but not where it was going. We needed real-time testing information. The idea of using information on spread to power a more targeted approach to mitigation and reopening had merit, but we just didn't have the right data to support the strategy. Even once testing started to get into place, we didn't have enough of it and lacked adequate insight on how and where SARS-CoV-2 infection was heading. What factors created the conditions for spread? The CDC lacked an analytical framework that could be used to inform us on current outbreaks and future trends.[47]

So, on Sunday, March 29, before the fifteen days to slow the spread were set to end, President Trump urged governors to maintain the measures for another thirty days, until April 30, largely limiting to their homes the roughly 230 million Americans who were not essential employees. "Avoid social gatherings, especially those with more than 10 people . . . use pickup or delivery options instead of eating at restaurants or bars. Work from home if at all possible.

Do not go to work if you feel sick," the plan directed. "The better you do, the faster this whole nightmare will end," President Trump said in announcing the extension of the nationwide order.[48]

I had been on the phone with White House officials that Sunday. They were worried that Trump would lift the measures entirely, only to have the epidemic explode. I was told that the president was ultimately persuaded by modeling he'd been shown from the Institute for Health Metrics and Evaluation, a project created at the University of Washington. It showed that deaths could reach a million or more if the spread wasn't brought under control.

Earlier in the day, Anthony Fauci told CNN's *State of the Union* that the timing for lifting social distancing restrictions would depend heavily on the availability of new tests that can give results in fifteen minutes. These antigen tests, he said, would allow health officials to identify infected patients, isolate them, and do contact tracing to locate the people they interacted with.[49] By that point, COVID had infected more than 130,000 Americans and killed nearly 2,500. Hong Kong had recently installed vending machines in its subway stations that dispensed rapid COVID tests right to consumers. In the US, the rapid antigen tests that Fauci envisioned wouldn't be available until the fall of 2020. And they wouldn't be sold directly to consumers until the spring of 2021.[50] We needed another strategy to control transmission. Without the ability to diagnose infection and to isolate those who were contagious, as soon as you close things down and then reopen them, the virus rebounds. That's exactly what happened.

The application of public health requires public trust. When the tactics to confront a threat don't line up with people's perception of the risks, that trust is eroded. That's what happened with COVID. We had adopted a model for mitigating a pandemic that would involve a flu and failed to see how COVID would spread through features that had some critical differences from flu, and in turn frustrate some of the tactics we'd use. Even after we learned more about the novel coronavirus and its mode of transmission, and even after we developed more reliable ways to track its transmission, we had the chance to fine-tune our tactics, and mostly didn't.

Given how little was known about COVID at the outset, it was reasonable to base our early assumptions and tactics on the flu blueprint we'd written in 2005. But this doctrine wasn't revisited as more data became available about the unique features of SARS-CoV-2. Experts were trying to protect Americans, and we can't blame them for being wrong in the absence of good information. The question is whether there is an effective process for rapidly gathering new evidence and then reevaluating our tactics as new data emerge.

Science isn't a set of unchanging truths handed down by a government agency. One of the virtues of the CDC's process is that the agency is able to move fast in a crisis. At the CDC, you can whip up public health recommendations and get them out of the agency in twenty-four hours if necessary. The CDC recognizes that the available information and science will likely change, and they'll have to revise the recommendations. However, often updating prior recommendations isn't a priority for them and they sometimes lose track of the stuff that needs to be revisited.[51]

While the CDC's guidance is nonbinding, it often has more force than many regulations, but its composition has much less transparency. There are no opportunities to provide comments or expert input, such as through public meetings. Moreover, the CDC isn't always clear on when the science behind its recommendations is unsettled, making it harder for people to identify which advisories are more fungible. The CDC also doesn't always identify the underlying science of its recommendations. That's how we ended up not knowing the exact basis for the six feet of distancing.[52] Among other steps, the CDC ought to have a public vetting process that involves outside experts. This way, they also are less likely to end up with some of the problems they had during the pandemic by putting out bad advice because it was being edited by political appointees.

When trying to contain a pandemic, it's essential to focus on the precautions that are likely to make the biggest difference. The public is willing to follow sensible, evidence-based directions. But experts can only ask people to sacrifice so much before resistance starts to form, given the potential social and economic hardship. If the CDC expects Americans to follow its guidance, it will have to be more transparent and get the public invested in how these decisions are made.

Ultimately, the overreliance on the model for responding to flu, and the late and incomplete efforts to readjust that position as new science emerged that showed that COVID wasn't behaving like flu, caused us to make costly mistakes. These challenges should have been clear early and should have prompted more of a concerted effort to better tailor our response.

These distinctions were evidenced by the limited success of our mitigation at stopping COVID, compared with its striking success to virtually eliminate the 2020–21 flu season. New strains of flu often emerge in the Southern Hemisphere during its winter (May through September) and then migrate to the Northern Hemisphere for its winter (November through March). But the numbers from South America, Africa, and Oceania told a remarkable story. In Argentina, labs that enrolled in the FluNet Global Influenza Surveillance System—a network of reference labs that sample for flu as a way to track its global spread—recorded 4,623 cases of flu in 2019 but just 53 in 2020. In Chile, there were 5,000 cases in 2019 and 12 in 2020; and in South Africa, the network's labs detected 1,094 cases in 2019 and just 6 in 2020.[53] In New Zealand there was a "near extinction" of influenza.[54] With so little flu virus migrating, a similar scenario played out in the US during our fall and winter. By the end of January 2021, the CDC had recorded only 1,316 positive flu cases in its surveillance network, compared to 129,997 they had recorded over the same time frame in 2019.[55] The mitigation we put into place was designed to deal with a pandemic flu, not COVID, and it worked much better against its intended viral target.

This was another measure of how a plan first envisioned for

confronting a pandemic flu was not the right blueprint for mounting a response to a coronavirus. A key distinction between COVID and flu was how many new cases could arise from each index case. Flu spreads from one person to two or three people, because droplets don't carry well, and people are most contagious only after they start showing symptoms. By contrast, the novel coronavirus can spread far more broadly, because of asymptomatic and aerosol transmission. Much more of the total spread of SARS-CoV-2 was accounted for by so-called superspreading events, which made mitigation less effective and strongly suggested that, under the right conditions, the virus was spreading through aerosols and not just respiratory droplets. Suppressing routine, day-to-day interactions wouldn't break as many transmission chains if the types of settings that were driving much of the COVID spread were a small number of outlier events. These were typically social gatherings when people tended to gather inside in poorly ventilated spaces, where aerosol spread was optimized, and then let their guard down.

The superspreading events were personified early in the crisis, by episodes where the introduction of a single infected person into a confined, indoor space led to large outbreaks. In Boston the virus was introduced, perhaps by a single individual, to an international professional conference held by the biotechnology company Biogen with about 175 attendees, more than 99 of whom would end up testing positive.[56] The transmission chains ignited by this event would be implicated in up to 300,000 downstream infections, according to one estimate published in the journal *Science*.[57] Other superspreading events would be traced to a funeral in Albany, Georgia, with more than one hundred attendees. That gathering would leave the local community, a small rural county in Georgia, with one of the highest cumulative incidences of COVID in the country.[58] In Chicago, the CDC would report on a cluster of sixteen cases and three deaths tied to two family gatherings—a funeral and a birthday party.[59] In Ohio, an infected man went to church in mid-

June, leading to illnesses in ninety-one people, including fifty-three who were at the service. Eighteen of those individuals would go on to infect other people, setting off a long chain of transmission that ignited a regional outbreak. "It spread like wildfire, wildfire. Very, very scary," Ohio governor Mike DeWine said in unveiling an analysis of the event done by his state's public health authorities.[60] In March 2020 in Mount Vernon, Washington, sixty-one people attended a choir practice and sang in a large room, distanced from one another as a precaution. They were provided hand sanitizer and left the doors open to improve airflow. But fifty-three of them were later confirmed or believed to have contracted COVID as a result of the gathering, and two died.[61]

There was early evidence that SARS-CoV-2 wasn't spreading like a flu, but the CDC and HHS took a long time to adjust the national approach based on this information.

Studies showed that about 70 percent of patients infected with SARS-CoV-2 wouldn't spread the virus any further, but about 5 percent of infected individuals without masks could account for as many as 80 percent of all subsequent cases, mostly the result of superspreading events.[62] In thwarting these kinds of events, widespread mitigation can still work; it just works less well than it might in relation to influenza. However, the biggest shortcoming wasn't the limitation of these tactics, but the inability to implement case-based interventions after the mitigation was lifted. We didn't have the tests or the public health infrastructure to diagnose people, identify contacts of those who had the infection, and voluntarily quarantine those who may have been exposed to the virus. In most cases, we didn't even try. These case-based tactics are why New Zealand kept the virus out of their country and why the nations of the Pacific Rim got control of their outbreaks. When we finally lifted the stay-at-home order, we failed to transition from community-wide mitigation to tracking down individual cases because we didn't have the testing and contact tracing systems to support it. So, the virus flared after we reopened. It was a continuous cycle of infection. We never escaped.

There were other critical distinctions between COVID and flu when it came to the application of the 2005 plan. Perhaps the most crucial was the perceived role of children in intensifying community spread. The flu-based strategies were focused on kids as major amplifiers of infection within communities. That's why closing schools was such a key part of the 2005 plan.

However, in the case of COVID, closing schools didn't have the same measure of benefit, at least not in the first year, when most schools remained shut—much of the evidence suggested that children didn't appear to be significant drivers of community spread, especially when careful measures were taken in schools to prevent outbreaks.[63] One widely cited study from the *Journal of the American Medical Association Pediatrics* reviewed thirty-two studies comprising 41,640 children from different counties as well as 268,945 adults. The analysis also included eighteen studies (including three based in schools) in which scientists had traced the contacts of infected individuals to follow chains of transmission. The researchers found that younger children were roughly half as likely as adults to become infected, although children older than fourteen may be just as likely as adults to catch the virus.[64] That study tracked with another analysis that was released by the CDC around the same time, which evaluated 277,285 cases among children aged five to seventeen whose illness was diagnosed from March to September. The CDC also found that COVID incidence among adolescents aged twelve to seventeen years was about twice that in children aged five to eleven years.[65]

The new and more contagious variants that would start to spread in the winter and spring of 2021, such as B.1.1.7, the strain that originated in the U.K., and then B.1.617 which was first found in India, seemed to affect a higher proportion of children than the original strain. It wasn't that B.1.1.7 was disproportionately targeting children more than adults; it was more contagious and virulent to everyone, children included. Data from about twenty thousand people infected with B.1.1.7, including three thousand children under ten, showed that while everyone was more impacted by the new strain, relative to the adults, the young kids remained about half as likely as adults to transmit the new strain to others.[66]

A study published in *Nature* provided some understanding of why young children seemed less likely to develop COVID and transmit the infection. The authors found that children produced lower levels of antibodies in response to coronavirus infection when compared to adults. It was an indication that children might have been clearing the infection more quickly and easily, and therefore weren't generating the same robust immune reaction to the virus.[67] If they cleared the infection more quickly than adults, they may have had fewer opportunities to spread it to others.[68] Some studies pointed in the opposite direction—that children, and especially older children, were just as likely to contract the infection. However, the balance of the research seemed to agree that when it came to SARS-CoV-2 infection, children and adolescents were less likely than adults to become symptomatic or have severe outcomes.[69]

Some data also suggested that infected kids would be less likely to bring the virus back into their homes. One study done by researchers at the Geneva University Hospital used contact tracing to identify when household contacts became infected by other family members and who was the source of the initial introduction into the household. Of the thirty-nine different households that the researchers examined, in only 8 percent was a child the suspected index case, with symptom onset preceding illness in an adult in the household. In all of the other households, the study found, the child developed symptoms after an adult became sick, or they both developed symptoms at the same time. So, in these cases, the child was less likely to be the source of the initial introduction. Taken together with some similar research, the authors concluded that children most frequently acquire COVID from adults rather than transmitting it to them.[70]

The data on COVID and kids was by no means a lock. Other studies drew opposite conclusions, and the role of children in catching and spreading SARS-CoV-2 remains, even now, the subject of study and debate. While kids clearly serve as vectors for spread, it was generally agreed that they didn't play nearly the same role in transmission as adults, and certainly not in relation to the role that kids played in spreading influenza. There were caveats: one study found that while children were about a third less susceptible to coronavirus infection than adults, when schools were open, the kids' social milieu changed. The authors found that children had about three times as many contacts as adults, perhaps offsetting their virological advantage.[71]

None of this is meant as a justification to forgo precautions in schools. Schools aren't inherently safe in the setting of community spread of SARS-CoV-2, but they can be made safer through proper measures. We still needed to take steps to prevent outbreaks in classrooms. That included the use of masks, distancing, hand washing, improving air handling and filtration, and, especially, keeping students in strict social pods: what one epidemiologist described as the "full Harry Potter."[72] However, if these precautions were taken, then the real-world experience showed that the risk of transmission could be substantially reduced, and the social and educational benefits of keeping kids in school, especially grade-school kids, could offset the risks.[73]

But as with the original pandemic plan first crafted in 2005, the issue of closing schools would remain perhaps the most contentious measure taken in response to COVID.

For some of these same reasons, I remained cautious on closing schools, recognizing the social, educational, and emotional impact on children from being out of school. On March 9, speaking to the editorial board of *USA Today*, and addressing the outbreak in Seattle, I said, "I think that you want to think about reactive school closures because closing schools has its own impact on social lives and an impact on public health that could be adverse and could potentially even worsen the epidemic. . . . So broad preemptive school closures, I personally wouldn't advocate that."[74]

Most schools would ultimately close in the spring. In view of the uncertainty we faced and the mounting death and disease, it was a reasonable precaution to take as the epidemic first gripped the nation in March and April, and we didn't understand its risks and how it could be stopped. But then, as we learned more, we should have adjusted our tactics.

In an op-ed titled "Schools Can Open Safely This Fall," in the summer of 2020, I wrote in the *Wall Street Journal* that "the debate over schools has been swept up in a political maelstrom. Reopening schools will draw more controversy if people believe their school district was forced into opening. I've talked to Republican and Democratic governors about their strategies. . . . Their approach is appropriately varied to local conditions. The main risk is transmission inside school buildings, but there are ways to reduce the chance of a big outbreak."[75] Data looking at the experience in schools that remained open during the COVID pandemic seemed to support the premise that schools could open safely, if they took proper precautions. One survey that looked across the national experience found that in communities where schools were opened during the fall surge, there was no increase in cases and hospitalizations in the local community relative to areas that chose to keep their schools closed.[76] The experience in states supported these observations. Some of the best data came from North Carolina, where schools were open. They adopted measures like masks, distancing, and podding, and didn't see outbreaks, or evidence that schools were contributing to community spread.[77] The lessons from some rural parts of Wisconsin were similar.[78]

For many districts, the physical distancing requirements were what kept schools shut. In many towns, local school boards felt compelled to adhere closely to the CDC's recommendations, and they simply didn't have enough space in their schoolhouses to keep students six feet apart. The federal government was working off a model for flu, and it was the wrong framework for COVID.

It's unclear how much insight the CDC had into the impact that this single requirement was having, and the Department of Education seemed oddly detached from trying to troubleshoot the challenges. President Trump tried to jawbone the schools into opening, holding White House events to target districts that remained shut. Members of the White House staff told me that the president's economic team were telling senior principals, including Trump, that the US economy couldn't begin to recover until children were able to return to school and their parents could return to work. The president took to Twitter to rebuke some governors for allowing schools to remain closed. It was emblematic of a policymaking process that tried to use pressure and persuasion as its principal tool. The president was right to focus his attention on trying to get schools opened for a host of reasons, and principally the important role that in-class instruction plays for children. But the White House had failed to hone in on a root cause for why many districts remained shut.

The CDC's guidelines were the single greatest obstacle. I was speaking to the White House over this time period, and some officials there didn't fully appreciate how much impact the six-foot requirement was having on efforts to reopen schools in the fall. They didn't connect the lines between their policy goals, the parts of the pandemic plan that impacted those objectives, and the actions of the CDC that frustrated these outcomes. It was a breakdown in policymaking, and the way the pandemic playbook was implemented, that would plague other aspects of our response.

CHAPTER 13

THE INFORMATION DESERT

In a crisis, we often don't have time to wait for deep analysis. We're forced to make the most effective decisions we can on the best information we have available at the moment. In a moment of public health crisis, that information is always going to turn on some complement of raw data and conjecture.

However, the CDC isn't in the business of providing this kind of real-time, actionable, but often notional information. It's counter to the agency's culture. While the CDC does provide some near-real-time data streams like their flu surveillance program, their best reporting comes from longer-term investigations that are often wrapped into their Morbidity and Mortality Weekly Reports, the CDC's epidemiological digest that provides deep analysis of disease trends.[1]

After an especially devastating flu epidemic struck the US in the winter of 2018, it took the CDC a year to establish just how deadly it had been, and start drawing inferences on why it was so lethal.[2] I was running the FDA during that 2017–18 flu season and, during that winter, I wanted to share our concerns with the public, and warn them about the unusual severity of the flu, and our agency's belief that the vaccine might not have been protecting people as well as it had in past years. The FDA wanted to urge people to be cautious and take added precautions to avoid infection.

The CDC tried to stop my communications, appealing to the Office of the Secretary of Health and Human Services to intervene and block me from issuing my warning. The CDC was arguing that it, and not the FDA, should be commenting on the severity of the flu. They said it was the CDC's job to discuss disease trends with the public, and they weren't ready to draw any firm conclusions about that particular flu season. The CDC was still gathering and analyzing their data.

I released my warning anyway, except that instead of issuing it as a public health advisory commenting on the flu season, I fashioned it as a safety update on the vaccine. The FDA regulated the flu vaccine, not the CDC. The CDC couldn't stop me from commenting on an FDA-regulated product and the FDA's belief that the vaccine might not be protecting patients from flu as well as intended.[3]

Because we didn't know for certain the reason behind the rising death and disease from flu, and whether reduced effectiveness of the vaccine was definitely the culprit, the CDC didn't want to say anything that was potentially wrong. But at the FDA, I wanted to warn the public based on what we knew, while explaining that we weren't certain. The CDC wanted to wait and do more investigations. Later analysis would show that the vaccine was only about 25 percent effective against the predominant strain of H3N2 influenza A that was circulating that season.[4]

It isn't a question of competency or public health commitment. CDC officials are deeply devoted to their public health mission. It's just a question of institutional style, structure, and culture.

The CDC's reporting is reflective and aims to provide definitive analysis. It isn't geared to providing the sort of real-time information that's often early, imprecise, and incomplete, but an essential currency in helping inform policymaking in a crisis. The CDC's output is more akin to the journal articles issued by an academic department of epidemiology than a battlefield report issued by the Joint Special Operations Command. Providing timelier and more actionable reporting would require a reengineering of the CDC's commitments, capabilities, and ethos.

During COVID, the general refrain, quite appropriately, was that our policy decisions should be guided by, and shouldn't get ahead of, the science. However, what slowed action, and left our decisions less well informed, was that this common refrain is actually a lot more complex when applied to our response to COVID. It wasn't the scientific data that the CDC didn't want policymakers to ignore or second-guess; it was the CDC's interpretation of that data.

The problem was that the CDC's analysis was never timely or complete. The CDC's edicts, contained in careful journal papers published in the Morbidity and Mortality Weekly Report, were often thorough, data driven, well researched, and late. The agency issued recommendations on the risk of coronavirus transmission through touching contaminated surfaces (fomites), measures to improve safety in schools, the role of respiratory aerosols and droplets, and countless other aspects of COVID's pathogenicity and spread that took many months to generate; and were often published long after decisions based on these important parameters had to be made by patients, providers, and policymakers.[5]

The resulting information desert that plagued real-time decision making impacted more than policy, it was also felt in clinical practice. Even after we were months into the COVID epidemic, there was no reliable information on doctors' collective experience in treating COVID patients. We had not tracked and reported what treatments and interventions medical providers were relying on and the outcomes of patients who received different forms of care. The best information linking different approaches to care with outcomes was coming from Italy and China. These countries had earlier experiences with the virus, but they also did a much better job than our CDC in tracking and collecting clinical experience, and sharing timely, bottom-line data with providers.

For clinical data, the CDC largely relied on information derived from death certificates, but reporting from these documents was usually delayed by one to two months. That's why a lot of the early clinical data that informed doctors in the US and helped them determine which patients were most likely to have the worst outcomes was coming from other countries. Nations in Europe derived this information from their electronic health records in a near real-time fashion. They thus had better information about patient experiences to guide clinical practice.

We also lacked timely, reliable epidemiological data derived from effective contract tracing and used to inform policymakers on the settings where spread was most likely to occur. These data could have helped us better target mitigation to the highest-risk venues. Studies show that people will generally follow public health directives for about two weeks, and then compliance will break down. Without better ways to focus our interventions on regions where there was the greatest spread, we eroded public trust and squandered the social and political capital needed to maintain consent.[6]

Bars were closed basically on a hunch that they were a significant source of spread. There was some contact tracing to support this policy, but the data were imperfect. Early in the epidemic, we left many schools and day care sites open for essential workers, but we didn't systematically monitor whether those settings became sources of spread, or which interventions helped reduce the likelihood of outbreaks. So we missed a critical chance to collect data that could have informed future policymaking on the issue of opening schools. Lots of other gaps could have been plugged, risks reduced, and hardships avoided with timelier and more complete information.

Could the quarantine period after a COVID exposure have been shortened?

Almost a year after the pandemic began, the CDC shortened the recommended quarantine for those exposed to the virus from fourteen to ten days, and then seven days if a person tested negative for the virus seven days after exposure.[7]

Was six feet the right minimum distance?

The 2006 pandemic playbook had discussed the use of three feet of separation. The six-foot rule led to situations in which people congregating indoors, in spaces that were poorly ventilated and created easy conditions for spread, but who were kept six feet apart, felt they could take their masks off if they remained outside a six-foot circle.

Was fifteen minutes of exposure the right measure to determine when someone who was in contact with an infected individual might have contracted the virus?

It seemed like an arbitrary judgment and led to manipulation of the agency's advice. The CDC eventually changed its guidance from fifteen minutes of sustained exposure to fifteen minutes of cumulative exposure, but how could you measure your cumulative exposure to a person?[8] It was almost as if the CDC was treating exposure to the virus like exposure to radiation and trying to measure a cumulative dose. It was reported that some establishments moved people around at the fourteenth minute to avoid passing the fifteen-minute regulatory threshold.[9]

As sociologist Dr. Zeynep Tufekci noted in *The Atlantic*, "None of this made any practical sense. What happened at minute 16? Was five feet okay? Faux precision isn't more informative; it's misleading."[10] The evidence behind these recommendations was always shaky, and the measures hard to implement. The CDC had to start somewhere, and so some of the initial recommendations were based on imprecise evidence, but the agency should have been more transparent on just how weak some of the data were. People could have made more informed judgments about where to apply limited resources, focusing on the guidance that had the greatest chance of reducing risk or the strongest scientific foundation. Good epidemiological surveillance, conducted by the CDC, could have informed more precise recommendations, and the CDC could have provided us with more information about the conditions that contributed the most to spread, improving our ability to keep people safe. Without reliable and actionable data, and good systems for collecting and reporting information, we lacked the infrastructure to try to build a new evidence base for a novel pathogen. So we largely worked from what we knew about flu, which in many important respects didn't apply to the crisis we faced from COVID.

Some of the CDC documents were subject to revisions or delays by political officials at HHS and the White House, and the general refrain was that this interference made the CDC reluctant to advance other guidance or degraded the impact of the recommendations it issued. But there were plenty of matters the CDC opined on, that flew well below the radar of its political interlopers, where the CDC had failed to release relevant and timely information. Moreover, the CDC's approach to these efforts didn't change much once President Biden took over. Two of the CDC's most senior career officials left the agency within the first six months of Biden's term, in part, I'm told, over friction with the new administration.[11] Frustration with the agency, it seemed, had bipartisan appeal. It's perhaps convenient, but self-serving, to blame all of the CDC's faults during the Trump term on political interference into its work. The CDC wasn't just slow to develop this evidence, it also didn't offer it in practical terms that made it actionable.

Still, there were White House and HHS political staff who wrongly believed that more information would confuse or alarm the public and drive people to take decisions that conflicted with the Trump administration's reopening goals. So the CDC was, at times, intentionally stymied. In other cases, the White House and HHS lost confidence in the CDC and, not knowing how to reform the agency, they moved instead to suppress its work, isolate its leadership, and usurp some of its responsibilities.

However, what I also saw were political officials who misread the practical value of providing consumers with better information, and the benefit in leveraging the CDC to help gather and report it. Reliable information about risks would help advance the policy goals of reopening schools and businesses, because the absence of information created uncertainty, and uncertainty bred indecision. People chose to take no action at all if they didn't have data that could help them correctly calculate, and lessen, the risks of the actions that they wanted to pursue.

Take the issue of opening schools. The White House wanted schools to be reopened in fall 2020. But political officials feared that more information about transmission in schools and the conditions that led to outbreaks could frighten additional schools into staying shut. So these same political officials stymied efforts at the CDC to put out more prescriptive guidance to schools outlining the steps that schools could take to reduce the likelihood of outbreaks.

This political posture probably had an effect that was opposite its unfortunate purpose, causing more schools to remain closed. The schools didn't have enough information to guide safe decisions to reopen. At best, these political efforts were a misreading of the value that information could play in supporting action in the setting of uncertainty. Did masks lower the likelihood of spread in classrooms? Did distancing help? Was keeping students in distinct social pods effective? These were critical questions that needed to be answered. If we had data to guide these actions, more schools would have had a framework to know how to both stay open and reduce the risk of outbreaks. Secretary of Education Betsy DeVos said it wasn't the responsibility of her department to collect and report this information.[12] That was probably true, although the education department could have led that effort. However, the obligation to collect and report these data certainly belonged to the CDC.

National pandemic strategies going back to 2006 included schools as part of national disease surveillance. There were requirements for the reporting of school-based data to local health agencies who would then provide it to the states. The states, in turn, are committed to providing that information to the CDC. However, staff in the White House hesitated to systematically track this school-related data and share it with the public, and the CDC seemed to struggle with collecting this kind of bottom-line information, anyway, even if the White House had encouraged it to do so.

During a fast-moving crisis, in the absence of good information, people tend to be more conservative, and less willing to try something perceived as risky. When we can discharge uncertainty and properly handicap a danger, we can help people embrace reasonable risks.

I learned at the FDA that people are often willing to confront risks that they can adequately measure for themselves, but balk when forced to embrace risks that seem open-ended, ambiguous, or hard to measure. We learned that timely and complete reporting on drug side effects was reassuring to patients. They needed to have confidence that, if there were risks associated with a drug, the FDA would unearth this information and promptly report it to patients.

Armed with good information, patients were able to make informed choices and assume risks that they felt were reasonable for their individual circumstances. To help support this informed patient decision making, the FDA made substantial investments in recent years to develop more information about the real-world use of medical products, especially about their safety, and to provide this information directly to consumers in a regular and timely way. This was a major focus of many efforts I undertook while serving as the agency's commissioner. The same principles apply across public health challenges. In the setting of COVID, more data about how and where COVID spread occurred in schools would have provided more certainty to school administrators on how to lower the chance of outbreaks. In the absence of good information to inform these decisions, facing uncertainty, many cautious districts chose to close schools instead. By the end of March 2020, as the US epidemic was getting under way, 94 percent of American schools were closed, and the majority of them would remain shut for the duration of the year.[13]

Dr. Christopher Murray was the director of the Institute for Health Metrics and Evaluation and the architect of a model of

COVID spread that was closely followed by the Trump White House. Writing in the *New York Times*, just as the fall surge was gaining explosive momentum, he said that federal agencies, including the CDC, had been telling him since March that the government was compiling bottom-line, county-level data on COVID cases, hospitalizations and deaths, the timing of social distancing mandates, testing, and other factors that could provide insights on how policy actions were affecting how fast and wide the virus would spread. This kind of data would have been invaluable in helping to establish more-targeted measures. As Murray wrote, "This information can provide insights into how combinations of public health mandates—masks, social distancing and school closures, for instance—can keep the virus spread in check. But the government, inexplicably, is not sharing all of its data. Researchers have asked federal officials many times for the missing information but have been told it won't be shared outside the government."[14]

The *New York Times* had to sue the CDC under the Freedom of Information Act to obtain data on COVID cases tabulated by race and ethnicity. It was basic information that could help focus resources on communities that were being hardest hit by the virus, to help save more lives. The information would eventually prove that Black and Latino Americans were being excessively harmed by COVID in a "widespread manner that spans the country, throughout hundreds of counties in urban, suburban and rural areas, and across all age groups."[15]

The dominant narrative over this time period remained that the White House pressured the CDC to subdue certain reporting. The record shows political actions certainly played a role in the suppression of some critical information. But there was another problem afoot. The CDC didn't have all the pertinent information in the first place, or the ability to collect it in a reliable fashion that would enable timely reporting. The agency paid local and state health officials to report bespoke feeds of data that was typically collected in a format that made it inaccessible to anyone other than the CDC.

The agency used proprietary forms that required healthcare providers to input the information into specialized data streams that were for the CDC's exclusive use. It slowed the collection of information and increased the chance for errors, since many providers had to extract data from other systems and separately transpose it onto the CDC's forms. There was also no natural market for this information, so it was not in anyone else's normal work stream. It was being gathered and shared only for the CDC's consumption. And since it fell outside the normal systems for collecting healthcare information, with all the compliance rooted in those tasks, there was no embedded audit function. Because the data collection was done separately from other routine healthcare reporting, it also meant that systems for collecting it were generally outdated.

It was possible to derive the same information by culling the data from our existing electronic health records, a process that would have also provided a more natural audit trail and more quality control over the information. The CDC could have taken the role of being an aggregator of existing data feeds rather than a proprietor of unique reporting streams that were distinct from other pools of healthcare information. The CDC took significant pride in its proprietary data feeds, however, and clung to its model, even though the agency's approach had many obvious shortcomings.

The challenges played out in how the CDC reported on COVID hospitalizations. It turned out that the CDC wasn't actually providing data that was being reported to the agency on real hospitalizations, but was instead posting an estimate derived from a predictive model. In other words, the number of hospitalizations that the CDC was reporting each day weren't people who were actually being admitted to the hospital but were hypothetical patients being modeled off a small sample that the CDC was collecting.[16]

The CDC reported on the number of hospitalizations each day as if it were tabulating these totals, but it wasn't. Making matters more uncertain, the data that underpinned that modeling exercise was being lifted from a system that had been built years earlier as a way to monitor the prevalence of hospital-acquired infections like urinary tract infections and hospital-acquired pneumonias, not for tracking hospital admissions. This system provided reports to the CDC from one thousand participating hospitals, sometimes through faxes.[17] When COVID struck, instead of asking the hospitals to submit data on hospital acquired infections, which is what they'd customarily report through this system, the CDC asked them to also start reporting on COVID infections. The CDC then used these data to estimate the number of COVID admissions for the entire country.

So the numbers on COVID hospital admissions that the CDC reported to the public were not actual data on admissions that were taking place in the nation's six thousand hospitals but estimates off the smaller one-thousand-hospital sample set. That's a pretty small sample to use when you're trying to model the entire national experience. It's even more flimsy when you consider how COVID was being experienced. It wasn't advancing uniformly across the nation, but through regional pockets of dense epidemic spread. COVID wasn't hitting every city with the same ferocity. As you moved across the country, some communities were being devastated by the virus, while others were largely spared, and the CDC's sampling left a lot of regional gaps. The entire northern half of California didn't report any data at all to the agency feeds. To fill these gaps, the CDC crafted algorithms to estimate what the missing data might show. Nonetheless, the agency argued that, for making projections, their modeling was better than actual reported data.

The shortcomings with the way the CDC would collect its data, and then use these samples to try to render a national estimate through modeling, showed itself in other settings, including how the agency reported data on seasonal flu, where the CDC also uses models to try to make estimates of the annual number of hospitalizations. This is why, when the CDC reports flu data each year, its estimates have such a wide range, why they have a big "confidence interval." For example, in the 2018–19 flu season, the CDC estimated that there were 97,967 people between the ages of fifty and sixty-four who were hospitalized for flu.[18] That number sounds fairly precise, right down to the last sixty-seven people. However, it turns out that the confidence interval for that data, the range of potential estimates, was actually between 69,808 and 167,123 hospitalizations. So that means it might have been 70,000 patients. Or 100,000. Or 150,000. Or perhaps 143,294.

The CDC couldn't tell you for sure. That's a pretty big variance to tolerate for estimating something that's seemingly as reportable as the number of Americans admitted each year to a hospital and diagnosed with influenza infection. This would seem to be a figure that could be collected and reported with greater precision. How hard could it be for the CDC to have hospitals report the precise number of admissions that they have each winter with a diagnosis of flu?

As it would turn out: not very hard.

Medicare has data on influenza hospitalizations for those over the age of sixty-five. Insurers will have this data for their beneficiaries, and they'll often report these trends. However, for the CDC, aggregating this information seemed to be too difficult. Deborah Birx convinced the Coronavirus Task Force to direct money to the CDC to modernize its reporting of the COVID hospital data, but the CDC said no. Birx had previously worked at the CDC for nearly a decade, running the agency's global HIV/AIDS division. After that, she had continued to collaborate with the CDC in her role as the US global AIDS coordinator, an ambassador-level job inside the State Department. Now, prodding the CDC to reform the way it collected hospital data, she found it was hard to get the agency to change approaches.

The problems with reliably collecting and evaluating the COVID hospitalization data were deeply rooted in the CDC's institutional approach, but it was also a product of the outdated systems that the agency worked from that made it hard to change its methods. Several years earlier, during a measles outbreak, state health officials were tracking it with pen and paper and reporting to the CDC on fax machines.[19] A CDC official told the journal *Science* that during the COVID crisis, compiling the actual total number of COVID hospital admissions was simply not possible.[20]

The CDC's inexact modeling finally became a significant liability when the drug remdesivir was made available for the treatment of hospitalized COVID patients. It was the first treatment authorized by the FDA that could reduce the severity of COVID symptoms and the length of hospitalization. There was some evidence to suggest it could reduce the likelihood of death for hospitalized patients.[21] The drug was by no means a panacea. It appeared at the time to be a weakly active antiviral, but there was reason enough to believe it had a treatment effect for sick, hospitalized patients who otherwise had no effective options. However, the drug was initially in very short supply, and doses needed to be carefully rationed to the sickest patients. The federal government had bought the entire available supply, and HHS needed to know where to ship its limited doses, to make sure that the scarce medicine would get to those hospitalized patients who might benefit most from the treatments. However, the CDC didn't have actual data on who was currently hospitalized for COVID, just estimates built off a model. They couldn't tell HHS where real patients were, only hypothetical patients that were being extrapolated from an algorithm.

Birx said that the government couldn't ship scarce doses of the valuable medicine to treat estimated patients that were hypothetically hospitalized according to a mathematical formula. So she gave hospitals an ultimatum. If they wanted to get access to remdesivir, they would need to start reporting real data on the total number of COVID patients that they admitted each day. Hospitals quickly started to comply, reporting actual data on their total daily hospitalizations to a new portal that Birx had helped set up inside HHS. Rather than try to cajole the CDC into fixing its reporting system, Ambassador Birx and Secretary Azar decided to re-create that structure outside the agency. They had concluded that getting the CDC to change its own scheme, and abandon its historical approach to modeling these data, would have been too hard.

Under the new reporting system, 95 percent of US hospitals soon provided 100 percent of their daily hospital admission data. In an unfortunate twist, the CDC declined to work with the new data, worrying that since it wasn't their data, they couldn't assure its provenance and couldn't fully trust its reliability. As one senior HHS official put it to me, the CDC "took their ball and went home."

The new system was housed at HHS, with the data collected by the healthcare analytics company TeleTracking Technologies; the software technology company Palantir helped build and manage the database. It was not without its own challenges. *Science* magazine would report that data published by the new federal portal, dubbed HHS Protect, initially suffered from systematic discrepancies from the data that was being separately reported directly by the states on state-run websites.[22] Over a short time, these inconsistencies narrowed considerably, and in October, Medicare issued a regulation mandating that hospitals report the COVID hospitalization information to HHS.[23] The new reporting system grew to be highly reliable, and the data began to be posted publicly. The cofounders of the COVID Tracking Project, one of the most authoritative and closely watched enterprises to report bottom-line information about the pandemic, would later say of the TeleTracking information this way: "the data set that we trust the most—and that we believe does not come with major questions—is the hospitalization data overseen by the Department of Health and Human Services. At this point, virtually every hospital in America is reporting to the department as required. We now have a good sense of how many patients are hospitalized with COVID-19 around the country."[24]

Yet when HHS initially took the task of reporting the hospitalization data away from the CDC and set up the new system, many in the media would almost reflexively cite the incident as support for a dominant narrative that the CDC was seeing its role obstructed and reduced by the political leadership at HHS and the White House.[25] US senator Patty Murray sent HHS an oversight letter in which she argued that the new system was wasteful and "duplicates existing CDC work."[26] It was alleged that the Trump administration had taken the hospital reporting away from the CDC and given it to TeleTracking with a political not a public health goal in mind—political officials wanted to fudge the data to give a false rosy picture of the pandemic. Or so the narrative went. While the CDC was certainly subject to some deeply unfortunate and ultimately damaging political intrusions into its work, this wasn't one of those instances. Some of the frustrations with the CDC's execution had merit, and the CDC's method for reporting COVID hospitalizations was one of those moments.

The problems with the way the CDC collected and disseminated information in the crisis were systemic—the data and analysis that the agency released often weren't practical in the sense that it was collected and analyzed in a way that would make it immediately actionable by consumers, providers, businesses, and policymakers, which would be the natural constituency for the agency's findings. The CDC just didn't have this mind-set. Its reports were often geared to answering important bottom-line questions about core science, and its constituency was other researchers and academicians. The CDC didn't generate much data for real-time policymaking. The CDC's orientation was always that its information defined the policy, through its findings and recommendations, rather than feeding into the decisions that would need to be made by others. The CDC was the final word, so it didn't need to tailor its data to feed into someone else's analysis.

That worked because the way the CDC's data and analysis fit into the broader mosaic of public health information was always fairly well defined and understood. In the COVID crisis, the role and constituency for the CDC's information changed. Suddenly, every policymaker, every consumer, every business, and every healthcare provider was looking to the CDC to define the rules of engagement. Carter Mecher memorably called the CDC "a peacetime institution in a wartime environment."[27]

One illuminating episode occurred with a CDC report published in the summer that tried to trace the source of infection among a group of people tried to isolate the kinds of activities that were contributing the most to spread. The CDC report concluded that people who had dined out in the prior week were more likely to contract COVID than those who had not eaten at a restaurant. The finding was used to draw a conclusion, widely reported in the press, that dining out increased a person's risk of infection.[28] News articles covered the report with headlines such as "Dining Out Increases COVID-19 Risk More than Other Activities, CDC Report Finds," and "CDC: Dining Out Tied to Coronavirus Infection."[29] However, it was summertime when the survey was done, and the participants were not asked perhaps the most relevant question: Did they eat at an indoor dining establishment or an outdoor one?

Similarly, the same survey didn't distinguish between bars and coffee shops when assessing the risk of contracting COVID in those kinds of establishments. Bars and coffee shops were grouped in the same bucket, even though the settings and the behaviors in each establishment (fewer people linger in a Starbucks) and the conditions for spread could be very different. When pressed on why the agency had grouped bars with coffee shops, the CDC said that it judged the two settings to present comparable risk for transmission.

In fairness, reporting real-time information in the setting of a crisis was never an explicit part of the CDC's mandate. Nor was the task of providing actionable information for policymakers. It's hard for a bureaucracy to self-organize around a complicated new mission, especially in the setting of a crisis. As Dr. Yuval Levin of the American Enterprise Institute observed, these stumbles shouldn't have come as a surprise to those who had worked closely with the CDC in the past. Levin had been a senior White House official in the Bush administration overseeing the public health portfolio for the Domestic Policy Council at the time that Bush was leading the effort to shape a more durable pandemic response plan. In an article for *National Review* in the winter of 2021, Levin recounted efforts to enact the Pandemic and All-Hazards Preparedness Act.

The law required the CDC to "establish a near real-time electronic nationwide public health situational awareness capability through an interoperable network of systems to share data and information to enhance early detection of rapid response to, and management of, potentially catastrophic infectious disease outbreaks and other public health emergencies that originate domestically or abroad."[30] As Levin observed, "the simplest way to describe the CDC's response to this binding legal mandate was that it just ignored it. It did nothing." The GAO noted in 2017 that the CDC still didn't have an effective tool for public health surveillance in a crisis. The CDC and HHS had drafted a plan to bring their systems into compliance with the legislation, but "the actions identified in the implementation plan did not address all of the requirements defined by the law," the GAO concluded, and "as of May 2017, HHS had made limited progress toward establishing the required electronic public health situational awareness network capabilities."[31]

During COVID, it became evident that the CDC didn't have the basic tools of data management for public health decision making. They didn't have the advanced data analytics needed to do the evaluations that were required, and they didn't have the right integration for electronic capture of information from health records. They didn't have a large group of data scientists and modelers. They outsourced most of these analytical tasks to academic partners.

And the problems weren't just on the back end in how the CDC captured and analyzed data, but also on the front end, in how they developed the raw information, particularly in cases where the agency had the primary responsibility for generating a body of evidence. Take sequencing: the CDC wasn't sampling and sequencing the virus at any meaningful scale, but they should have been. So, when new variants surfaced, the CDC was slow to identify their spread. Novel variants arose in New York and Los Angeles, and the CDC identified these strains long after they had already become epidemic, and in the case of Los Angeles, probably after the strain, B.1.427, had already raced through the population and contributed to the city's devastating winter surge.[32]

In spring 2020, Deborah Birx had asked the CDC to establish a contract to engage outside labs to do more widespread sequencing as a way to better understand how the virus was being spread. By evaluating the sequence of different strains, and merging these findings with good analytical data, you could enable earlier identification of dangerous new mutations. New variants could be correlated with information about how easily they were spreading, and how likely they were to make people sick. This would reveal when dangerous new strains had evolved.

Using sequencing, much as Trevor Bedford had shown, you could also trace how the virus was being passed through a local population based on the mutations it would acquire as it underwent successive generations of spread. Relying on these genetic signatures, you could then figure out who was an index case and how they went on to infect other people. Armed with this information, it was possible to identify the most significant sources of spread in a community—was transmission most likely to occur in a bar, a restaurant, or a local Lions Club meeting? Sequencing patient samples could help uncover chains of transmission that were leading to local outbreaks.

Ambassador Birx's idea was to focus first on Mississippi, because it had diverse social compartments. Its population was dispersed across rural markets, small communities, and cities. Birx felt that the state would provide a good test bed to see if sequencing could be used as a way to track where infections originated, and then how an outbreak spread across a population.

Were cities the focal point for spread into rural communities, or vice versa? By regularly sequencing thousands of cases, you could see where superspreading events were occurring and how single introductions of infection were leading to wider community transmission.

However, the CDC didn't want to turn over the sequencing work to outside labs; they wanted to keep it in-house, which limited the agency to sequencing hundreds of cases a month when it was necessary to sequence many thousands of samples to understand the patterns of spread. The larger effort never got under way, and it proved to be a major, missed public health opportunity.

When dangerous new variants emerged in the winter of 2020–21, we were doing a lot of sequencing, but it was distributed across the country, and it was mostly done in private labs or by academic teams. There was no organized, systematic process in place for sequencing viral samples at a large enough scale, and aggregating the information, so that we could detect the entry of the new variants. There was a lot of sequencing work going on in various US labs tracking new mutations, including the more contagious B.1.1.7 variant that had originated in the UK, but the CDC wasn't collating and reporting this information.

The agency started to track these new variants, but it declined to recognize the sequencing data that it wasn't doing itself, or even to include it on the agency's website. This included the sequencing work being done by the public health labs, which the CDC also bypassed. From the CDC's standpoint, the problem was that not all of the sequencing was being done according to the same protocols, and if outside labs weren't following the same approach as the CDC, then the agency didn't believe the data could be reliably pooled. So the CDC simply set it aside and didn't include it in the agency's analyses or national estimates on spread.

Looking at the CDC's website, you would see only a fraction of the sequencing work being done across the country. Worse still, by the spring, the CDC's website only updated data on a monthly basis. So the sequencing information they publicly reported reflected analytical work that had been conducted weeks prior to its posting. The CDC's data were as much as a month out of date. As a consequence, to get an accurate snapshot on the spread of the new variants, you'd have to go to the websites maintained by different states and cities. New York and Los Angeles were posting regular updates on the sequencing work being done by each of those cities. There were also some researchers who were posting to Twitter the sequencing they were doing. The CDC wasn't only doing less sequencing, it was reporting only a fraction of the sequencing that was under way. In contrast to the US, the UK was able to detect the more contagious B.1.1.7 variant because it was sequencing about 10 percent of all patient samples. By the time the delta variant, B.1.617, emerged in May 2021, the UK was sequencing 60 percent of all patient samples.[33] By comparison, the CDC had set as its initial goal getting five sequences from each state per week. When B.1.1.7 arrived in the US, the CDC was sequencing only 0.3 percent of patient samples. This figure probably undercounted the total amount of sequencing going on in the US because it was based on what was being reported into central repositories (like GISAID, a global initiative set up in 2008 to provide open access to genomic data on influenza strains). Since there was no organized reporting for COVID sequencing data in the US, we weren't even capturing what was getting done.[34] The *Washington Post* estimated that the US ranked forty-third globally in its percentage of cases sequenced.[35]

The CDC's historical role is to provide careful analysis and shape our insight into more fundamental questions of public health. It's a retrospective mind-set. In a crisis, what's needed is rapid sharing of information and quick analyses that can inform real-time decision making. It's a much more prospectively focused challenge and requires a forward-looking mind-set.

An analysis that suggests, at the outset of a crisis, that spread is primarily through aerosols can be far more impactful at informing our decisions and response than waiting for a report to provide a more definitive answer to the same question after you're twelve months into the pandemic. By then, a lot of mistakes might have been made based on faulty assumptions. The CDC's carefully crafted analyses take time to develop, time that may not be available to policymakers in a crisis, where even partial reporting is better than making the same decisions in an information vacuum.

Ultimately, the mandate to collect and report clinical information in a fast-moving crisis may need to reside with a different agency, one that has a national security mind-set. The CDC can take on some of these functions, but its mission and capabilities would need to be profoundly reformed. Epidemiologists Dr. Caitlin Rivers and Dr. Dylan George advocated the creation of a centralized system for disease forecasting. The idea is to have the epidemiological equivalent of the National Weather Service.[36] Such an agency would be tasked with developing sophisticated disease modeling to help guide public health policy. It could provide the base on which to rebuild the nation's data infrastructure for pandemic monitoring and preparedness. Congress ultimately set out to create such an effort inside a newly constituted component of the CDC that will be properly staffed, resourced, and more importantly clearly programmed by Congress to execute this new mission.

Wired magazine referred to this concept (in analogy to Philip K. Dick's novel *Minority Report*) as "federal PreCrime for pandemics. Precognitive epidemiology. Make up whatever sci-fi words for it you want; the fact is, one thing the COVID-19 pandemic proved is that pandemics can happen and certainly will again. Building a place to develop the sophisticated models and simulations that can give a hint of when and where an outbreak will hit, and give guidance on how to stop it . . . well, that sounds like a pretty good idea." The center, *Wired* observed, "would also become a central place to gather all that data, via public health surveillance and lab work—the equivalent of ocean buoys and satellites—and for dissemination of that information to local public health workers. Right now, lots of the most important data is siloed among different researchers and labs. The result is, disease modelers have to wheel and deal to get access to data, and production of models responsive to emergent problems is ad hoc. Meanwhile, if you want to know whether you need to double up your masks today, there's no National Epidemic Center web page where you can check the forecast."[37]

During COVID, the lack of reliable, real-time information on what measures were working to help contain the spread meant that many of our efforts remained subject to debate long after careful evidence could have substantiated their value or disproved their merits. Worse still, in the absence of definitive evidence on what worked, it left a void where critics were able to put forward their own cherry-picked evidence, or try to discredit the value of certain mitigation steps that they rejected and there wasn't enough definitive data to fully contest these intrigues.

In the absence of good information, bad information was able to guide the debate. Perhaps nowhere did this play out with more conflict and misfortune than in the debate over masks.

I was involved in some of the early efforts to coax the Coronavirus Task Force to endorse the use of masks. Other nations, including South Korea, had mandated the wearing of masks. Initially, task force members were reluctant to embrace a similar recommendation. One of their early concerns was that guidance to wear masks would send a mixed message about the benefits of social distancing. At the time, the task force was encouraging people to stay at home. Some members told me that they were concerned that issuing a call to wear masks would be interpreted by people as an indication that it was safe to go out, so long as you wore a mask.

It could, they worried, confuse the public.

There was also long-standing ambivalence at the CDC toward the use of masks. One task force member told me that the CDC raised concerns that masks would end up encouraging consumers who wore them to touch their face more, and in turn make them more likely to spread infection through fomites. One instance: the CDC initially told a major airline that their flight attendants couldn't wear masks because the CDC was concerned that the flight personnel did not know how to properly fit the masks, and it would lead to more touching of their faces and could ultimately increase spread. The concerns were based, in part, on the CDC's flawed premise that more of the early transmission was being driven by droplets and contaminated surfaces rather than aerosolization. One senior airline executive told me that their internal data showed that after their flight personnel started to wear masks, the incidence of coronavirus infections among staff fell sharply.

Masks had never been viewed as a standard part of the response to a pandemic. In fact, the much-discussed pandemic plan that the Obama administration had provided to the incoming Trump team didn't take up the issue of widespread masking as a potential approach to containing a pandemic, or even mention masks a single time.[38] I worked with a group of experts at Johns Hopkins University to craft a report we issued through the American Enterprise Institute (AEI) on March 29, 2020, on how to safely reopen the economy. (The report was titled "National Coronavirus Response: A Road Map to Reopening.")[39] We recommended universal masking. A lot of the heavy lifting in writing the report was done by Caitlin Rivers. She is a skilled scientist, with the rare gift for being able to translate scientific goals into prose and condense complicated public health objectives into policy goals that had enough coherence and unfussiness to be easily adopted. In my experience, this kind of practicality, this skill of being able to convert policy into interpretable narrative, is what often separates good policymakers from great ones. In crafting our proposal, we wanted to make sure there was enough supply of masks to support the objective. So we talked to the chief executives of the major medical product distributors. They told us that the supplies were still severely limited, and hospitals were struggling to maintain enough masks for their medical personnel. Ultimately, we issued a proposal for high-quality cloth masks.

President Trump was asked about our recommendation at a March 30 press conference.[40] "Scott Gottlieb, your former FDA commissioner, wrote a roadmap for recovery after coronavirus. . . . The roadmap suggests that everybody wear a mask in public. Is that something that the Task Force thinks is a good idea?" a reporter asked Trump. "I saw his suggestion on that," the president replied. I was told by Marc Thiessen, a *Washington Post* columnist and colleague of mine at AEI, that Thiessen had sent Trump a copy of the report, through the president's personal assistant, to be printed off and shared with the president. "So, we'll take a look at it. For a period of time, not forever," Trump said. "I mean, you know, we want our country back. We're not going to be wearing masks forever, but it could be for a short period of time. After we get back into gear, people could—I could see something like that happening for a period of time, but I would hope it would be a very limited period of time. Doctors—they'll come back and say, 'for the rest of our lives, we have to wear masks.'"

The president took the idea back to his advisers. The task force had already been debating the measure and was on the cusp of issuing a recommendation. Four days later, on April 3, the CDC issued a new recommendation that Americans wear cloth face coverings in situations when they were in public and couldn't socially distance. President Trump announced the directive at a task force briefing, emphasizing that the guidance was "voluntary" and saying that he wouldn't be wearing a mask himself, undercutting the message while unveiling it.[41]

The data showed that masks were not a panacea, but they could help reduce spread.[42] One study later showed that weekly increases in per capita mortality were four times lower in places where masks were the norm or recommended by the government, compared with other regions.[43] Another study published around the same time looked at the effects of mandates for mask use issued by governors in April and May. The authors estimated that these state policies reduced the number of new COVID cases by up to two percentage points per day.[44] Other studies would go on to report similar findings. Masks were not a solution, but a higher quality mask worn properly could reduce risks.

The central premise behind our call for masking in the AEI report that laid out a road map to reopening the economy was our view that widespread adoption of maks could reduce asymptomatic spread. If people were infected, asymptomatic, and in settings where they might transmit the virus but they had a mask on, the mask would reduce the chance that they could spread the infection. Wearing a mask was in many respects an act of civic virtue, a way to protect your friends and neighbors if you were one of those individuals who was unknowingly infected and at risk of becoming a superspreader. Many cloth masks weren't intended to protect you from getting infected, although a high-quality cloth mask would provide a person with some protection. To secure more robust protection, a person would need a higher quality N95 mask. The idea of cloth masks, simply put, was to protect others from you, not you from others, a distinction I made often.[45]

But the whole premise of mask wearing quickly became a faux protest, an expression of the disdain that some people felt for government directives that brought us shutdowns, closed schools, and other restrictions. Sensing that political zeitgeist, the president and his staff couldn't resist the political urge to support these impulses and undermine their own guidance. People who wore masks inside the White House were mocked by senior staff. Members of the task force also didn't model the behavior early on, even the doctors. They would get tested when they entered the White House, but then wouldn't wear masks once they were inside. It was one of our greatest missed opportunities; especially our inability to get higher-risk individuals N95 respirators that could offer them better protection. Like the dispute over other forms of mitigation, the debate over masks had echoes in the tensions that surfaced during the 1918 Spanish flu, when antimaskers had gone as far as to fabricate a telegram from the US surgeon general, falsely claiming that he didn't believe laws requiring masks were effective at reducing the pandemic flu.[46]

However, in the setting of COVID, it was unfortunately the president who did the most to turn face masks into a political flashpoint, creating an ideological movement that ultimately cornered the administration into a dogmatic position that proved harmful to itself and harmful to the nation.[47] It may have been a misreading of the politics: antimasking didn't become a political movement until many political leaders made it so. In fairness to the White House, the CDC gave political leaders mixed guidance on masks during the early months of the pandemic and was itself initially resistant to their widespread use.

The president could have found a middle ground on masks. His message could have been: We don't need mandates. We're adults. We control our government; our government doesn't control us. However, we're going to act responsibly and wear masks. We can do this ourselves. Personal responsibility; not government control. He could have couched it in a way that appealed to his political supporters while appealing to the nation to take more public health precautions. He could have encouraged people to wear masks. I had continued to urge the White House staff to support the use of masks, and I asked myself many times, why didn't the president?

Most thought his resistance was to satisfy his political base, some of whom saw mask mandates as a breach of their liberties. Others said it was the president's innate disdain for government control, his contempt for the "nanny state." It was, in my estimation, some combination of all of those things. However, I heard another theory from those close to the president: Trump just thought the masks made people look funny.

When people insisted on wearing masks around him, the president would ask them to stand outside of the camera shot. As one person close to Trump told me, "the president thought that masks made you look weak." White House staff had been prodding the president to wear a mask in public as a way to reinforce their use and sought the optimal occasion where it would be hard for the president to refuse. They found one in early July when Trump would wear a mask for a trip to Walter Reed National Military Medical Center to visit with wounded service members.[48] However, he had agreed to do so only after finding a mask that he believed looked sharp: dark blue and emblazoned with the presidential seal.[49]

Our challenges in collecting, analyzing, and reporting information reinforced the shortcomings in our pandemic planning. As we learned, the tactics designed to mitigate spread of flu didn't work as well against a novel coronavirus. Flu was not a bad starting point for our efforts, considering how little we knew about COVID. However, we didn't learn fast enough about what was and wasn't effective. We were caught in the fog of a viral war. We lacked the data to reveal how the virus spread and how our tactics were working against it. Then, we lacked the information and analyses to tell us how to adapt our response.

All of these problems showed just how poor our systems were for confronting a pandemic: our information systems, our analytical tools, and our ability to modify our policy response.

The emergence of HIV had proven that viruses with characteristics that are both highly novel and highly lethal can arise unexpectedly.[50] When it came to new and deadly forms of coronavirus, MERS and SARS-1 should have prompted us to plan differently. The appearance of these pathogens triggered efforts to develop vaccines and therapeutics that targeted this class of viruses. However, that work was largely shelved when the immediate risk seemed to recede. A group was formed inside the CDC to work on coronaviruses, but its efforts were dwarfed by the much larger group dedicated to influenza, and much of the coronavirus effort dissolved.

All of our focus was on flu. The next pandemic, it was firmly believed, would be triggered by a novel influenza. It's likely that the next pandemic will indeed be from a new strain of influenza, and it will be a lot worse than COVID. However, we need to be prepared for the unknowns. That starts with the analytical systems to identify and characterize a new pathogen. It could be a new strain of influenza with features that are irregular. Or another coronavirus that's even deadlier and more contagious than SARS-CoV-2. Or, instead, something else that we never anticipated.

CHAPTER 14

HARDENED SITES

The family had just returned to South Korea after a short trip to China, where an unidentified respiratory disease had started to spread, and they brought the pathogen home with them. One of the parents and a child soon became seriously ill with pneumonia. Before they developed symptoms, they were in close contact with friends and other relatives. A spark of transmission had been lit. The mysterious disease began to spread uncontrollably inside South Korea.

A team from the South Korean Centers for Disease Control and Prevention activated a plan to isolate the respiratory virus, identify its source, and deploy a diagnostic test to screen other people for the novel pathogen. They managed to get control of the outbreak before it exploded.

The entire incident was just a drill, a virtual scenario that South Korea ran as a tabletop exercise in December 2019. It was part of preparedness efforts first implemented after the country's dangerous experience with MERS in 2015. A month later, South Korea would confirm its first case of COVID-19.[1] "It was blind luck—we were speechless to see the scenario become a reality," said Dr. Lee Sang-won, one of the South Korean CDC's experts who led the drill. "But the exercise helped us save much time developing testing methodology and identifying cases."[2]

South Korea's rapid response to COVID was not a coincidence. The country's public health system was well prepared. Like the US, South Korea identified its first case of COVID in mid-January. However, the two nations would experience divergent trajectories. South Korea was prepared to deploy widespread testing from the very beginning, and would identify, isolate, and largely eliminate its outbreak. Even before South Korea had its first cluster of cases, it had developed a battery of diagnostic tests and established high-volume testing facilities. It never ran out of testing supplies or protective equipment. They had stockpiled the materials for years.

South Korea's first positive COVID case was identified on January 20. Its daily confirmed cases in the first wave peaked at 851 on March 3, and the country flattened its initial curve without the kinds of strict shutdowns that the US was forced to implement.

South Korea learned hard lessons from SARS-1 and MERS. The country never faced a major risk from SARS-1, but the contagion prompted it to implement major changes in how it prepared for outbreaks. The most visible product was the creation of the South Korean CDC.[3]

As for MERS, in 2015, South Korea developed the largest known outbreak of the disease outside the Arabian Peninsula, mostly the result of a single traveler returning to South Korea from a trip to four countries in the Middle East.[4] Most people who contracted the virus became infected through very close contact with someone already sick with MERS. The virus didn't spread nearly as easily as SARS-CoV-2, but it was also far deadlier, claiming the lives of 30 to 40 percent of the people who became infected. In total, the country experienced 186 confirmed cases and 38 deaths, much of which was the result of patients spreading the infection in local hospitals where they were being treated. In May 2015, a single superspreading event in the 1,900-bed Samsung Medical Center shut down the entire hospital and was responsible for eighty-two cases, including thirty-three patients, eight healthcare workers, and forty-one visitors.[5]

The WHO said the medical center was slow to react to the crisis, taking weeks to implement adequate quarantine measures and get "up to speed."[6] In one instance, an infected man spent two nights in the emergency room without being diagnosed, where he transmitted the virus to at least eighty people before being quarantined. "During that time, it is clear this person was coughing a lot and was also mobile in the emergency room," the WHO later said.[7] Another individual implicated in igniting a large chain of transmission was a person employed in the hospital's emergency department as an orderly, reported to have worked for days after developing symptoms. This person was estimated by health officials to have come into contact with more than two hundred people while possibly contagious.[8]

In response to the MERS outbreak, the South Korean government established hundreds of high-capacity screening sites and leveraged its private-sector diagnostic firms to create a substantial supply of testing components to be stored for a future crisis. The South Korean playbook also directed the creation of an elaborate system for identifying and tracing infections imported into the country and relied on close collaboration between public health authorities and private companies capable of quickly developing and distributing tests to screen for a new virus.[9]

So, on January 27, one week after South Korea identified its first case of COVID, officials from the South Korean CDC convened representatives from twenty local diagnostics companies in a hastily arranged meeting held in the main train station in Seoul and asked them to begin developing COVID tests immediately.[10] One of the country's top infectious disease officials promised the companies that they would receive close cooperation from regulators, and swift authorization. After MERS, South Korea's FDA had established a fast-track approval process to streamline the clearance of tests in a crisis.[11] Though there were only four confirmed COVID cases in the country at that point, the South Korean CDC understood the need to move fast.[12]

Two South Korean manufacturers had already begun developing tests for COVID. One of the manufacturers, Kogene Biotech Co., got clearance for its test four days after the meeting.[13] A week later, Seegene Technologies, a major global manufacturer of molecular diagnostics based in Seoul, received authorization for its own kits. Clearance for three more tests followed shortly after.[14] South Korea guaranteed the manufacturers that the government would reimburse them for the tests, reducing the financial risk. It wasn't by chance that South Korea had companies that were ready to go. The country had invested money in building up its local diagnostics industry after its experience with MERS, specifically to create the capacity it would need in a future crisis.

The South Korean CDC provided the manufacturers with early access to SARS-CoV-2 viral samples, making it easier for them to develop new tests, and also made available standardized assays that manufacturers could use to independently validate the accuracy of their tests. This capacity was baked into their pandemic planning, and manufacturers with the capacity to develop and mass-market test kits were prespecified.

To help advance the development of these tests, after MERS the South Koreans also created a regulatory model that had their CDC and FDA working together to advance new tests in the setting of public health crises. Under the approach, the South Korean CDC would do its own clinical evaluation of new diagnostic kits to validate their accuracy using patient samples that the agency would have access to. It would then send the data it generated to the South Korean FDA to do an independent review and issue the equivalent of an emergency-use authorization. The key advance was the commitment by the South Korean CDC to use patient samples to do the independent evaluation of new tests developed by commercial manufacturers. This accelerated the development and validation of the tests.

When SARS-CoV-2 started to spread, in the first week two domestic manufacturers had produced 700,000 test kits. The tests were able to run on equipment that had been standardized across South Korea's labs, so they could be sent to any of the country's 120 facilities. Each test could be turned around in under a day.[15] However, to speed the kits to market, the South Korean FDA had only put the tests through a preliminary review. Regulators knew that the rapid development process left gaps in how the tests had been validated. So the South Korean FDA continued to evaluate the tests after they reached the market. The South Koreans acted with speed but not neglect, in the words of Dr. Lee Hyukmin, head of the Korean Laboratory Society.[16]

"Of course, a kit that's approved in one week isn't as good as one that goes through a year of clinical trials," Lee said. So, in the early days, he said, the government actively spot-checked the results that labs were generating to make sure that the tests were performing well.[17]

Tests were only part of the challenge. South Korea also needed a way to safely get thousands of people screened. Dr. Kim Jin-yong of the Incheon Medical Center, one of the doctors who treated South Korea's first positive COVID case, was concerned that even with proper ventilation equipment, rooms at indoor testing sites had to be aired out for thirty minutes between patients, meaning only two tests could be conducted per hour in each room. So he suggested a new approach: drive-through testing sites. Patients would stay in their vehicles while they were tested. Kyungpook National University's Chilgok Hospital in Daegu (the country's hardest hit region) installed South Korea's first drive-through testing center on February 23. In comparison to indoor testing sites, the drive-through centers had the capacity to screen up to six patients per hour, because they didn't require a lot of time for disinfection and ventilation between patients, and they could be operated by a minimum of four personnel.[18]

The model was replicated at Yeungnam University Medical Center, another hospital in Daegu. In less than a week, the cities of Goyang and Sejong also established drive-through sites. By April 1, about eighty were in operation.[19] The sites were kept away from hospitals, which helped South Korea avoid what happened in Italy, where testing was largely done inside hospitals, causing sick patients to swarm the facilities. In Italy, hospitals ended up being a major source of spread.[20]

South Korea's testing capacity reached 20,000 tests per day in the opening weeks of its outbreak. That was for a country of just 51 million people. It equates to about 130,000 tests a day in the US. It's a level of testing that the US wouldn't reach for about four months into our epidemic.

Even though the US and South Korea would each report its first confirmed cases of COVID on the same day, South Korea would record daily testing totals over the next two months that would dwarf those in the US. This wasn't because the South Koreans had more infection, but because they were actively looking for the virus, diagnosing cases, and then tracking and tracing infected patients to contain the spread. In South Korea, they were finding their cases and isolating those infected with the virus. In the US, over the same time period, SARS-CoV-2 was spreading unchecked.

Broad testing enabled the South Koreans to use test results to support an elaborate system for tracking down new cases and those who might have been exposed to the virus. Some American policymakers who were skeptical of taking more aggressive steps in the US to contain spread said that South Korean–style interventions couldn't work here. They attributed South Korea's success to the country's ability to mine electronic data on people's movements and interactions, enabling tracing and testing on a scale that would have been unobtainable in the US.

The South Koreans employed a workforce of hundreds of epidemiological intelligence officers to track down those who were infected and get them into isolation, or find those who might have been exposed to place them in quarantine while they waited out the incubation period. To support these efforts, South Korea records almost every credit card and bank transaction on government databases. Health agencies were able to legally access this information and combine it with closed-circuit television footage and data from mobile phones to track people's movements.[21] South Korea's response relied on a high level of public trust and a willingness to yield elements of privacy that would have never flown in the US.[22] However, to be successful, we didn't need to come near this level of intrusion. Tracking and tracing didn't need to identify every case, just some of them. In the US, even if we could find a fraction of the infections, and convince people to voluntarily self-isolate, it could delay the start of the epidemic, or reduce its severity.[23]

Following the country's experience with MERS, South Koreans had accepted the need for this kind of real-time tracing, and the government amended its laws to allow health authorities to collect the same sort of information as law enforcement. The law expanded the country's authority to quarantine people who, it was believed, could be infected, and stipulated the public's right to be informed of exposures, requiring the government to send text messages.[24]

"Other countries simply don't have the capacity to do these thorough investigations on patients," said Dr. Jung Ki-suck, a former director of the South Korean CDC. "We had a smaller absolute number of cases than other nations, but more importantly, the social norm, where people are okay with their privacy being infringed for the wider public interest, allowed comprehensive investigations, which is just unimaginable in western countries."[25]

In the US, critics of more widespread testing and tracing said that it wouldn't be successful without the obtrusive measures that South Korea had employed, and that those tactics amounted to a major invasion of people's privacies. So why try. They said the South Korean government would have a harder time carrying out such effective contact tracing if it couldn't use electronic information.[26]

A key to the success of South Korea, and other Asian nations, identified by healthcare advocate Andy Slavitt, wasn't just the surveillance tools that they used, but social cohesion that made people more willing to engage in collective actions to advance societal interests.[27] It's worth observing that even the South Korean government faced domestic challenges. It was criticized in the South Korean media for its surveillance tactics and the private information it leveraged to carry them out.[28] To be sure, the US efforts to adopt case-based interventions, as a way to isolate the infection and slow its spread, were stymied by cultural resistance to the testing and tracing that would be required (many critics felt that even basic tactics would invade privacy, as public health workers tried to hunt down sick contacts and encourage people to go into quarantine). However, case-based interventions to limit spread didn't require the kind of extreme monitoring that South Korea employed. We should firmly reject the more intrusive tactics the South Koreans adopted, but we can still learn from their success in deploying screening tools. Widespread testing alone could have accomplished a lot, even without the active surveillance.

The Italian town of Vò, in the hard-hit Lombardy region, showed how. Local officials were able to contain spread by combining extensive testing with isolation or quarantine for those infected or exposed to the virus. They accomplished this without the cell phone tracing and the other electronic tools that South Korea used to augment its contact tracing efforts. Starting on March 6, researchers in collaboration with the University of Padua and the Red Cross tested all three thousand residents of Vò, including people who weren't showing any symptoms. This allowed local officials to isolate infected patients and eradicate the virus from the town within two weeks.[29] In the US, by the time we were fully aware of the epidemic spread, it was largely too late to rely on case-based interventions. The US infection levels were so high, and testing was so scarce, that tracking and tracing was never effectively employed during the epidemic's first wave. We simply had too many cases to identify and isolate new infections, and public health departments had been badly underfunded for many years and didn't have the personnel and resources they would need to do contact tracing at the required scale.[30] By the time we turned to these tactics it was too late. The infection was already everywhere. In the end, American health authorities were hampered from using case-based interventions and the conventional tools of public health, not because we lacked South Korea's surveillance capacities, but because we couldn't deploy a diagnostic test.

By early March, America was staring into the abyss. However, it still wasn't too late. On March 12, I appeared on CNBC's morning show *Squawk Box* and said we still had time to get control of the virus and avert the worst outcomes. There were still things we could do to slow the spread and avoid a national epidemic. But time was short, and we needed to take immediate action.[31] America, I said, was at a precipice. We sat at the edge of two alternative fates. On the one hand, we still had the chance to contain the virus and mirror some of the experience of South Korea, which was getting control of its outbreak and was already showing sustained declines in new cases. If not, we risked suffering the outcome being faced by Italy, where the epidemic was spiraling out of control. Right before my CNBC appearance that morning, I had posted to Twitter the points I planned to make during the television segment, airing my main message before I shared them with CNBC's audience. This was something I would do before important television appearances, where I wanted to deliver a careful message. I often used Twitter as a way to get feedback on what I planned to say and pressure-test my main points.

In the series of tweets I posted that morning, I observed: "In U.S. we face two alternative but hard outlooks with #COVID19: that we follow a path similar to South Korea or one closer to Italy." I also wrote, "We probably lost chance to have an outcome like South Korea. We must do everything to avert the tragic suffering being borne by Italy."

I continued, "It starts with aggressive screening to get people diagnosed. . . . While testing capacity expands it's not evenly distributed to places most needed, we're far behind current caseloads. Too many people still can't get screened. So, we can't identify clusters and isolate disease. . . . Social separation works. Every day we delay hard decisions, every day leaders don't demand collective action, the depth of epidemic will be larger. We must act now. We have narrow window to avert a worse outcome. The virus is firmly rooted in our cities. We're losing time."[32]

When I got off the television set, my iPhone lit up. President Trump had retweeted one of my messages.[33] Later in the day, the president called and asked me to come to Washington to discuss my outlook with him. No doubt, he had seized on the hopeful part of my message, that we still had time to be South Korea. The next day I was in the Oval Office laying out the steps I believed we had to take to achieve South Korea's success. I advocated a major effort to expand testing, to use contact tracing to determine where the infection was, and to encourage people to self-isolate to prevent further transmission. We would need to take targeted steps to reduce social interactions as a way to limit spread in places where widespread community transmission was already under way. However, the key was getting much more testing in place, I said. That was going to be pivotal in addressing the crisis. We had a narrow window to secure the experience in South Korea, where the infection was coming under control. Instead, we ended up being Italy.

How could we have realized a better American outcome?

Part of the answer lies in having the right capacities. South Korea showed the importance of having in place the infrastructure that was required to respond to a pandemic. This was true especially when it came to the equipment needed to deploy an enormous amount of testing. However, having enough test kits and test supplies was just one half of the equation. We also needed an installed base of labs and equipment that could run the tests and the ability to deploy testing sites that could quickly screen a massive population.

Some lessons I learned from the FDA's response to the devastation caused by Hurricane Maria two and a half years earlier influenced my views on how to build these needed capabilities.

On September 20, 2017, Hurricane Maria made landfall in Puerto Rico and overwhelmed the island. Maria was a category 4 storm when it hit Puerto Rico and traversed the entire long axis of the island, bringing wind gusts of up to 113 miles per hour. Puerto Rico's power grid was completely destroyed by the storm, leaving millions without electricity.

I was commissioner of the FDA when this tragedy struck, and the death and devastation inflicted on the island's residents was our urgent concern. The FDA helped in the effort to restore essential services and public health. However, the impact on the island's manufacturing sector also had broad implications that put at risk the public health of our entire nation. Puerto Rico was home to a large concentration of advanced drug and medical device manufacturing facilities, and these sites were shut down as a result of the destruction. The disruption risked the supply of critical medical products for all of the United States.

The clustering of these facilities in Puerto Rico was not by chance. It was the result of a policy enacted in 1976 that gave drug and device makers special tax incentives that made it economically favorable for them to locate their manufacturing on the island and then ship their products to the US mainland. It was part of a coordinated effort to move life-science manufacturing to the island as a way to support its economy. As a result of these efforts, the manufacture of medical products became a key part of Puerto Rico's economy, and the island became responsible for the production of about 10 percent of all the prescription drugs consumed in the US.[34]

The FDA was working closely with the Puerto Rican government and manufacturers to assist in the hurricane recovery efforts. On Friday evening, September 22, I got word that a key dam had become vulnerable and might burst. The resulting floodwaters could have taken out a number of warehouses that had stocks of important cancer medicines and other therapeutics. If the drugs were lost, we feared it could trigger shortages of some lifesaving medicines.

I worked with the FDA's chief of staff, Lauren Silvis, to secure permission to move the drugs out of the warehouse by air, but the airport was partially closed, and obtaining landing rights was difficult. We didn't have a lot of time. I phoned Andrew Bremberg, who at that time was head of the president's Domestic Policy Council, at the White House, to ask whom I should call for help. He connected me with Chris Krebs, an acting undersecretary at the Department of Homeland Security. Krebs would go on to head the Cybersecurity and Infrastructure Security Agency, a stand-alone federal agency that is an operational component under DHS oversight, and he would gain international attention when his efforts to affirm the integrity of the 2020 US presidential election put him in public conflict with President Trump and the president fired Krebs from his job. When I reached Krebs late on the night of September 22, he was at his brother-in-law's wedding reception. I explained the problem. "Let me take my kids home and call you," he replied. He called me thirty minutes later and spent the rest of the evening helping me secure the permissions we needed to move the products off the island. Thankfully, the dam ended up holding.

The crisis that night underscored some of our deeper vulnerabilities. We had located the manufacturing for some of our most important medicines to a small number of geographically concentrated sites. In some cases, we didn't have an adequate plan B if our existing supply chain were disrupted. This wasn't an issue particular to Puerto Rico. Across the drug industry, critical products are often made in just one or a few plants. A disruption in a single manufacturing site can cause a cascading series of shortages. After 9/11, we paid some attention to these risks. There was a focus on situations where a single manufacturing site was responsible for the production of most, if not all, of a critical medical product. Back then, the concern was that a terrorist attack on one of these sites could trigger a public health disaster. So, in situations where such risks existed, we started to quietly approach drug makers and encouraged them to build additional facilities. A deliberate attack on any one of these sites could cause drug shortages that could lead to mass casualties.[35] One area of our focus at the FDA was the supply chain for insulin. It turned out that much of the country's supply came from a single manufacturing site. That's not the case anymore, but back then, we were worried about the risk that a deliberate attack on that site could disrupt the majority of the nation's insulin supply. So the FDA worked closely with the drug manufacturer to help support the company in its efforts to build an additional manufacturing facility.

But that work to build redundant manufacturing sites for key medicines was a limited affair, and ultimately, attention faded and the production of a lot of critical medicines became geographically consolidated again. A week after Hurricane Maria, I traveled to Puerto Rico and visited with our FDA staff on the island. Many were without power and supplies, but they were working to keep the agency's office operating because they were dedicated to the agency's mission and knew how critical it was to help make sure FDA regulated products on the island, like food, blood, and medicines were safe and available to support the recovery efforts. Puerto Rico was the location of a large FDA field office where about one hundred of our staff worked. The agency's presence reflected Puerto Rico's importance to America's drug supply chain. Before leaving for that trip, working with Lauren Silvis, the two of us had called the chief executives of each of the companies that maintained a substantial manufacturing facility on the island. We wanted to hear from the CEOs firsthand how much of an impact the storm had on their plants, the consequence to their operations, the public health implications of their outages, and what steps the FDA could take to help them resume critical functions. Our highest priority was to help the people of Puerto Rico. However, for the facilities that manufactured critical drugs, where shortages could have devastating nationwide consequences, we were also working directly with the Federal Emergency Management Agency to help the plants get back online. In some cases, we were helping plants obtain the fuel, generators, and diesel storage tanks they needed. Chris Krebs became a huge help to the FDA and to the people of Puerto Rico in achieving this mission.

When I called each of the CEOs, the status reports varied. Some of their facilities had been badly flooded or damaged by the high winds. All of the plants had their electricity knocked out and were running on backup generators powered by diesel fuel—if they were operating at all. Most had only partially restored their functions. In many cases, the manufacturing sites were running enough of their key infrastructure like air-conditioning systems in order to protect their facilities, but they wouldn't be able to resume substantial production anytime soon. Each of the sites I surveyed had been greatly impacted. There was one exception: Amgen.

The biotech company operated a large manufacturing facility in Puerto Rico, where it produced the drug Neupogen. The medicine is principally used to help the body make white blood cells, chiefly for patients who are receiving cancer chemotherapy. It works by stimulating the production of new white blood cells in the bone marrow. Neupogen is also vital to national security because it can be used to save the lives of people who have been exposed to radiation by helping the body reconstitute bone marrow.[36] It's for its clinical application in a radiological or nuclear emergency or incident that Neupogen has become a critical part of our national defense, a therapeutic hedge against an act of terrorism.[37] If a dirty bomb or other radiological weapon were detonated in a populated city, there would be a surge in demand for Neupogen as a tool to help rescue people harmed by the radiation. For this reason, the US government has added Neupogen to the Strategic National Stockpile, to ensure there will always be an uninterrupted supply of this critical medicine.

When I got Amgen's CEO, Bob Bradway, on the phone, he described the status of his facility. Located in a remote part of the island, it was operating virtually uninterrupted. The manufacturing site was so hardened that it could probably have churned out Neupogen almost indefinitely, even if it were kept off the electric grid for many months. Staffing the facility was his urgent concern, along with providing for the safety and well-being of his workers and their families. But his plant's physical structure was fully secure. He had ample generators and enough diesel fuel stored in massive tanks to operate his manufacturing site for long stretches of time.

After a day spent on the phone with CEOs who were piecing together shoestring strategies to keep their plants limping along at partial capacity, Amgen stood out. I asked Bradway why his company had invested enormous resources in building a facility capable of withstanding a major disaster. He said that Amgen had made a guarantee to the federal government that there would never be a disruption in the supply of the drug. To help fulfill that promise, it had created a domestic network of hardened manufacturing sites and distributed these facilities around the nation. Amgen also maintained some excess inventory at its plants so there would always be a "stockpile" in the event of a crisis. The drug was a biologic, so it couldn't be stored indefinitely. It wouldn't have made sense for the US government to buy a whole bunch of it and put it in a warehouse. Instead, reserves were kept at the manufacturing sites. When Neupogen came off the production line it didn't get immediately shipped out. Some of it went on a shelf at the plant, held in a careful storage unit, while product that had been manufactured weeks earlier was shipped out. This was a vendor-managed inventory, maintained on behalf of the federal government. Through this approach, a small but constant stockpile of Neupogen was being continuously refreshed.[38]

Amgen's efforts provide an instructive lesson for how we can harden ourselves in preparation for a pandemic and secure the capacity to expand diagnostic testing in a crisis. COVID showed that testing is critical to identifying and controlling the epidemic spread of a new pathogen. The approach that Amgen took to guarantee a continuous supply of Neupogen could also apply to securing an adequate supply of testing: build the residual capacity we need into the system.

The backbone of our testing preparedness is the network of about one hundred public health labs spread across the nation. But these are mostly small labs that do specialized surveillance work. At full tilt, their available capacity to screen for SARS-CoV-2 at the height of the crisis peaked at about ten thousand tests a day—enough to do initial investigative work of a slowly spreading pathogen, and to offer a critical tripwire to a new pathogen, but not enough for a fast-moving epidemic.[39] Even if we invested money to expand the capacity of this network, these labs couldn't maintain enough idle volume to be able to provide sufficient surge capacity in a crisis. However, they offer a critical starting point for building a system that can guarantee better preparedness for the future.

The seeds of a more robust system had already been planted years earlier. Leveraging these public health labs, the US invested in a lab response network following 9/11, but it was largely a bespoke effort to evaluate bioterrorist threats. It wasn't meant for the large-scale screening needed to keep up with a galloping epidemic. Writing in *The Lancet* in 2011, Dr. Ali S. Khan, the director of the CDC's Office of Public Health Preparedness and Response, observed that "Large-scale and unpredictable natural, accidental, or intentionally caused disease outbreaks and environmental disasters need many of the same routine surveillance, laboratory, risk communication, and other core public health systems. The flexibility of the Laboratory Response Network shown during the anthrax attacks, for example, has also played a key part in validating BioWatch results, and for responses ranging from severe acute respiratory syndrome (SARS), monkeypox, West Nile virus, and H5N1, to investigations for ricin and saxitoxin poisonings and numerous exposures to mercury."[40] To support the national testing needs that would be required if an epidemic ever got under way, this system needed to be expanded on.

To keep up with the needed testing volume, such capacity has to be already in place. We need to create a reserve capability that encompasses the full spectrum of testing platforms and consumables that we'll need in a crisis. It can't just be mothballed somewhere, either. It needs to be kept operational. This reserve capacity should be thought of as similar to the way that Amgen created those hardened facilities for the continuous production of Neupogen, making sure there was always a supply of the medication available, and that we had the ability to ramp up production in the event of a crisis. We should take the same approach when it comes to testing capacity, and create facilities that have excess volume that can be expanded in a crisis.

The best way to secure this capability is to leverage the nation's existing labs. Under such a scheme, the government could contract with the large commercial labs and ask them to create some additional testing capacity in their existing facilities. This excess capacity would always be available in the event of a crisis. Other smaller labs could also provide some of this volume, but the three largest commercial labs (LabCorp, Quest, and BioReference Laboratories) with their enormous throughput and facilities, could become the core pillars of this new system.

Under this approach, the federal government would subsidize these labs to build excess capacity inside their existing sites. The labs would also be paid to keep a fresh stockpile of testing consumables (reagents, swabs, etc.) to support a surge in testing demand. Demand for lab services already fluctuates by season, and there's always some standby capacity available to handle a natural surge in the demand for testing that occurs each winter. We can build on that concept to increase the size of that reserve. Here is how it could work:

Right now, the commercial labs might maintain about three

thousand testing machines in their largest reference sites. They typically operate these facilities at 80 percent capacity for maximal productivity. Their present focus is to operate these facilities for maximum efficiency. That helps keep costs down. Instead, there needs to be an equal focus on operating these sites with greater resiliency. Some surge capacity needs to be built into the system as a hedge against a major crisis.

This could mean, for example, operating five thousand machines but running each of them at 50 percent capacity. Running more machines at a reduced volume isn't maximizing their efficient use. You're leaving a lot of expensive machinery operating below capacity. However, that excess volume is now available to be tapped in the event of a public health crisis. Addressing a surge in the demand for testing then becomes a function of just running the existing machines a little harder. In this way, the surge capacity is baked into the infrastructure and can be lit up in the event of a crisis. It's kept hot. The federal government can pay the labs for the additional cost of maintaining this excess capacity, as part of our national security preparedness. Asking labs to run more machines at a reduced capacity is preferable to building a giant lab and mothballing it. This kind of sophisticated equipment needs to be changed out every five to seven years, so it must be operating to make sure it's up to date, and it needs to be in the hands of a lab that's able to properly staff these facilities and implement those regular upgrades.

It's also critical that we distribute this infrastructure around the nation. Diagnostic labs need to be close to population centers to maintain their efficiency. Testing large volumes of patient samples is as much a challenge of logistics as it's a question of having enough testing machines and consumables to process the samples. One of the challenges during the peak of the COVID crisis was just the physical processing of the samples. Getting the patient samples shipped around the country to a small number of reference labs created bottlenecks. National reference labs were strained to take in and process the many thousands of additional packages that were arriving every day by FedEx, filled with swabs. To reduce testing delays, it's important to have the testing sites spread around the country, so that they're closer to the points of care. State public health labs can fill a key role in a crisis, helping to direct this traffic in patient samples across different private and public labs, enabling states to make maximum use of existing supplies and testing capacity.

As for the hardware, different testing platforms are needed to test for different kinds of pathogens. For detecting new viruses, labs need to maintain PCR and sequencing machines. Labs also need equipment to perform antibody tests. The platforms used for such tests can be especially important for detecting infections found primarily in the blood, like the Zika infection.[41] To have a proper infrastructure that prepares us for the full range of potential threats, we need to support a diverse base of complementary testing equipment.

We also need to think about the front end—the system for collecting patient samples. In the case of COVID, we couldn't fully leverage the large footprint that commercial labs like Quest and LabCorp maintain because those facilities were designed primarily for collecting blood samples. COVID was a respiratory pathogen that required nasal swabs. So, we had to create new testing sites that were able to handle huge volumes of swabs. South Korea, following its experience with MERS, built and maintained a lot of the infrastructure they'd need in order to erect a massive screening program for a respiratory pathogen. That's why it was able to establish those drive-through sites so quickly. Initially, at the outset of a crisis, the testing is likely to get done at these kinds of large testing centers. This was the experience with COVID. But over time, to give more people access to timely screening, the goal must be to move testing into a distributed model where it's done at medical offices, at retail locations like pharmacies, and at home.

There were other obstacles to putting in place a better testing infrastructure that will need to be solved as part of a new pandemic blueprint. One was the way that the CDC claimed intellectual property rights over the tests it designed. The agency's scientists are allowed to file patents on assays they develop for public health purposes, including in emergencies. The CDC can then seek royalties on the tests.[42] However, this practice can complicate the CDC's ability to quickly hand off tests to commercial manufacturers. That's what happened with COVID. Commercial kit makers worried about how the CDC's claims of intellectual property over the tests could impede the ability to market the tests, and get reimbursed. Negotiating the terms for licensure can eat up valuable time in a crisis. That's precisely what occurred during COVID.

The CDC's assertion of intellectual property rights over the COVID tests would be more understandable if the agency had allowed commercial manufacturers also to compete to develop tests. But the CDC exerted tight control over the access to the viral samples that test makers needed and didn't allow labs to go forward with their own tests without first getting permission from the agency. If labs were eventually given permission to develop tests, they would be required to follow the CDC's protocols. Through these regulatory measures, the CDC basically owned the entire market, essentially preventing manufacturers from developing their own inventions. Anyone who wanted to make a lab test for COVID had to follow the CDC's test design, but to use that blueprint, they had to first secure a license to the agency's intellectual property.

During COVID, critics said that the CDC paid inordinate attention to its intellectual property in the agreements it negotiated with manufacturers. Companies seeking to make the test kits described extended negotiations with the CDC that stretched for weeks as the agency made sure that the contracts protected its inventions. Some test makers balked at the provisions. In other cases, companies tried to maneuver around the CDC's intellectual property claims, by modifying tests to avoid infringing the CDC's claimed designs. The FDA told the CDC on February 20 that the commercial manufacturers were telling regulators that companies couldn't deploy the CDC test because of "IP limitations." Following up on this, the FDA pressed the CDC simply to license its design to manufacturers that could mass produce testing kits that could be commercially distributed.

In one episode, in an attempt to protect its intellectual property, the CDC balked at providing its quality-control procedures to New York State so the state's labs could start using the protocol to validate the tests they had developed. The FDA provided the procedures to New York anyway, without notifying the CDC that they had been shared. On another occasion, the CDC asked the FDA to change the authorization for the CDC's tests to make sure that the regulatory documents contained language that would protect the CDC's inventions and allow the agency to collect royalties on its diagnostics. The CDC wouldn't fully relent on some of its most burdensome intellectual property claims, particularly those governing the process for validating the reliability of tests, until March.

These practices should be revisited. They don't serve an identifiable policy goal, especially in the setting of a public health emergency, where speed to testing is paramount. The CDC isn't pressing its patent claims to make sure that the resulting tests are more accessible, nor is the agency asserting property claims to make the tests more affordable to patients. It's largely just guarding its turf.

Part of the agency's posture reflected its belief that in the past it had been burned when it fashioned innovations, only to see private companies profiting off its inventions. An episode that stood out in the minds of some CDC officials occurred in the early 1980s, when the CDC and drug maker Chiron entered into a research collaboration aimed at developing a test for hepatitis C, but then several years later, Chiron received exclusive rights to the test and a lucrative commercial monopoly.[43] After academic scientists criticized the CDC for letting Chiron get the monopoly over the test, the agency challenged Chiron's patent and proposed an agreement that would have assigned the CDC a half interest in the patents and a ten-year royalty of 3 percent of net sales. Chiron declined to go along and, following a protracted legal fight, ultimately prevailed, agreeing in 1990 to a much smaller concession to the CDC and to Dr. Daniel W. Bradley, who had spent twenty-three years at the CDC and whose research to isolate and clone the hepatitis C virus, the CDC argued, was central to the development of the test. Under the settlement, Chiron agreed to pay Bradley $67,500 a year for five years "in recognition of his contributions to the field" of hepatitis C research.[44] Shortly before the five years ran out, in 1994, Bradley (who had since left the CDC for a job at the WHO) filed suit against Chiron, claiming, once again, that Chiron had used his research and then shut him out of a patent resulting from the discovery. Bradley would lose on appeal. The experience with Chiron left some bitterness at the CDC that, along with some similar episodes, made the agency more aggressive in trying to assert intellectual property rights over its work.[45] That indignation carried through into its work on other pathogens, including SARS-1, when the CDC asserted ownership of the SARS-1 virus and its entire genetic content after CDC researchers had helped map the pathogen's genome. Rather than try to profit from the patents on SARS-1, the CDC said its aim was to prevent other companies and researchers from monopolizing the field the way Chiron had with hepatitis C.[46]

Part of the challenge to America's efforts to rapidly field a test in a crisis is that the CDC can effectively maintain a monopoly on the development and design of tests through its control over the access to viral samples and the power to determine which labs and companies are authorized to manufacture a test and start screening for a new pathogen. This creates the potential for a conflict of interest if the CDC can compel commercial manufacturers to forgo their own development work through the agency's control over test creation and then require commercial manufacturers to pay royalties to the CDC for the right to use the agency's test design—and submit to the CDC's commercial terms.

At the very least, the terms that the CDC seeks should be standardized and settled in advance with a handful of manufacturers who are then prepositioned to take a warm handoff from the agency in the event of a crisis. This would ensure that the transition from the CDC to commercial manufacturers happens smoothly and quickly in a crisis. In such a framework, the commercial test kit makers would function largely as contract manufacturers to the CDC. During COVID, some manufacturers that might have been able to copy the CDC's test design, and produce kits in large quantities, instead initially sat on the sidelines, worried that they would infringe the CDC's intellectual property claims, fail to secure access to its assays, or be unable to reach commercial terms with the agency.

Going forward, we need a hot base of industrial testing capacity. This is what South Korea did. Its government offered contracts to its manufacturers to get them to focus on COVID before cases even arrived. It also guaranteed its test manufacturers a return on their investment, so there were economic incentives in place at the very outset. The equipment needed for deploying hundreds of new testing sites and running samples was already in place.

In the US, we turned to our commercial labs and test kit manufacturers late and didn't coordinate their efforts in any meaningful way. The result was hundreds of different labs and manufacturers entering the market, often with custom-made solutions that couldn't be deployed widely because they would work only on specific instruments or only in certain labs.

This disorganized approach used up unnecessary supplies on tests that could never achieve sufficient scale to provide a meaningful volume of screening. It diverted the resources of regulators who should have focused their efforts on a more curated set of manufacturers capable of producing tests that could be used in the largest number of labs and on the greatest number of patients. Ultimately, once commercial manufacturers were able to get into this market in late March and April 2020, the scattershot nature of the early development efforts meant that we had fewer accessible tests than if that development work had been more carefully orchestrated.

A pandemic strategy should include a plan for leveraging the expertise and capacity of a defined set of labs and test kit manufacturers. They should be contracted with in advance, and their help should be enlisted early in a crisis, to ensure that the needed testing is in place if an outbreak unexpectedly turns into an epidemic. These partners should be given the resources they need (such as viral samples) to develop kits that can be run on the largest possible installed base of testing instruments. The CDC should have these commercial firms on standby, with some dedicated resources that are financed in advance and ready to be ramped up when needed.

We also need a mechanism to buy test kits in advance of an epidemic, and not wait until a new pathogen becomes widespread. The federal government needs to be willing to start stockpiling these kits when a new virus first surfaces, so we'll have them if an epidemic erupts.

In the US, very little money was allocated to diagnostics, even once Operation Warp Speed got under way. Early on, BARDA made some individual investments in certain tests. However, many of these were novel, point-of-care diagnostics and at-home tests that would be needed later to solve specific needs. Most of the US government funding would be focused on vaccines.[47] Meanwhile, the US manufacturers capable of designing tests for COVID were reluctant to jump into this market. The FDA's professional staff had been working with commercial manufacturers and labs starting in January, but there was no leadership from HHS or from the FDA commissioner to encourage these developers into the market. Secretary Azar looked to the CDC to solve the testing shortage problem. During January and February 2020, the HHS leadership was mostly preoccupied with the travel restrictions and repatriating Americans from other nations. The commercial manufacturers and labs, in turn, were frozen by a belief that the CDC had this covered and would soon flood the market with its own kits. The industry's difficult experience trying to market tests for Zika skewed their response to COVID. When the Zika outbreak emerged, device makers had pivoted quickly into the production of new tests only to get financially burned when the threat dissipated, and the government wouldn't reimburse some labs and manufacturers for the tests that they had produced and performed.[48]

To get the manufacturers and commercial labs into the market early, senior officials needed to guarantee the purchase and payment for their tests. We needed to anticipate the threat, and the officials at HHS needed to create a market where one didn't yet exist. However, nobody predicted how quickly the epidemic would explode, how much spread was already under way, and how much testing we would need. Even if these needs had been anticipated, there was no natural home for mounting such a testing response. South Korea reformed its CDC after MERS and gave it responsibility for creating a testing market in a crisis. In the US, HHS turned to the CDC, and the CDC didn't have the policy orientation, operational experience, or industrial expertise to pull off an effort on this scale. As a consequence, we never had enough testing to keep up with the initial spread, and we lost control of the pandemic at its very outset. It made the need for safe and effective therapeutics and vaccines even more urgent.

CHAPTER 15

EVIDENCE IS HARD TO COLLECT IN A CRISIS

He had a low-grade fever. Then some muscle aches. However, at first it wasn't so bad. Not bad enough to keep him from traveling. He got on a commercial airplane from Saudi Arabia, where he was working as a healthcare provider, bound for Chicago, and then took a seventy-minute bus ride before arriving at his home in Indiana. Three days later, his nagging symptoms abruptly worsened: he grew short of breath and developed a high fever. He was evaluated at a local hospital, where a chest X-ray showed a right-lower-lobe pneumonia, and a CAT scan would show that the infection had progressed to both of his lungs. As his condition deteriorated, he was admitted to the Community Hospital in Munster, Indiana.

These events took place at the end of April 2014. The man admitted to the hospital was America's first known case of a novel coronavirus known as Middle East Respiratory Syndrome, or MERS. During the course of his trip through the healthcare system, fifty-three healthcare workers were in contact with him before his diagnosis emerged. Probably hundreds of others were exposed to the virus during his journey from Saudi Arabia.[1]

A few days after the Indiana man was admitted to the hospital, a second case of MERS emerged in America, another healthcare provider who worked in Saudi Arabia and had similarly traveled by a commercial airline from the kingdom via London, Boston, and Atlanta, to his final destination in Orlando. He had first started to feel sick on the flight, but it wasn't until a few days later, on May 9, that he visited a local emergency room and was admitted.[2] The Florida Department of Health confirmed his diagnosis on May 12.[3]

Once again, hundreds of people had been exposed along his route.

The two episodes showed how easily a novel virus can journey across the world and infect people along the way. MERS had first emerged in Jeddah, Saudi Arabia, two years earlier, with a cluster of infections in June 2012. It would continue to spread across Asia, Africa, and Europe before reaching America.[4] By December 2015, a report from the WHO identified 1,621 cases in twenty-six countries, with 584 deaths.[5]

The MERS virus never obtained the ability to spread easily between people; it was largely transmitted through close contact with its animal hosts, usually camels. When human transmission occurred, most of the cases were confined to healthcare settings, where providers would have sustained contact with patients, often without taking adequate precautions to protect themselves.[6] But the virus proved to be dangerous, and it should have been a provocation to take the threat of novel coronaviruses more seriously. MERS arrived just a few years after SARS-1 had already sparked a dangerous epidemic, infecting more than eight thousand people and ultimately spreading to twenty-nine countries, killing at least 774 people worldwide. Most of the cases of SARS-1 would occur over an eight-month period in 2002 and 2003, but the virus would linger for another year.[7] It was clear that coronaviruses were on the march.

The emergence of SARS-1 changed our conventional understanding of this class of viruses. It showed that novel coronaviruses could emerge, evolve to infect humans, and become lethal killers.

MERS underscored these risks. There were ample reservoirs for new and potentially more serious strains. Coronaviruses could be found in bats, whales, pigs, birds, cats, dogs, and mice.[8] The SARS-1 epidemic had its origin in a bat virus that had spread first to civets, an animal sold in live food markets that served as the virus's intermediate host, before spreading from civets to people. For MERS, the intermediary host animals were camels.[9] It was now clear that this once benign class of viruses had pandemic potential.

In virology and in nature, it has often been said that what is past is prologue.[10] The proven potential for evolutionary change meant that, as a category of viruses, coronaviruses could pose a substantial risk to human beings. A novel and dangerous strain that was already lurking in an animal reservoir could adapt to human biology and emerge into human circulation for the first time.

This is how SARS-1 came about. It was an existing virus that transferred with just minor modifications, from an animal reservoir in bats and then to its eventual arrival in the human population.[11] Making matters more ominous, the reservoir of potential coronavirus threats was deep. One research team traveled to twenty different countries to survey animals where the viral species is known to hide. They sequenced the viruses they found and identified about one hundred different clusters of novel coronaviruses, more than 90 percent of them in bats.[12] In a single site known as the Shitou Cave, located just outside of Kunming, the capital of Yunnan Province in southwestern China, the researchers found a coronavirus resembling SARS-1 with attributes that more or less proved that it had the ability to infect human cells. It would provide earlier evidence that SARS-like coronaviruses with the potential to hop from bats to humans were circulating in southern China. In 2013, the scientists published their findings in the prominent journal *Nature*. It should have provided a grave warning to the world.[13]

Historically, advances in public health preparedness have come as a result of major crises. In the case of infectious disease, these advances often followed an epidemic or the fear of one. A lot of the progress we made in improving pandemic surveillance, for example, grew out of concerns over the H5N1 bird flu. When it came to novel coronaviruses, investments were made after SARS-1 and MERS, but were transient. Once the immediate threat seemed to subside, so did the interest in improving preparedness.

In 2004, the National Institute of Allergy and Infectious Diseases awarded $104.7 million in annual funding for coronavirus-related research. However, that funding fell to $14.9 million by 2010. It climbed after MERS, but declined again after that threat receded, and by 2019 it was $27.7 million.[14] The private sector also made investments, but they, too, didn't get much support. The biotechnology company Regeneron fashioned a therapeutic antibody that could target MERS and might have been turned into an effective drug. The new treatment worked by mimicking the antibodies that our bodies would produce naturally in response to a MERS infection. Regeneron had tested the drug successfully, but it wasn't advanced into human trials. After the immediate risk of MERS seemed to fade, policymakers lost interest in these countermeasures.

Then there was work to develop a vaccine to MERS, much of it coordinated by the Coalition for Epidemic Preparedness Innovations (CEPI), a global foundation established to finance independent projects aimed at the development of vaccines against emerging infectious diseases. CEPI is run by Richard Hatchett, one of the original architects of the mitigation tactics included in the 2005 pandemic plan set in motion by President Bush. CEPI's work is focused on the WHO's "blueprint list of priority diseases,"[15] which include coronaviruses. Launched in 2017 with $460 million from the Bill and Melinda Gates Foundation, The Wellcome Trust, and a group of nations, including Norway, Japan, and Germany, CEPI is the closest thing we have to a collective international effort to develop countermeasures to global threats.[16] CEPI is one of the few bodies that remained committed to these risks, but many lost interest in CEPI's coronavirus-focused efforts after the perceived threats from SARS-1 and MERS seemed to recede.[17]

The traditional planning for pandemics had always envisioned the US developing countermeasures against specific pathogens. When it came to threats that could spring from nature, there was an almost single-minded focus on flu. When it came to pathogens that could be harvested from a lab and deliberately turned into bioterrorist weapons, much of the emphasis was on anthrax and smallpox. These pathogens consumed most of our attention.

The strategy was to stockpile countermeasures (therapeutics and vaccines) to these specific threats, so we would have them on the shelf, in a warehouse, to be employed immediately in the event that one of these dangers emerged. What SARS-1 and MERS showed was that we needed to prepare for the unexpected. Few predicted that a novel coronavirus would emerge that was such an efficient killer as SARS-1. Then it happened again with MERS. The initial policy reflex was to try to develop specific medical countermeasures to these two threats, but when these viruses seemed to recede back into nature, so did the bulk of the efforts to counter them with drugs and vaccines.

What we needed instead was sustained investment in broad capabilities that could counter a range of similar dangers—not specific countermeasures, but general approaches to designing and developing drugs and vaccines that could be employed against an array of adjacent risks. Among the diseases on the WHO's priority list are: Crimean-Congo hemorrhagic fever, Ebola virus disease and Marburg virus disease, Lassa fever, Nipah and henipaviral diseases, Rift Valley fever, and "disease X," the risk that a pathogen will emerge that's currently unknown to science.[18] SARS-1 and MERS may have retreated, but their very emergence proved that new dangers could arise from unexpected places. We needed platforms for designing drugs and vaccines to target whole classes of viruses in the event that something new and dangerous suddenly appeared.

We didn't make these sorts of investments, not after SARS-1, not after MERS, and certainly not the way we should have. We remained too focused on flu and not enough on broad capabilities to protect us against unexpected risks. We nonetheless were fortunate, in some respects, that when COVID emerged, we were right at the cusp of a major inflection in science that enabled us to derive fully synthetic drugs and vaccines based largely on information about a virus's genetic sequence, and to be able to quickly develop and mass produce therapeutic antibodies and genomically derived vaccines. We would swing each of these platforms into action to develop countermeasures to COVID. While the Defense Advanced Research Projects Agency, an agency of the Department of Defense that develops countermeasures to national security threats like bioweapons, had laid some of the foundation for using these technologies in the setting of a pandemic (through its work on developing antibody drugs and nucleic acid–based vaccines to address emerging infectious diseases), it wasn't by deliberate design that we had these capabilities to use against COVID.[19] We got lucky that the technology had advanced in the private sector to a point where it could be quickly pivoted to these national security endeavors.

Our ability to use these nimble platforms to develop specific remedies for COVID would be a powerful proof of principle for the value of having broad capabilities that could be used to counter a diversity of unexpected threats. Going forward, to reduce the risk posed by future pandemics, the lesson was clear: It's too hard to guess where the next threat is going to come from. It could be a novel strain of flu, a new coronavirus, or something that mutates in an unexpected way to pose a completely novel risk. We're just as likely to guess wrong if we try to guess at all. What matters most is having core capabilities that can allow us to quickly fashion countermeasures against a broad selection of potential threats and then manufacture drugs and vaccines in the quantities required to protect us.

Antibody drugs provided one such capability, and they could have had an even greater impact in reducing COVID deaths during the fall and winter surge, giving us a potent tool to control the pandemic. However, developing these drugs wasn't the biggest problem.

A lot of the technology for engineering them was already on the shelf; even if the programs against MERS had never gone forward, the science for quickly fashioning antibody drugs against novel viruses had advanced considerably. We didn't develop these therapies as quickly as we could have, however: we were slow to conduct the clinical trials to prove that they worked, and then we were slow to deliver them to patients.

Getting the clinical trials done quickly should have been an achievable task. These trials should have been a high priority, and yet they struggled to enroll patients. Too many doctors steered their patients to competing trials evaluating arguably less promising medicines. There were literally thousands of clinical trials under way with various drugs that doctors hoped could help treat COVID, though most of these drugs had dubious value. It was a therapeutic free-for-all.

It's already difficult to run clinical trials in a crisis. Doctors don't have time to enroll patients and follow rigid protocols when they're overwhelmed by the daily demands of delivering medical care in a pandemic. Moreover, patients are often reluctant to enter randomized trials in which they might receive a placebo when they're sick with a potentially fatal condition. If they're going to take their chances with an experimental drug, they want to make sure they're getting the actual medicine and not a sugar pill, and it's often the drugs that show the least potential that are tested in less rigorous, open-label studies where everyone is promised the medicine.

Yet in a public health crisis, it's critical that drugs that show the most promise are prioritized and advanced more quickly. The enrollment and completion of clinical trials needs to favor the drugs with the most potential to improve outcomes. An analysis by the then head of the FDA's drug center, Dr. Janet Woodcock, would find that about 90 percent of the COVID trials run during the pandemic were designed in a way that would never yield actionable results that could change medical practice.[20] That meant that an awful lot of patients entered trials that offered little hope of generating evidence that could improve the care of patients. Given the difficulty of running clinical trials in a crisis, this was a terrible amount of wasted effort and opportunity.

To steer enrollment to studies that were most likely to yield actionable results, what we needed was something akin to the Defense Production Act for clinical research. The DPA was enacted in 1950, at the start of the Korean War, to help mobilize resources in support of the war effort. The law gave the federal government the authority to require businesses to accept and prioritize contracts for materials judged to be necessary for national defense, regardless of whether the businesses would lose money on the transaction.[21] The DPA would be invoked dozens of times in the setting of COVID to prioritize contracts aimed at mobilizing the production of drugs, vaccines, and medical equipment.[22]

However, nothing similar existed when it came to the conduct of clinical trials. Federal regulators at the FDA had no way to prioritize the conduct and completion of certain trials over others, or to steer enrollment toward higher-value studies. Researchers at the NIH could use their funding mechanisms to support high-priority studies, but they had limited ability to direct academic sites to prioritize enrollment in some clinical trials over others. With the right authorities, the two agencies could be traffic cops on clinical trial enrollment in the setting of a public health crisis. The time needed to conduct studies, and access to patients willing to participate in trials, is a scarce resource in a crisis. Such authority was needed to make sure we turned over cards on the treatments that were most likely to save lives.

We also needed to make sure trials were designed with an eye toward simplicity. If the trials are too elaborate, take too long, and require the collection of too much complicated data, then overwhelmed providers will simply decline to participate. During a pandemic, when doctors are overworked by the need to deliver crisis levels of care, they simply don't have the time to participate in complicated trials that require them to input a lot of extra information on all of their patients. We need to keep it simple. We need trial designs that are crisis proof.

One of the few successful efforts was the British RECOVERY trial, an acronym for Randomized Evaluation of COVID-19 Therapy. RECOVERY was a practical trial that assigned patients to different COVID treatments at random and limited the data it collected to just the most important measures of disease severity and outcomes. This approach made it easier for providers to participate in the study.[23] It was the RECOVERY trial that showed that the steroid drug dexamethasone helped reduce COVID death rates by as much as 30 percent in patients who needed oxygen or mechanical ventilation. This was a profound impact from a simple, cheap, and relatively safe intervention.[24]

Conducted in hospitals across the UK, by the summer of 2020 RECOVERY had enrolled more than eleven thousand COVID patients who were randomized to receive usual care or one of six treatments.[25] The trial's simple design was a stark contrast to the more elaborate trials pursued in the US. Instead of focusing on the limited number of variables that were most relevant to determining whether a drug was delivering a benefit, as RECOVERY did, many US trials tracked dozens and sometimes hundreds of different variables and required providers to input reams of sometimes marginally useful information. The more elaborate US trials just took too much effort to run. The more cumbersome trials also made it harder to enroll patients, and the conduct of these studies were often confined to a smaller subset of academic sites capable of administering the complicated trials. This created additional barriers for underserved communities that didn't have as ready access to these specialized healthcare sites and, in turn, lacked access to promising experimental treatments through clinical trials, potentially worsening outcomes for these communities.

In the end, most of the key answers on which drugs were providing a benefit to patients came not from the thousands of trials under way in the US, but from the single British RECOVERY trial. Yet US public health authorities didn't even participate in RECOVERY or support its model, viewing it as insufficiently rigorous because it didn't collect data on all of the variables that they believed might influence COVID outcomes. Instead, the NIH supported its own clinical trial network, the Accelerating COVID-19 Therapeutic Interventions and Vaccines (ACTIV), which took a more traditional approach to how the clinical trials were conducted, relying on carefully randomized studies that collected a lot more information to evaluate different therapeutics. After a year, not a single therapeutic would be authorized by the FDA based on information derived from ACTIV.

The development of evidence proving the substantial benefit of steroids in treating COVID, and the efforts of the RECOVERY trial to uncover these findings, simultaneously reveals the virtue of the RECOVERY approach, and the reason why American regulators didn't support the trial.

The RECOVERY trial was referred to as a practical trial because it had a simple design where it looked at just the key variables that could help doctors tell if the drug was working. Usually a clinical trial might track dozens or even hundreds of factors that could impact a drug's effect on a patient. The idea is to monitor everything that might affect the outcome you're looking for, so you can control for these exogenous factors that might influence outcomes and be mistakenly construed as an effect of the drug. But in order to make the RECOVERY trial more efficient and easier to run in a crisis, it didn't require providers to track the full complement of clinical variables that a regular study would evaluate. For example, doctors weren't required to monitor the level of sugar in patients' blood. We know that when patients are treated with steroids, their glucose rises. Some might argue that we need to know if there's a correlation between the magnitude of that rise and the clinical response to steroids. The architects of the RECOVERY trial would say: Why monitor a variable when we know what it's likely to show? Moreover, we know that doctors are going to treat high blood sugar. The critical data point is whether the steroids improve survival. Knowing how the change in blood sugar might impact that outcome is secondary to the key question we're trying to answer.[26]

In normal times, America's decentralized clinical trial system serves us well as the competition for products and scientific findings helps spur research and investment. In a crisis like COVID, this same competition for drugs, ideas, and scientific findings really hurt us. In the setting of a fast-moving public health emergency, there's no time to waste. We squandered far too much time running studies on dubious therapies, often with trial designs that were inadequate to answer the question of whether a therapy worked. We needed more practical studies like RECOVERY and more central organization around the conduct of research in a public health crisis. We need to support deliberate efforts to steer patients into the trials that are more likely to provide benefits and yield actionable evidence that can improve treatment and outcomes.

The controversy surrounding use of hydroxychloroquine to treat COVID would illustrate the significant risk of allowing critical scientific questions to linger, where the desire for a swift therapeutic solution and the need for public health rigor collided to produce some unfortunate challenges. Initially, there was some belief among doctors, based on small studies, that hydroxychloroquine might provide a benefit. So the drug was heavily used in the first several months of the pandemic. In April, the majority of hospitalized patients in New York were receiving the therapy. However, subsequent research didn't support the drug's hoped-for benefits. Yet more than one hundred studies with the drug would be under way by the summer of 2020, even though by this point its claimed benefits had been fully disproved.[27]

A devotion had formed around the drug's perceived benefits, which was initially promoted on Twitter by some investors and tech executives, and these postings got picked up in mid-March on cable news and were advanced prominently by Fox News host Laura Ingraham, who featured multiple segments about the medicine and emailed the White House about the drug.[28] Within days of Ingraham's initial segment on Fox News, President Trump made his first mention of hydroxychloroquine from the White House podium.[29]

I didn't believe the president was being well advised on the drug and research evaluating its use in COVID. A group of outside doctors had gained access to the Oval Office and advanced their anecdotal experiences as proof of its purported efficacy. Some of these meetings were attended by FDA commissioner Stephen Hahn, whose involvement lent further credence to the speculative therapy.[30] The drug soon became a political ambition for those who believed it could bring an end to COVID, and it continued to be promoted heavily by a handful of TV commentators. During a single two-week period between March 23 and April 6, hydroxychloroquine was mentioned on Fox News nearly three hundred times.[31]

President Trump called me on March 18 to ask my opinion about his desire to authorize the drug's use. The president would have been superseding the advice of the FDA's professional staff if he were to issue an authorization for the drug to be used in the treatment for COVID. I was on the speakerphone while others were gathered in the Oval Office around the president's desk. Secretary Azar had said that Trump had the authority to issue the order and that Azar would implement it. I expressed my serious concerns about the idea, joined by Anthony Fauci, who was conferenced into the call after the meeting had already started. Hahn was in the Oval Office during the conversation, but remained silent through the discussion. Fauci and I voiced similar worries, arguing that such a move by the White House, to force a premature action on the drug, would undermine the regulatory process at a moment when the American people needed to trust the independence and rigor of the FDA's decision making. It was clear that there were a lot of critical actions that the FDA would be drawn into on other therapeutics and vaccines. The agency's scientific independence would become a crucial currency. More important, by that point, it was becoming more obvious that hydroxychloroquine didn't work in treating COVID and might harm patients.

The president was cautious and seemed to agree that the White House shouldn't intervene. However, it wouldn't be the last word. The president continued to promote the drug, and pressure to find other ways to broaden access to the medicine would continue among his staff and at HHS. The next day, at a press conference, the president said that the drug was soon going to be made widely available and "given out to large groups of people."[32] Ten days later, on March 28, Hahn would issue an authorization for the drug's emergency use as a hospital-based treatment of COVID.[33] Privately, those around the commissioner said that the decision to issue the EUA was driven by pressure from officials at HHS, who in turn were under pressure from the White House to make 30 million doses of the drug (donated to the strategic stockpile by the drug maker Novartis) available to patients.[34] Since the donated tablets were intended for foreign markets, and weren't an FDA-approved formulation of the drug that could be legally sold in the US, the EUA was needed to authorize their prescribing and allow them to be legally distributed in the US.

At the same time, there was a second reason why HHS officials wanted the EUA. I was told by White House staff that Larry Ellison, the executive chairman of the tech firm Oracle, had proposed to the president that Oracle would lead an effort to develop an app and a database that could track information about the drug's use and evaluate its treatment effects in patients.[35] Inside the White House, proponents of the drug saw the database as a potential way to develop data that would confirm the drug's purported benefits. It could, in the eyes of some who supported the scheme, serve as an end run around the FDA. However, to go forward with its plan, HHS officials told the FDA that Oracle wanted protection from liability, and that's where the EUA came in. The EUA would extend indemnification to Oracle under the terms of the PREP Act, which provides immunity from liability for claims relating to or resulting from administration or use of countermeasures used in a public health emergency.[36] So, the White House was leaning on HHS to put the Oracle app in place, and HHS, in turn, was pressuring Hahn to issue an EUA.

However, an EUA might also instigate more prescribing of the drug at a time when evidence was mounting that it didn't work. On May 18, Trump would become a patient himself, declaring that he was taking hydroxychloroquine as a prophylactic to prevent infection after he had come into contact with the virus. In his announcement he said that he had received "so many letters" from constituents who said it worked. "I want the people of this nation to feel good," Trump said.[37]

Yet the next day, May 19, scientists at the FDA's drug center completed an analysis of the hydroxychloroquine's side effects, based on reports from physicians who were treating COVID patients with the drug and had spotted problems. The analysis found that there had been nearly four hundred adverse health events linked to the medicine, including eighty-seven deaths. Fourteen cases were identified with ventricular arrhythmia, a known side effect of the medicine, of whom seven died.[38] The scientific data was shared with the White House, where it set off conflict.[39] The White House still wanted the drug broadly distributed. Emails showed that Hahn ultimately approved a plan to allow doses to be sent from the federal stockpile to community pharmacies, even though the FDA's authorization restricted the drug's use to hospitals.[40]

A month later, on June 15, after more data showed that hydroxychloroquine was providing no benefit while introducing additional risks, the FDA revoked its prior action authorizing the drug's use in the treatment of COVID. "The agency determined that the legal criteria for issuing an [Emergency Use Authorization] are no longer met," the FDA said in announcing the action. "Based on its ongoing analysis of the EUA and emerging scientific data, the FDA determined that chloroquine and hydroxychloroquine are unlikely to be effective in treating COVID-19 for the authorized uses in the EUA. . . . Additionally, in light of ongoing serious cardiac adverse events and other potentially serious side effects, the known and potential benefits of chloroquine and hydroxychloroquine no longer outweigh the known and potential risks for the authorized use.[41] The FDA's safety report, the one from May 19 detailing the adverse events, was posted to the FDA's website on July 1.[42]

I care deeply about the FDA's mission, the dedication of its staff, and its central role in addressing the crisis. It was personally hard for me to watch what were, in my estimation, the challenges Hahn faced in not being able to safeguard some of those principles with HHS and the White House and to protect the FDA's prerogatives to make independent decisions grounded in science. When I was in the role, we had our own fair share of challenges, and efforts by others to use political power to infringe on some of the agency's decision-making authority. We strenuously opposed an effort by the Department of Defense, which wanted to seize from the FDA authority for overseeing the approval of medical products that would be used for soldiers.[43] We made our case to Attorney General Jeff Sessions, who wanted the FDA to affirm that it didn't have authority to regulate drugs that would be used to provide anesthesia to ease the suffering of prisoners who were being put to death.[44] We had wrestled with the Department of Agriculture, which wanted to usurp from the FDA some of the agency's authority to regulate the safety of food products.

If the FDA had relented to such positions, I believed it would have undermined some of our core authorities, sacrificed aspects of our public health mission, and put patients at potential risk. In each of these episodes, we waged a spirited defense of the agency's prerogatives, we "spent political capital" in advancing our positions, and we prevailed on the issues. The FDA wielded immense authority over areas that touched people's lives in profound ways. There were always political forces looking to encroach on the agency's authorities. And in the setting of a global pandemic, those political pressures were especially profound, and the fault rests with those who tried to pressure Hahn to take actions that the agency couldn't support through its objective review of the science. I'm sure in the actions Hahn took, he tried to limit harm to patients and to the agency's public health mission and its independent decision making. However, the hydroxychloroquine episode was a cautionary tale for many reasons.

Foremost, it reinforced why it's critical to produce meaningful evidence in a crisis. A false perception of the drug's benefits was able fester because we didn't firmly settle these questions early, through the conduct of good studies. The drug was widely embraced in April, when there was some belief that it could provide benefits, and these hoped-for benefits seemed to outweigh its known risks in a setting where patients had few options. As evidence accumulated that the drug wasn't helping patients, and as other treatments emerged, doctors stopped prescribing it, but at that point, the drug had become a cause célèbre. Far too many people continued to be inappropriately exposed to the medicine, and perhaps harmed by it or denied more effective options. The false narrative had persisted for so long, and had become so hardened, it eventually obscured the evidence that it wasn't working. We failed to produce an early and conclusive study to frame accurately its risks and benefits and put some of the original speculations to rest. When that firm evidence finally arrived, the perceptions were already deeply embedded, and the data didn't break through. A year later, in June 2021, when a single, small, retrospective study came out by doctors from a hospital in New Jersey—suggesting that their experience with the drug showed that it might benefit a select group of very sick patients when used in high doses and in combination with the antibiotic azithromycin—supporters of the president seized on it to say he was right all along.[45] The hardened views on the drug had persisted.

The COVID response continued to be hampered by the challenge of fielding properly crafted studies that could decisively answer key clinical questions, both before the hydroxychloroquine bubble inflated and burst, and well after. By December 2020, there were 251 studies examining the benefits and risks of different antiviral drugs that, it was hoped, could act directly on SARS-CoV-2 to block its replication, but only twenty-four were randomized, adequately controlled studies that were designed to yield a definitive answer about a drug's effectiveness. Similarly, there were 225 studies with drugs that targeted the intracellular environment as a way to disrupt the function of the virus, but only nine were designed in a way that they would produce clear answers. There were 737 studies with drugs designed to suppress the hyperactive immune response that some patients developed, but only 31 were properly designed to generate a decisive conclusion.[46] Yet while all these studies were under way, many of which wouldn't yield useful answers, the trials with the antibody drugs, the most promising medicines in development, were delayed because of difficulty enrolling patients. A lot of eligible patients were being siphoned away by competing studies that were evaluating inferior drugs.

The episode involving hydroxychloroquine was instructive for another reason. It revealed the importance of holding the line where core principles were at stake. Decisions in a public health crisis should be based on clear evidence to support these convictions. I was especially focused on what was happening at the FDA, having worked there for many years, and I was concerned about any perceived intrusion into the agency's scientific autonomy. Commissioner Hahn faced strong pressure to open up parts of that science-based process to the scrutiny of officials at HHS and the White House, and having crossed that line on hydroxychloroquine, he would be at pains to reclaim that independence later. In Washington, one lesson I had learned by watching others be subject to political forces, and by coming under such pressure myself, is that once you show you can be influenced or controlled, it invites more people to try to push you around. Bend your principles a little, and you get pressured later to bend them a lot.

There was an even broader lesson here. Regulatory agencies like the FDA have politically appointed leadership that are subject to Senate confirmation for a reason. The role of a politically appointed chief is to help align an agency with the political objectives of Congress and the executive branch, and to be accountable to both. However, there's a mutual concession in this arrangement. A politically appointed agency head has a twin obligation to educate political leaders about an agency's prerogatives and to maintain the delicate line between an agency's core mission and the political goals of elected officials. As the American Enterprise Institute's Yuval Levin noted, "Political appointees in federal agencies are not ideological compliance czars but essential two-way conduits between the sources of public accountability and the sources of professional expertise in government. They represent the agency's views and priorities to the president's senior team at least as much as they convey the president's views and priorities to agency staff."[47] As the head of an agency, I knew that there were times when I would need to call balls and strikes, and that my decisions might be in conflict with someone's political objective. However, I always had a chance to educate leaders in the process, building support where I could or seeking concession where I had to, but I had an obligation to support the agency's findings and affirm its mission.

These considerations would play out in one of the biggest challenges that the FDA confronted in the COVID crisis—bringing a safe and effective vaccine to the market. The pressure and scrutiny that Hahn, and the FDA, faced around these decisions was historic and unmatched in any other modern period. It was certainly in excess of anything I confronted while I was commissioner. I am therefore reluctant to second-guess the decisions they took. However, we should reflect on those challenges to learn from them where we can, because we'll face a crisis again one day.

Following the hydroxychloroquine episode, Hahn would face criticism for his actions related to another drug for treating COVID: convalescent plasma. This is a treatment in which antibodies recovered from patients who have successfully battled COVID are infused into other patients who still have an active infection. The belief is that the antibodies that helped one person recover can transfer some immune protection to the person who is still sick.

At a White House press conference on August 24 announcing the authorization of convalescent plasma as a treatment for COVID, Hahn would overstate its benefits while standing next to the president, using "a deeply misleading statistic to claim that a treatment the agency had just authorized for treating the coronavirus would save 35 lives out of every 100 people who get the treatment," National Public Radio would write. "That false claim has brought withering criticism from scientists, in news articles and on Twitter, who argue that it was a gross exaggeration of the benefits."[48] A day later, Hahn apologized for the unfortunate misstatement, saying on Twitter that he had used a poor choice of words in describing the clinical trial results. "What I should have said better," he wrote, "is that the data show a relative risk reduction not an absolute risk reduction."

However, right or wrong, fair or not, critics would accuse Hahn of furthering the embellished description that the White House put forward of the drug's usefulness in comments Hahn shared at that press conference with the president. A year later, internal FDA emails were obtained by reporters and would reveal that, contrary to his apologies and corrections, Hahn had planned the remarks in advance and opted himself to use the 35 percent figure.[49] It didn't help the circumstance that subsequent studies called into question the therapy's benefits, including one by the NIH, framed as a definitive evaluation, that was stopped early for futility, because the treatment wasn't providing any benefit.[50] By May 2020, the British RECOVERY study would demonstrate in a trial involving 11,569 hospitalized patients that plasma delivered no significant reduction in mortality, no significant increase in the chance of being discharged alive at twenty-eight days, no significant reduction in the need for mechanical ventilation, and no significant reduction in death.[51] (The use of plasma was largely curtailed by the spring of 2021 owing to persistently lackluster result from clinical studies, but not before more than 722,000 units of plasma would be distributed to hospitals owing to the federal push, with the US government paying $647 million to the American Red Cross and America's Blood Centers to support the collection and distribution of the blood products, and Hahn appearing on roadside billboards around the nation to promote the therapy.)[52]

After Hahn's misstatements related to plasma, the medical community turned their attention to the COVID vaccines that were in development and asked whether Hahn would now be able to hold the line and assure the FDA's independence on the review of these products. "Hahn's mea culpa comes at a critical moment for the FDA which, under intense pressure from the White House, is responsible for deciding whether upcoming vaccines are safe and effective in preventing COVID-19," wrote ABC News.[53] The review of the COVID vaccines would be among the most important medical product decisions that the FDA would make in modern times. People wanted to feel confident that the commissioner wouldn't be cowed on these decisions, or worse still, have the decisions taken away from the agency and made by political officials in HHS. To many, these fears didn't seem far-fetched. They were being fueled by President Trump's repeated references and tweets where he castigated the FDA and said he wanted to get the vaccine approved before the election. The president was explicit in linking the authorization of the COVID vaccine to the timing of the election.

Ultimately, the FDA's professional staff would put down a marker of their own, laying out in a guidance document the parameters that would govern their review of a COVID vaccine.[54] It was an audacious and important effort to stipulate objective parameters, led by the career scientists in the FDA's biologics center. Through this process, by every indication, Hahn showed support for the FDA's staff in these efforts. After the episode involving convalescent plasma, FDA colleagues said he saw the need to change course, and

Hahn let the professional staff take steps to insulate the process they would use to evaluate a COVID vaccine from outside pressure.

The FDA guidance document would establish a clear and immutable benchmark. However, the guidance would also be issued well in advance of any chance to see the interim data from the ongoing clinical trials. Nobody knew at that time how effective the vaccines were going to be. If you had asked me, I would have guessed 60 or 70 percent based on the earlier data that I had reviewed for the Pfizer vaccine. Before the data from the vaccine's pivotal study was available, I didn't expect the vaccine to be 95 percent effective, and I suspect few others did.

In spelling out far in advance the criteria it would use to judge the vaccines, the FDA had ceded its discretion to make those judgments after it had a chance to see how effective the vaccines might be. When that interim data finally arrived, showing that the vaccines were far more effective than many expected, the FDA would be largely locked into the guidance that they had already issued where, as a practical matter, the agency's criteria meant that they would have to wait for the conclusion of the trial, and couldn't base an authorization on an interim analysis of that data. I couldn't help but think: Would the agency's bar have been set differently had the FDA constructed its framework with knowledge of how well the vaccines would pan out? Would the FDA have used its discretion to authorize the vaccines sooner for the highest risk patients, based on an interim analysis, if it knew how effective those shots would be?

At the peak of the winter surge, we were losing as many as seven thousand patients a week in nursing homes to COVID. About thirty thousand long-term care facilities were impacted by COVID with outbreaks. Less than 1 percent of America's population lives in long-term care facilities, but by the end of 2020, this population accounted for about 40 percent of all COVID deaths—nearly 175,000 in total.[55] During the worst week of infections, the week of December 17, there were 72,586 COVID cases diagnosed in nursing homes.[56] (The nursing homes that did a better job at controlling spread were the ones that were affiliated with hospitals, probably because these facilities had more experience in implementing strict infection control procedures.) Dialysis facilities were another site where vulnerable patients congregated and where the impact of COVID was especially brutal. Dialysis patients suffered a higher rate of hospitalization for COVID disease than any other group in the Medicare program.[57]

If a vaccine could have been available just four weeks earlier, perhaps based on the first interim analysis of data available in early November, and initially authorized just for use in nursing home patients, it might have translated into a potentially meaningful benefit to those at highest risk from the virus. Yet, in reacting to the political pressure that Hahn faced and the agency's understandable concerns that this pressure could erode consumer confidence in the regulatory process at a key juncture, the FDA's staff felt justifiably bound to lay down a scientifically sound and objective marker on the criteria they would use to evaluate these products. Owing to that new criteria, it would take an additional one to two months to authorize the first vaccines based on a more complete data set.[58]

The political pressure by the White House to hasten vaccine approval may have had the opposite effect on its commissioner, instigating a series of actions that ultimately might have delayed those products from reaching patients sooner. In seeing the political pressure on its process and recognizing a need to issue a clear marker on the criteria it would use to evaluate the vaccines, the FDA's professional staff would also have to give up some of its later discretion to set those criteria based on what the studies actually showed when they were unblinded.[59]

I always felt the FDA was at its best when it was in firm control of its own process and decision making, when it could make independent decisions based on the close evaluation of data. The FDA is diligent in explaining its decisions when it does use this discretion, including public discussion with an independent advisory board. These were key features of the process and elements that I tried to affirm as I aimed to advance the ability of the FDA's professional staff to call it as they saw it. Sometimes, that meant they would articulate their standards in advance, in guidance, and sometimes, the process was best served if the FDA's staff reserved judgment and was able to make a decision later based on the evidence, adjusting their thinking based on what the data showed. The White House thought the FDA would take more risk if they were put under pressure. The reality was quite the opposite: the agency would be more able to do a careful weighing of risks and benefits, and perhaps embrace some additional uncertainty, if it had the independence to evaluate the available evidence. I can't help but think how the process might have played out if the FDA had felt more firmly in control of these events, and if Hahn's unfortunate entanglements with hydroxychloroquine and plasma had never occurred.

CHAPTER 16

GETTING DRUGS TO PATIENTS

President Trump started to feel sick on Wednesday, September 30, just a day after a presidential debate. The first sign of illness in his ranks had come that evening, when his close aide, Hope Hicks, developed symptoms on a trip back from Duluth, Minnesota, where she had accompanied Trump to a campaign rally. The next morning, Hicks was tested. The result came back positive. Only the president and a small circle of senior staff knew about Hick's test, but several aides who had been around Hicks that week were seen wearing masks, a rare sight inside the West Wing that tipped off others that something was amiss.[1]

President Trump would continue with his full schedule on Thursday, including his plan to attend a fund-raiser at his club in Bedminster, New Jersey, that afternoon. When he returned to the White House, he took a rapid COVID test that came back positive. A confirmatory PCR test was sent off at the same time, but it would take longer to return a result. In between the two test results that evening, the president called in to Sean Hannity's show on the Fox News Channel and seemed to foreshadow his diagnosis.[2]

"You know it's hard when you're with soldiers and with airmen, you're with marines and the police officers, I'm with them so much," he said, seeming to invoke a sequence for how he might have contracted the virus. "And when they come over to you, it's very hard to say: stay back, stay back. It's a tough kind of situation, it's a terrible thing, so I just went for a test, and we'll see what happens, I mean, who knows. . . . It's very, very hard when you are with people from the military or law enforcement and they come over to you and they want to hug you, and they want to kiss you because we really have done a good job. They get close and things happen."[3] Trump's comments would seem to betray the diagnosis that he was anticipating.

White House aides should never have let the president be put at risk of contracting COVID. In an Oval Office meeting the previous Monday, the president had complained privately to senior aides about an event that had been held inside the White House on Sunday, where guests attending a ceremonial gathering had gotten close to the president. After Trump was diagnosed with COVID, there was a refrain in the media that his infection was the consequence of his own cavalier approach to the virus, a narrative that was overly simplistic. Though the president appeared, on many occasions, to outwardly minimize the threat that COVID posed—downplaying its severity, eschewing the use of masks, giving short shrift to other precautions—inwardly he was cognizant of his own personal risk, and privately he was being reassured that the precautions adopted by the White House would keep him safe.

When I met with the president in March, he showed me the sanitizing hand wipe he used, a Sani-Cloth by PDI. This wasn't an ordinary wipe. The cloth was imbued with isopropyl alcohol, like hand sanitizers in common use. However, it was also impregnated with ethylbenzyl ammonium chloride and dimethyl benzyl ammonium chloride, two powerful agents that are often used as hospital-grade disinfectants. The PDI Sani-Cloth is typically sold to healthcare facilities for wiping down contaminated surfaces. I told him that with repeated use, it could be irritating to his skin. He said he had a problem with it only one time.

It was shared by those who knew him that the president was

conscious of the risks from contagion. Years earlier, *Newsweek* had described a series of interviews Trump had given on *The Howard Stern Show*, where Trump wryly described himself as a "germaphobe" who "washes his hands frequently throughout the day and needed to drink through a straw because he wants to avoid contamination."[4] It seemed clear to me that the president didn't want to get COVID and that he was led to believe that the White House testing protocols would keep him from being infected.

At 12:54 a.m. on Friday, October 2, the president tweeted the news of his diagnosis: "Tonight, @FLOTUS and I tested positive for COVID-19," Trump wrote. "We will begin our quarantine and recovery process immediately. We will get through this TOGETHER!"[5]

After the tweet, the president's condition would deteriorate, distressing his staff and his doctors. An attorney from the White House Counsel's Office called FDA commissioner Stephen Hahn, to ask the agency to authorize the use of an antibody drug being developed by the biotech company Regeneron that targeted COVID but hadn't yet been authorized by the FDA and wasn't commercially available. The White House would tell Hahn and one of his FDA colleagues only that the drug was for "two senior administration officials," presumably the president and first lady.[6]

Regeneron had already told the White House that it would provide expanded access to its drug under a compassionate use investigational new drug application. This is a pathway that the FDA uses to grant a company permission for an experimental drug to be used on a specific, named patient, outside of clinical trials. The pathway is typically reserved for circumstances where patients have a serious or life-threatening condition, have exhausted conventional therapies, and where an experimental treatment may offer the best and only option.[7] Regeneron immediately shipped a package of doses to the White House.[8] The next morning, White House officials also called the pharmaceutical company Eli Lilly to request a similar antibody drug that Lilly was developing. Lilly told Trump's doctors that the president could also access its drug as part of an individual patient investigational new drug application, so long as the White House filled out the requisite paperwork to authorize the expanded access. That was the last time Lilly would hear from the White House on the matter.

The antibody drugs being developed by Regeneron and Lilly to target COVID were built on the same clinical premise as the medicine that Regeneron had used to successfully develop an experimental antibody drug to treat MERS. In the weeks just prior to the president's diagnosis, both firms had released promising data showing that their antibody drugs could help reduce the viral load in patients infected with SARS-CoV-2.[9] It was the strongest signal yet that these medications might work to reduce the symptoms and consequences of COVID.

Later on Friday, the president's physician announced that Trump had received the drug from Regeneron as an infusion administered inside the White House.[10] He had also received an infusion of the drug Remdesivir.[11] However, his medical condition deteriorated over the course of the day, and by the afternoon, he was requiring supplemental oxygen. White House staff made the decision to transfer him to Walter Reed Army National Medical Center, fearing that if his condition continued to worsen, he might not be able to walk to the helicopter under his own power.[12]

President Trump may have been an optimal candidate for an antibody drug. The antibodies seem to be most effective when they're used early in the course of COVID and for patients who aren't mounting an effective immune response on their own. Data released later by the president's physician, Dr. Sean Conley, showed that Trump didn't have detectable antibodies of his own when he was first diagnosed with COVID.[13] That could have been because his antibodies were measured early in the course of his infection, before his body had had enough time to mount a response. Or, as seemed possible, it could have been because the president was among the subset of patients, usually older individuals, who don't mount a robust initial immune response to the virus, putting them at more risk. It's for these patients that the antibody drugs seem to work the best. What happens in these situations is that the virus replicates largely unchecked because patients don't develop antibodies to interrupt its progress. By the time their immune systems kick in, a lot of virus has accumulated. Faced with a high load of virus, their bodies will then overreact to the infection and dump a whole lot of immune cells into the bloodstream.[14] This is the immune system becoming overcharged all at once, the "cytokine storm" that can damage organs.[15]

The clinical circumstance was described elegantly by the celebrated author, oncologist, and hematologist Siddhartha Mukherjee. If you're able to mount a robust immune response during the early phase of the virus, then "you control the virus and have mild disease," he wrote, and "if you don't, you have uncontrolled virus replication in the lung that results in misfiring of the immune response that fuels the fire of inflammation leading to severe disease."[16]

Making matters more concerning was the early onset of the president's symptoms and their rapid escalation. This also suggested that the president might be among a subset of patients who, the data seemed to suggest, might be at heightened risk and could derive more benefit from the antibody drugs. This is medically speculative, of course—we will never know for sure. It's a judgment based on the data that his doctors shared about his diagnosis and symptoms, and the fact that he didn't appear to trigger an early antibody response on his own based on his blood results. It's possible that the Regeneron drug may have been more important in helping treat his condition than was fully appreciated. Almost a year later, data would emerge from the RECOVERY trial demonstrating that hospitalized patients who don't mount their own, initial antibody response to COVID could cut their risk of death 20 percent by using the Regeneron antibody drug.[17]

This is also, paradoxically, why patients in these circumstances (including, possibly, the president) might not develop enduring immunity to COVID after taking one of these antibody drugs. Such patients might have been on a worse clinical course because they didn't mount an early and effective immune response. Then, once they are administered the antibody drug, it's basically like giving the patient an intact immune system in a bottle. It supplants their own immune response. So, for such patients, their bodies may never have had to develop a robust, innate antibody response of their own. These patients may be no worse off in terms of their long-term immunity to COVID than if they never had the infection. However, they may not be much better off, either. Since their own immune system was never sufficiently activated by the virus, they may have never had an opportunity to develop as much lasting immunity to the pathogen. So, the bulk of their durable immunity was what was transferred to them by the antibody drug. Once those infused antibodies wear off, in one to two months, so might any residual protection against subsequent coronavirus infections. (President Trump was later vaccinated against COVID in January 2021.)[18]

A few days after the president became ill, former New Jersey governor Chris Christie announced that he too had contracted COVID. Christie was likely infected by the same chain of transmission that sickened Trump, one that might have had its origin at a White House event on September 26, which included an indoor reception, to mark the nomination of Judge Amy Coney Barrett to the Supreme Court. The White House might have been able to uncover the chain of transmission among the people infected as part of the cluster if they had sequenced the individual viral samples.

Christie would also be treated with an antibody drug—the cocktail that Lilly was developing. Lilly's combination-antibody drug would later produce impressive results, substantially reducing the risk of hospitalization and death in high-risk patients with early COVID disease.[19] This cocktail included the same component that was in Lilly's single-antibody treatment, combined with a second antibody that was derived from a patient in Wuhan who had recovered from COVID. This second antibody component was licensed by Lilly from Shanghai Junshi Biosciences Co. in May. (Christie had initially sought Regeneron's drug but was offered the opportunity to join a clinical trial and refused, because he didn't want to take a chance of being randomized to receive a placebo.)[20]

By combining two antibodies that could target the coronavirus at different points on its spike protein, Lilly's combination proved far more effective than either antibody by itself. This cocktail, however, was in short supply. While Lilly would be able to make 1 million doses of the single-antibody drug in 2020, it could only produce 50,000 doses of the combination therapy for the rest of the year. Much more supply would be coming to market in 2021, but for the time being, few people would be able to get the combination drug. Christie was fortunate to be one of them.[21]

Antibody drugs like the products that Trump and Christie received could have been a bridge to a vaccine. They could be used as effective treatments in high-risk patients, especially when the drugs were administered early in the course of the disease. Moreover, these drugs could also be used as a prophylaxis to prevent infection in those at greater risk of contracting COVID and suffering a bad outcome. A single administration could provide a patient with a few months of protection.[22] This is how a similar antibody drug, developed by Regeneron for the treatment of Ebola, was approved by the FDA for use, and was successfully deployed to help control an epidemic.[23]

The antibody drugs that may have helped rescue the president, Chris Christie, and many others were a scientific bright spot. We were right at the cusp of a technological revolution in the creation of drugs and vaccines using fully synthetic processes where we were able to fashion them almost entirely from digital data derived from the coronavirus's RNA sequence. These capabilities helped demonstrate precisely the kinds of adaptable competencies that we need in order to combat a pandemic—broad proficiencies that allow us to quickly tailor countermeasures to the full range of viral threats, not only the species we anticipate, but also those we cannot.

In this one case, at least, we would have some of what we needed.

The story of how these COVID antibody drugs came about at once shows the benefits offered by nimble platforms that can be used to develop countermeasures to unpredictable threats, along with the opportunities that we missed by not securing these capabilities years in advance, as part of a comprehensive approach to pandemic preparedness that we never fully put in place.

To develop its drug, Regeneron used its VelocImmune mice—"magic mice," in the words of the company's CEO, Dr. Leonard Schleifer. The mice were genetically engineered to have an immune system that was a perfect replica of the human body. When exposed to COVID proteins, the mice developed antibodies to the virus and, since their biology was a facsimile for human immunity, the antibodies they produced were essentially human antibodies.[24]

Regeneron began work on its COVID drug right after the sequence of the novel coronavirus was loaded onto public servers. The company's researchers were able to use that information to synthesize a strand of genomic material that coded for the production of the coronavirus's spike protein. They introduced that sequence into their magic mice, which enabled the mice to start producing the spike protein and, in turn, antibodies to attack it.[25]

Regeneron had experience fashioning similar antibody-drugs that targeted MERS and Ebola.[26] When policymakers lost interest in Regeneron's drug for MERS, the company advanced an idea for a pan-coronavirus antibody that could be used if another coronavirus should emerge and threaten a pandemic.[27] Regeneron believed it was possible to develop a cocktail of antibodies that would be effective against any coronavirus. But the project couldn't attract investment from federal agencies like BARDA that would have had to fund these sorts of countermeasures. In 2015, the head of BARDA, Robin Robinson, told the publication *BioCentury* that his agency didn't have funding to develop coronavirus countermeasures. "We are funded for pandemic influenza and man-made threats like anthrax, smallpox, nuclear devices and chemical agents. If we had another emergency from another emerging infectious disease, we would need funding from another source because we don't have it. We don't have funding to work on coronavirus."[28]

Lilly decided to pursue its antibody drug on February 20, after the company's CEO, Dave Ricks, received a proposal from AbCellera Biologics Inc., a small drug-discovery shop with which Lilly collaborated. AbCellera had developed a "lab-on-a-chip" technology for sorting through blood samples for antibodies, giving the biotech company the capacity to isolate antibodies from the blood of patients who had recovered from COVID and to identify the antibodies that seemed most effective at neutralizing the virus. However, the small company needed help developing these antibodies into drugs and producing them at sufficient quantities to provide them as a therapy to patients. These were proficiencies with which Lilly had deep experience.

By April 10, Lilly had developed its lead drug candidate. It was just fifty days after it had started screening blood samples for the right antibody using their specialized chips.[29] The first patient was dosed with the new drug on May 30. By October, Lilly had filed a request with the FDA for an emergency use authorization to start distributing the drug, and it began putting the finished drug on pallets in preparation for its commercial release. From start to finish, it took Lilly eight months to go from concept to a commercially available medicine.[30] The company's previous record for discovering, developing, and commercializing an antibody drug—soup to nuts—was four and a half years. Lilly received FDA authorization for its COVID antibody drug on November 9, and Regeneron soon followed suit on November 21.[31]

Both companies took extraordinary steps to ramp up their manufacturing of these drugs.[32] Regeneron operated two major manufacturing facilities: one in upstate New York, and the other in Ireland. The Irish facility was newly built and had excess capacity because it was just coming online, but the company couldn't make the COVID drug in Ireland and expect to ship it back into the US; Irish authorities might claim the drug for themselves. So Regeneron made the decision to move the manufacturing of all their existing products out of the New York facility and into the Irish plant, to free up the entire American facility for the manufacture of the COVID antibody drug; pursuing this step before they knew whether the drug would work, long before it was approved by the FDA, and before they even knew the full extent of the threat COVID posed.

However, neither Lilly nor Regeneron would be able to produce enough supply in 2020 to meet the expected demand in the epidemic that was surging in the winter of 2020–21. Various officials in the federal government had been warned by me and others, at the outset of the pandemic, that the US needed to take steps much earlier, in April and May, to obtain more manufacturing capacity and have enough supply of the antibody drugs in time for the fall, in the event one or more of the medicines worked.

Frustrated by the lack of response, I had written a series of op-ed articles for the *Wall Street Journal* in an effort to instigate more action. On July 5, I said that these drugs had high odds of being effective given past experience with similar antibodies that targeted MERS and Ebola, but "the trick will be producing them at scale." I noted that "Operation Warp Speed focuses on finding and making vaccines. The government should set up a parallel effort for antibodies, as even a limited supply could save many lives." The federal government, I continued, "should work with drug makers to free up domestic manufacturing plants to start making more of these antibody drugs even before the FDA approves them."[33] On October 8, in an article I co-authored with former FDA commissioner Dr. Mark McClellan, I said, "It is essential to ramp up supply. Regeneron has said the company can have as many as 300,000 doses by the end of the year, and Lilly says it can churn out more than a million. That isn't enough. We estimate at least 12,000 of the new COVID cases each day would be indicated for the drug based on age and risk factors, even if there's no further surge of infection."[34] On *Face the Nation* on October 11, responding to a question by the show's host, Margaret Brennan, I lamented, "We would have needed to take different steps in April and May to ramp up manufacturing capacity to have the drug available in larger quantities right now. It's too late for this year. I think we could still take steps to do it for 2021. But we're stuck with the doses we have. . . . These drugs always looked promising, and they were always believed to be a bridge to a vaccine. We're not going to have it in the quantities we need."[35] And in a third article on December 13, I wrote, "The U.S. can buy a COVID insurance policy for next year. The federal government should scale up production of antibody drugs to make as many doses as possible."[36]

In the end, the US wouldn't be ready to produce these drugs in the quantities needed in the winter of 2020–21. (By the spring, the authorization for Lilly's single antibody treatment, and later its dual antibody therapy, would be revoked by the FDA, because the drugs were no longer sufficiently effective against the new variants that had started to circulate more widely.)[37] At one point, to try and maximize the available supply of antibody drugs in view of the limited capacity to manufacture these products, HHS had asked Lilly to forgo development of its cocktail and just produce the monotherapy; a decision that, if taken, would have proved fateful.[38] Operation Warp Speed had put most of its resources and focus on the development of vaccines—getting the clinical trials for the Moderna and Johnson & Johnson vaccines under way and supporting the development of about a half-dozen other vaccines that, by the summer of 2021, hadn't been authorized by the FDA. This effort took considerable time and resources. However, far less attention was focused on securing the domestic manufacturing required to churn out sufficient quantities of the drugs needed to treat the population, along with the systems, planning, and funding to distribute them to patients.

The shortcomings in manufacturing capacity were evidenced by a phone call that Lilly received from Operation Warp Speed when the company was on the eve of getting its FDA authorization. It turned out that Operation Warp Speed had identified a plant that could produce eight hundred liters of biologics. Did Lilly want access to the facility? they asked. It demonstrated the challenges we faced. The facility was too small to produce any appreciable supply. In the world of antibody-drug production, an 800-liter facility was tiny. To put it in perspective, the primary facility that Lilly was using in New Jersey to produce the drug was 86,000 liters—more than one hundred times larger than what the government was offering.

The US just didn't have the idle manufacturing capacity we needed.

The problem is that there aren't any big manufacturing plants sitting unused in the US. All the properly sized domestic facilities are busy making other drugs. Building new facilities was largely off the table. Under normal circumstances, it takes about three years to construct a facility to manufacture these types of drugs, and even a rushed effort would take eighteen months. Our federal government wasn't engaging in that sort of planning before COVID struck. Using disposable equipment like plastic containers to make antibodies was an option. Pfizer turned to these techniques later to augment the production of its vaccine, but it was hard to use this as the backbone of the process needed to produce the massive amounts of antibody drugs that were needed.[39]

The only way to get more capacity in the short term would have been to buy that capacity away from a company that was making another antibody drug. In other words, pay a company to stop the production of an existing drug and start making the COVID antibody. As I wrote at the time, "Most biotech companies prepare for disruptions in manufacturing by freezing and stockpiling enough of their most profitable drugs to last about two years. Why not pay these plants to use their capacity for antibody production? Companies can dip into reserves to avoid creating a shortage of other important medicines." That's what it came down to. The only way to secure more production capacity for the COVID drugs was to stop making other medicines and eat into stockpiles that companies had on hand.[40] That was the only fix that we had in the short run.

We didn't pursue this, or other stopgap measures. Instead, we had a limited supply of the antibody drugs when the winter surge arrived. It turned out, however, that our distribution system for the drugs was so poor that we didn't use even this small supply. States used only 20 percent of the 1.3 million doses that the federal government had on hand through the end of 2020.[41] The poor planning by HHS that went into securing production capacity for these drugs was eclipsed only by the even worse preparations to arrange for their delivery to patients. We had artificially depressed demand for these drugs as a consequence of poor execution on how they were distributed.

The entire episode exposed a critical weakness in our national readiness. We didn't have the capacity to produce domestic supplies of biologicals at a speed and scale sufficient to respond to a pandemic. And we didn't have an efficient system to distribute them to patients. It was always likely that, in the event that a new virus emerged, the first therapeutics that could be made available would be antibody drugs fashioned to attack proteins on the virus's surface. Yet we didn't have a plan on the shelf for how we would maximize their use.

That manufacturing capacity was another example of the broad capabilities that we needed to have in place in order to counter unforeseen viral threats, and it should have been operating as part of a hot base of preparedness. It needed to be built well in advance and kept operational in the event of a crisis. We couldn't just build it and mothball it, either, or try to keep it as part of a "warm base" of public health preparation. Such a facility needed to be kept fully functioning and embedded in some existing infrastructure. It needed to be kept hot. Like the volume to surge the supply of testing, we needed to have biological manufacturing facilities that were deliberately overbuilt so there was reserve capacity ready to be tapped in the event of a crisis.

We learned some lessons from our past experience, where we attempted to invest in the additional manufacturing capacity that we would need if a flu pandemic ever struck, only to see those efforts abandoned or lose their focus when the immediate threat seemed to subside. Many of these efforts had sprung from the planning that began in 2005 following concerns about the spread of the H5N1 bird flu. On the campus of Texas A&M University we established a facility to manufacture biologicals in cell cultures and then "finish" them—transferring them into vials ready for distribution to patients. During the first year of COVID, that Texas facility wouldn't provide any vaccine supply. Dr. Brett Giroir, who had helped secure the federal grant that enabled Texas A&M to construct the facility before he became the assistant secretary of health at HHS, said the facility would be prepared to quickly produce vaccines in a pandemic. "Once it's implemented, it really will solve the pandemic crisis," he said at the time.[42] We also built a vaccine manufacturing facility in Maryland that was run by Emergent BioSolutions, and another one located in North Carolina that used a cell-based manufacturing process to make flu vaccine.[43]

These and other capabilities were constructed as a hedge against a pandemic strain of bird flu, to give us the capacities we would need to manufacture a new vaccine.

All of these efforts ran into separate challenges.

The infrastructure was a sound investment. It was the contracting part that we got wrong. We invested in the construction of the initial capacity and then didn't keep it operating in a way that both maintained its readiness and guaranteed its availability for a crisis.

In the case of Emergent, the facility was never given an appropriate level of continued investment in view of its strategic purpose.[44] It was meant as a hedge for a future crisis; a way to make sure there would always be some manufacturing capacity that we could tap in a pinch. When the plant was conscripted to make the COVID vaccine developed by Johnson & Johnson, it ran into challenges getting the process up to sufficient scale and maintaining proper control over quality, while also trying to manufacture AstraZeneca's COVID vaccine in the same facility. At some point in the manufacturing process, components of the AstraZeneca vaccine cross-contaminated the Johnson & Johnson product, causing millions of doses of finished vaccine product to be lost as the government was trying to ramp up vaccinations in the spring of 2021. A government investigation had earlier concluded that over the years, Emergent was unable to attract and retain the specialized personnel that were needed to run such a facility, leaving it vulnerable.[45] Following an inspection of the Emergent facility, the FDA issued a report describing unsanitary conditions, peeling paint in manufacturing suites, and inadequate handling of medical waste.[46] The Biden administration made the decision to take supervision of the facility away from Emergent and put Johnson & Johnson in charge of the operation.[47]

To have the capability we need in the event of a public health crisis, the key is to make sure that the facilities we'll rely on are kept in continuous use and up-to-date. That requires us to contract them out for ongoing commercial work, and then maintain an option to call away their capacity in the event that a crisis emerges, and their volume is needed for another purpose.

The federal government can essentially contract for the right of first refusal on the available capacity of such a facility and secure the ability to commandeer the volume in the event of an emergency. The companies can be paid to maintain more frozen supply than they might have otherwise held, so they're prepared to turn over their facilities in the event it's needed, and then use up their stored drug. To maintain their readiness, facilities must be constantly operating. It's not just the specialized equipment that needs to be kept in a continuous state of use and readiness to remain up-to-date. The highly specialized personnel also need to be on hand to operate the plant. You can't maintain the necessary workforce if you build a plant and mothball it.

Another way to secure the reserve capacity that's needed as part of our national readiness is to build facilities that have extra lines and keep some of those lines operational and some on downtime. Products that are being made at these sites as part of their normal business can be rotated across the different lines, so all of them are kept up-to-date, but only some of them are being used at any one time. The rest of the capacity can be held in reserve.

To secure this capacity, the federal government can turn to a contract manufacturing organization that already does outsourced production for drug companies and pay them to maintain some reserve capacity for the government. The idea isn't to build a whole bunch of new manufacturing sites and keep them stagnant. We need to make sure the plants are properly staffed and maintained. If it's not operating, it might get atrophied or outdated. This is what happened to the equipment held in the Strategic National Stockpile, and it's what happened to some of the vaccine manufacturing facilities that were built following the 2005 pandemic plan. Instead, the contract manufacturers can be paid to operate plants that are oversized for their routine obligations. In an emergency, any one of these facilities could make its excess capacity available for the production of needed products.

The 2005 pandemic plan focused a lot of attention on the development of domestic sites to manufacture flu vaccine. It was widely recognized that we needed more capacity to make flu vaccine in the event of a pandemic with a novel strain of influenza, and we needed to make sure that these sites were based inside the US. But the effort stalled. "We never got anywhere pursuing the idea that we needed to develop more of a manufacturing base, back at the time when we were dealing with H5N1 and H7N9," Anthony Fauci told me. "We made investments in plants that would be able to manufacture flu vaccine, but we were always left with the problem:

you can't keep them cold, and if you just keep them warm, you lose a lot of money."[48]

To prepare for future threats, to have that capacity ready in the event of a crisis, we'll have to pay for that spare volume, much as we'll need to support the maintenance of some reserve testing capacity in the major diagnostic labs like Quest and LabCorp. We need to take new steps to make sure that the US will always have a surge capability in a crisis.

When it comes to making complex biologicals like antibody drugs and vaccines, we need to secure the needed manufacturing capacity in advance and build it into the nation's infrastructure. We need to create a hot base of preparedness that's kept in a state of continual operation, so it's always ready. We need to maintain that capacity in existing manufacturing sites and have it on full boil instead of simmer, prepared to be quickly repurposed to the production of new countermeasures.

CHAPTER 17

THE mRNA BREAKTHROUGH

On January 10, 2020, as soon as Chinese researchers posted the first genetic sequence of the novel coronavirus, labs across the US and Europe began working on a vaccine.[1] Within days, researchers were already developing constructs that could serve as the key starting point.

COVID came right at the moment when science was permitting the production of fully synthetic vaccines created wholly from data on the sequence of a virus—and not requiring actual samples of the pathogen itself. It was a computational exercise; the vaccine antigen (the component of the vaccine that elicits immunity against the target pathogen) could be initially designed almost entirely by software. We would get lucky when it came to the state of our technology. It would show the utility of having nimble capabilities that could be used to fashion countermeasures against a range of unpredictable threats. At the same time, the hunt for a vaccine would expose the capabilities we still needed to expand, especially the massive production capacity needed to quickly supply enough vaccine to protect the global population, and quickly distribute it to patients.

One key to acquiring immunity through vaccination is to develop an antibody-based protection, and it's this part of our immune system that was often discussed in relation to COVID.[2] Antibodies are produced when a key cellular component of our adaptive immune response called B cells come into contact with an antigen, which is typically a protein that sits prominently on the surface of a virus. To prevent infection and disease, our immune system creates antibodies that are tailored specifically to a virus's antigen, and able to bind immediately to an invading virus once it makes its way into our body. Antibodies can continue to circulate in our blood for many months, but not all antibodies are equally effective in neutralizing a virus.[3] The mechanisms of this neutralization process have been debated over the years, but they involve disabling the virus so that it can be destroyed by other circulating immune cells, including our T cells.[4]

The ability of a vaccine to stimulate the production of antibodies that are highly specific to a particular virus is one way vaccines protect against infection.[5] Researchers measure the level of these neutralizing antibodies in determining whether a vaccine is triggering an adequate level of protection.[6] But a good vaccine maximizes its ability to kindle all of our immune system's machinery, not only stimulating the production of antibodies but also spurring the activity of T cells and B cells that have memory, and are able to churn out new antibodies and other mediators of our immune response after they come into contact with a virus months or years later. The amount of priming that a vaccine delivers is sometimes referred to as its "immunogenicity." A vaccine teaches our immune cells to recognize specific regions on the viral antigen, called epitopes, that make the best targets for our cellular mediators to do their work.[7] This is how a vaccine confers protective immunity. However, there's always a risk that a vaccine can be too potent, or too indiscriminate, and activate too many parts of our immune system. If a vaccine overstimulates our immune system, it can cause our body to overreact and trigger side effects.[8] A good vaccine will strike a careful balance, maximizing the components of immunity that contribute most to providing protection from a virus, while minimizing the "antigenic load"—the number of viral proteins needed in order to stimulate a robust and durable level of protection.

To provide protective immunity, most vaccines will try to induce sustained and high levels of circulating antibodies.[9] However, activating the production of antibodies isn't the only way that we generate immunity through vaccination. There are two types of T cells that are especially important, and the function of each can be primed by a vaccine. The first type, called CD8 or cytotoxic T cells, recognize human cells that have been infected with a virus. These T cells are then able to bind to our infected cells and perforate their surface. Once they punch tiny holes, they inject a toxin that destroys these diseased cells before the virus particles inside are able to escape. The second type of T cells are called CD4, or helper T cells. These cells are "orchestrators, regulators and direct effectors of antiviral immunity," in the words of the journal *Nature*.[10] They are activated when our bodies are invaded by a virus that our immune system has already been trained to recognize. Once triggered, these CD4 T cells can play a "helper" function, principally by secreting chemical substances called cytokines that can amplify our immune response, recruiting other immune cells to the site of the infection.[11] Both of these types of T cells can linger in our bodies for years. By priming them, vaccines can offer more durable protection.[12]

All of the COVID vaccines that were put into development in the early months of 2020 used the same coronavirus surface protein as their antigen—the virus's spike protein. This protein sits prominently on the virus's surface and is the tool that the pathogen uses to gain its initial entry into our cells. The protein gets its name from its long structure and the way it latches onto our cells. Under a microscope, the part of this protein that actually hooks onto our cells looks like the spike on a long hook.

The spike protein was an attractive target for vaccine makers because of prior research with COVID's close cousin SARS-1. Once our bodies were exposed to this particular antigen, it triggered a vigorous immune response and activated a protective soup of B cells and T cells. The work with SARS-1 had also taught scientists how to make subtle genetic modifications to the spike protein that would make it far less inflammatory when it was injected into our bodies and therefore safer and more likely to linger. This could make it a more stable and effective ingredient for our vaccines.

In the traditional approach to making a vaccine, the antigens are cleaved off the virus that you're targeting, a tactic that requires scientists to isolate proteins on the surface of the virus. This typically entails growing the virus in large vats and deactivating it. The dead virus particles, or their surface proteins, then become the stock that goes into the vaccines. This is also the approach used by the two leading COVID vaccines that were made in China and marketed by the Chinese drug makers Sinovac and Sinopharm. The challenge with this approach is that it can be hard for the vaccine to deliver enough viral antigen to elicit a very robust immune response. The vaccine by Sinovac, called Coronavac, followed this strategy and was eventually shown to be about 50 percent effective, or less.[13] (By the spring of 2021, there were reports that some countries using the Chinese vaccine were giving patients a third dose to increase its efficacy, and China was pursuing a plan to start developing an mRNA vaccine to be used as a second booster for their original vaccines that were made from inactivated virus particles.)[14]

There were other challenges to the traditional approach China had adopted. It requires the virus to be manufactured in cell cultures so its proteins can be harvested. However, getting a novel virus to grow in a cell culture can be hard, and doing so in the quantities required to mass-produce a vaccine for a large population can require enormous facilities and may take a year or more to get to a proper scale. This method can also require major adjustments each time you want to adapt the process to a new virus strain. If the virus mutates and alters its surface proteins, you may want to modify the vaccine to reflect these new antigens and make sure that the immunity you're delivering is targeted to the new strain. However, you now have to get the new variants to grow in the cell cultures. It's not plug-and-play. Sometimes when you tweak the virus to reflect a new mutation, you find it doesn't grow as well in the existing cell cultures, and you need to start over.

Which leaves a lot of room for delays and errors. This method of using whole copies of the virus is how much of our seasonal flu vaccine is produced. The time it takes to complete this process is why scientists have to guess, months before the flu season, as to which influenza strains they believe will be the principal pathogens that circulate later that year. Twice a year, virologists, epidemiologists, and other experts meet at the WHO headquarters in Geneva to select the influenza strains that will go into the seasonal flu vaccine. They meet once to select strains that will be used in vaccines for the Southern Hemisphere, and then again to select the strains for the Northern Hemisphere. It takes a long time to get the chosen strains to grow in cultures and make enough viral stock to produce the finished vaccines. So, the entire process starts long before each year's flu season.

Making flu vaccines this way also requires the influenza strains to be incubated and grown in the chicken eggs. Newer approaches try to avoid some of these pitfalls. Rather than using the entire virus as antigen, scientists try to use just these key protein subunits found on the pathogen's surface. Using just the protein fragments, the vaccines can be engineered in a way so that they don't induce as many side effects like fevers. In the case of COVID, a downside of using viral subunits is that, in some cases, they were not as effective at triggering a strong immune response. However, there are ways to enhance its potency—for example, by mixing the antigen with a substance designed to boost its immune response, called an adjuvant.[15] This second approach was taken by some drug makers, including Sanofi and Novavax, for their COVID vaccines. By the fall of 2020, nearly seventy different experimental vaccines used segments of the coronavirus spike protein as a way to try to generate immunity. Most used recombinant technology, where the proteins are derived from synthetic genomic material manufactured in a lab, and then grown in cell cultures to produce the stock for the vaccine.

Another approach employed by some COVID vaccine manufacturers involved the use of a specially engineered virus to be harnessed as a vector for delivering the coronavirus genomic material. In this approach, a largely harmless virus is tweaked to render it unable to replicate and unlikely to elicit a strong immune response on its own. A snippet of genomic material that codes for production of the spike protein is then grafted into the virus, and the entire pathogen becomes a vector for delivering this gene to our cells. When the instructions are unraveled, our cells read the sequence and start making large quantities of the spike protein. Once this protein spills out of our cells, it activates an immune response by specialized B and T cells that recognize the protein as a foreign substance that needs to be eliminated. In this way, the viral vector is turned into a delivery vehicle for this part of the coronavirus's genetic material, and our own bodies become unwitting manufacturing plants for producing the spike. Instead of growing the protein in large bioreactors inside a manufacturing plant, our bodies are co-opted into becoming production facilities.

It's believed that using a viral vector to deliver the viral antigen is also more likely to provoke T cells to respond to the virus, in addition to antibody-producing B cells. There is some evidence that by teaching T cells how to attack the infected human cell that now expresses the spike protein, viral-vector vaccines can offer a longer-lasting immunity, protecting people not only from infection but also from symptoms of the disease. By the fall of 2020, there were about twenty vaccines in development that took this approach. The first program using this approach to make it over the finish line in the US was developed by Johnson & Johnson. But it would be another novel technology that would be the first vaccine to reach American patients.

In the early 1990s, as scientists were unraveling the genomes of different pathogens, they began to wonder if they could use this information to make vaccines directly from the sequence of a virus's genomic material. What would happen if they produced pieces of a virus's genetic material that coded for the production of its surface proteins and then injected these instruction sets directly into a person's arm? Would the recipient's cells take up this genetic material, read the sequence, and start manufacturing the proteins on their own?

The researchers' first effort relied on a small stretch of genetically engineered DNA called a plasmid. The strategy worked in animals, but the scientists couldn't get it to work well in humans. One of the challenges was that the plasmid would get chewed up by our immune cells before it could deliver its genetic payload.[16] So the scientists turned their focus to another kind of genetic material called messenger RNA, or mRNA. Unlike DNA, mRNA typically acts as a courier rather than a primary source of genetic information. Its role is to ferry instructions from DNA in the cell's nucleus to its machinery that's floating in the cell's cytoplasm, the biological soup enveloping the nucleus where the synthesis of protein occurs. There were some potential advantages of using mRNA as the construct for the vaccine. Compared to DNA-based vaccines, mRNA could be more readily taken up by our cells through a process called endocytosis and turned into the target proteins that you were trying to manufacture.[17] The mRNA strands were also a smaller strip of genetic material, unlike the long and bulky stretches of DNA, so it would be less likely to trigger a strong immune reaction and get destroyed before delivering its instruction set.

The idea of using mRNA in a vaccine was straightforward. If you could stabilize the mRNA, get it into a syringe, and put it in a formulation that could allow it to be taken up by our cells, then you could hijack the body's own protein-making machinery as a way to make copies of the viral protein. The mRNA could also be easier to manufacture in large quantities. Rather than relying on vats of cell cultures and big production facilities to make copies of the virus, making the vaccine construct with the mRNA approach could essentially be a chemical process with no animal or cellular components.[18] Once you had the mRNA isolated, our bodies would become manufacturing plants for the matching proteins, in a similar process to a viral-vector vaccine.

To make a vaccine against COVID, you want to teach the body to develop immune cells that target the virus's spike. With the mRNA approach, the idea is that you'd manufacture snippets of genomic material that code for these proteins rather than isolating the proteins themselves. The mRNA would be wrapped in liposomes (lipid nanoparticles) to stabilize it and allow it to evade immediate destruction by our immune cells. Then, once the mRNA is injected and gets taken up by our cells, its instruction set is read, and our cells start to churn out copious amounts of spike protein.[19] These proteins are solitary, meaning they can't assemble to form a virus themselves. But our immune system can detect these viral proteins and produce a defensive blend of immune cells in response to this perceived threat. The first time that mRNA construct was shown to be able to produce protective effects against an infectious pathogen was some work done in 2012, using the approach to try and fashion an experimental vaccine against influenza A.[20] The technology had other advantages. Because the vaccines were derived through the use of genomic information, the process for making them was fully synthetic. You didn't need access to the live virus to get started. And you didn't need chicken eggs.

By the fall, more than twenty-five COVID vaccines would be in development using mRNA as the starting point. Two emerged as front-runners and would become the first COVID vaccines to reach patients: one by Moderna, which had partnered with the National Institute of Allergy and Infectious Diseases, and the other by Pfizer, where I serve on the company's board of directors, which had teamed up with the German biotech company BioNTech. The process for making these two vaccines began as soon as the

SARS-CoV-2 sequence was published. In fact, Moderna never had the actual coronavirus on its premises. It never needed a sample, just the computational sequence of the virus's RNA. Once Moderna got the sequence, the entire process to construct a candidate vaccine took just two days.[21] And in six weeks, Moderna went from having the sequence in their computers to beginning the manufacture of a vaccine to start human testing, in collaboration with the Vaccine Research Center at the National Institute of Allergy and Infectious Diseases and the team led by Dr. Barney Graham and Dr. Kizzmekia Corbett.

I watched the Pfizer story develop from inside the boardroom. In January, when the first reports on the virus were published, the founder and CEO of BioNTech, Dr. Ugur Sahin, set out to see if he could use his company's mRNA platform to develop the construct for a potential vaccine. Pfizer had been collaborating with BioNTech on the development of an improved version of the seasonal flu vaccine. Among the advantages that mRNA might offer for that effort is a better ability to match the flu vaccine to the strains of influenza that circulate each season. Because the mRNA vaccines can be developed and manufactured in a short time frame, there's no need to guess far in advance which strain of influenza would circulate each fall. Using mRNA as the starting point, a company can develop the vaccine and begin making it closer to the actual flu season, when there would be more certainty about which strain was spreading. The mRNA approach could also provide a more reliable way to manufacture the vaccine in large quantities without relying on millions of chicken eggs. Using mRNA as the starting point, you might also be able to design a vaccine that's able to target both of the influenza virus's key surface proteins, HA and NA. An antigen to stimulate the production of antibodies that target NA was always hard to incorporate into vaccines made using the traditional approach involving chicken eggs, because the NA protein was less stable and could not be easily manufactured and placed in a syringe. But using as the vaccine stock the genomic material that codes for production of NA, rather than the actual protein, you could potentially solve for this challenge and produce a vaccine that can deliver an immune response against a broader complement of the influenza virus's surface proteins. Pfizer and BioNTech were planning to start clinical trials with an experimental flu vaccine in 2020.[22] Then COVID happened.

The novel coronavirus presented new challenges. However, some of the work the two companies had done on flu was transferable, giving them a head start on developing a vaccine for COVID. Sahin turned most of his research team toward the new effort.[23] Using information on the coronavirus's sequence, he initially designed ten vaccine constructs himself. These were short stretches of mRNA that coded for different variations on the SARS-CoV-2 spike protein. By the end of February, Sahin's research team would create ten more vaccine starting points, each slightly different, for a total of twenty different snippets of mRNA. Each one could serve as the initial building block for a new vaccine.[24] On March 1, with the effort showing promise, Sahin called Dr. Kathrin Jansen, the head of vaccine research and development for Pfizer, and told her that BioNTech was working on a vaccine for COVID. Did Pfizer want to jointly develop it?[25] Jansen said Pfizer was in. The approach "had all the hallmarks of speed," she told me. "And we already had invested so much, two years in analytics. So, in a way, we were already ahead in the game." The approach "just ticked all the right boxes," she said.

Earlier in her career, as the chief scientific officer at the biotech company Vaxgen, Jansen had helped develop a vaccine for prevention of disease caused by anthrax in those at high risk of exposure. It was shortly after 9/11 and the subsequent anthrax attacks. There was a palpable fear that terrorists would try to use bioweapons to carry out a mass-casualty event in the United States. At Vaxgen, Jansen was pursuing vaccines to these threats, having previously worked to develop vaccines at Merck. She was highly respected in the vaccine industry and credited with initiating and leading the program that developed Merck's vaccine for cervical cancer. Jansen's work on vaccines had helped reinforce the value of a flexible platform that could allow you to quickly engineer countermeasures as unpredictable new risks emerged. Just as few people were worried about coronaviruses before the emergence of COVID, nobody before 2001 had seen the immediate threat from anthrax, or the increasing focus on the risk of bioterrorism. Because it was hard to predict the next peril, an approach was needed for developing vaccines and therapeutics that could be easily adapted to new threats.

The mRNA technology had that dexterity.

When a new crisis arises, or a new pathogen emerges, scientists and policymakers are forced to quickly react to it, and they can lose the ability to more systematically "look at the disease landscape and ask themselves, how can we build an infrastructure," Jansen observed. "And how would you build such an infrastructure so you're not stuck working on any given pathogen when it's too late, but when you can actually create the framework to prepare for an emergency no matter what happens. . . . It really requires a change in how we look at potential pandemics and what we are willing to invest at a time when everything is fine, when there is no worry."

By April 12, the Pfizer and BioNTech teams had eliminated sixteen vaccine candidates. To carefully assess the remaining four, they tested them early in volunteers, in a study designed to compress all three stages of clinical development, from phase one to phase three studies, into one seamless trial.[26] It was an innovative and, in retrospect, a propitious decision. It was part of what Jansen called her "quick kill strategy." The researchers didn't know which of the four vaccine constructs would work best. But they also knew they couldn't take all four candidates forward, and so they had to figure out—as soon as possible—which one produced the strongest benefits with the fewest vaccine-related side effects. Given the extreme pressure to move quickly and the huge resources that it would take to run a pivotal study, they knew they would have only one shot at it. By evaluating the candidate vaccines in patients early, the scientists were able to get data that would let them better assess which one would work the best.[27]

Meanwhile, the manufacturing team at Pfizer was already building the custom-made systems to manufacture the vaccine in large quantities. Everything would be in place so that the company would be ready to start churning out the vaccine as soon as the decision came on which vaccine construct worked best in patients. This was part of a deliberate effort by Pfizer CEO Dr. Albert Bourla to have everything run in parallel. Classical vaccine development is highly sequential. You fully complete one step before moving on to the next one. Getting a safe and effective COVID vaccine authorized by the FDA and distributed to patients was a public health emergency. So Pfizer was doing everything at the same time as a way to shorten the process. The most obvious step was to ramp up the novel manufacturing infrastructure that was needed even before knowing which vaccine would be taken forward, and long before Pfizer knew whether the vaccine was going to work in patients.[28]

In selecting the four vaccine candidates, Pfizer had initially evaluated two different approaches. The first two constructs used a strand of mRNA that would code for the production of the entire spike protein, while the other two would use mRNA that coded for just a specific region on the spike, called the receptor-binding domain or RBD. It was believed that the RBD would make the most attractive target for human antibodies. It was the exact point on the spike where that protein latched onto the ACE2 receptors lining our tissues. There were a lot of these ACE2 receptors in our lower airways, explaining why COVID targeted its victims' lungs. Antibodies that target the RBD interfere with the spike protein's ability to bind to these receptors, effectively blocking its capacity to enter our cells, and neutralizing its ability to infect.[29] Of all the different regions on the long spike protein, it was the receptor-binding domain where our most potent, neutralizing antibodies would attach. Measuring the sheer volume of antibodies that COVID patients produced against their SARS-CoV-2 infections after they recover, about 90 percent targeted the RBD. So, a lot of a vaccine's effectiveness would be driven by its ability to induce antibodies to that very specific binding region on the virus's surface.

The idea of using mRNA had never before found its way into a commercially available vaccine. The tactic had been used in the development of vaccines for other diseases, which were in various stages of clinical development. Moderna was developing mRNA vaccines for Zika, cytomegalovirus, and flu. Another company, Curevac, was in phase one trials for rabies. In total, about a dozen vaccines had been put into human testing that involved the use of mRNA. Nine of them had been advanced by Moderna.[30] Pfizer had a preclinical program for flu in collaboration with BioNTech. The mRNA approach was novel—but it was not completely untested.

"When you look at pandemics, or other viral diseases [the mRNA approach] just opens the door to so much more than what we ever had access to, and we can develop vaccines so much faster than what was ever possible," Jansen said. "And now we're having to build the infrastructure for an RNA vaccine dealing with this pandemic, and of course that builds a foundation to deal with everything else" in the future. COVID would provide large-scale validation of the approach.

In evaluating the four different vaccine candidates, Pfizer focused on two of them early. Both generated robust immune responses. So the attention turned to their "tolerability profile," a description for the propensity of a vaccine to produce predictable but manageable reactions like sore arms, fatigue, and injection site pain. These features are described as a vaccine's "reactogenicity," and are related to the immune response that the vaccine stimulates. The RNA vaccines can also induce reactions like fever and chills in some individuals, and the key is to strike a careful balance: you want a vaccine that elicits a strong immune response to the antigens that you're focused on, but not so strong that it stimulates a lot of immune-related side effects.[31] This was a challenge, because the very properties that made a vaccine more likely to trigger the protective elements of our immune systems could also lead to more reactogenicity.

This is where the early clinical research that Pfizer did came into focus. By initially evaluating four vaccines' constructs, and quickly rejecting two while the remaining two were extensively tested in early studies, Pfizer was able to select the one that best struck that careful balance. On July 24, Pfizer and BioNTech agreed on a vaccine construct that integrated genetic material that coded for production of the full-length spike. Picking a vaccine that used the full length of the spike protein would turn out to be a fortunate decision. Based on the patient data that Pfizer had, the full-length spike offered the best compromise between robust stimulation of our immune cells and a lower incidence of vaccine-related side effects like low-grade fevers and injection site reactions. It would turn out later, by choosing to use the full length of the spike protein, the vaccine generated the broadest compilation of virus-neutralizing antibodies and activated T cells. It means that patients were likely to be generating antibodies against many different parts of the spike protein and its RBD, and not just one very specific region. So, as the virus mutated and changed the structure of its RBD (which happened in the winter, as new SARS-CoV-2 variants emerged), the vaccine would still induce antibodies against other regions of the spike protein that wouldn't undergo as much change. No single mutation would render the vaccine substantially less effective. The strip of mRNA that coded for that full version of the spike also contained two carefully crafted tweaks that were not in the sequence of the wild strain. Pfizer had adapted the mRNA sequence to make the resulting spike protein it produced more easily recognized by our B cells so it would stimulate a more robust immune response.

Pfizer began evaluating the vaccine in a pivotal trial on July 27, with the first volunteers receiving shots at the University of Rochester. Moderna's vaccine also entered its phase three study the same day. About three months later, on Sunday, November 8, I got an email around 2 p.m. "Board Call at 6 tonight" the subject line read, and the email went on to say, "Albert has confidential news to share regarding the vaccine," referring to Pfizer's CEO Dr. Albert Bourla. I was apprehensive. Based on the earlier data, we felt reasonably confident that the vaccine would work. However, you never know until the trial is done and you've seen the data. And even if the vaccine was proven to be safe and effective, we didn't have a clear sense of how beneficial it would be, just some hints from the earlier trials. Would the vaccine clear the FDA's hurdle for approval and prove to be more than 50 percent effective at preventing COVID's symptoms? Sensing the nervousness on the call, Bourla broke the tension right away. "I have news on the vaccine," he began, "and it's good."

The trial was blinded to the investigators and Pfizer. It's standard practice in trials for patients to be randomly assigned to receive an experimental treatment or an alternative meant as an inert or predictable comparator (a placebo), to see if the novel agent is working, and for the investigators and the patients to be blind to who received what. Only the data safety monitoring board (DSMB) knew who was getting the vaccine and who was getting the placebo. A DSMB is an independent group of clinicians and scientists who monitor the safety of patients while a trial is ongoing. They also continuously evaluate how well an experimental drug or vaccine might be working. They can intervene to stop the trial early if they find that the product is especially effective and should be declared to work and be given to everyone, or if it's judged to be ineffective and continued treatment with it is futile, or worse, it's judged to be causing harm.[32] The DSMB also reports the results when the trial reaches certain prespecified milestones.

In this case, the study was supposed to report preliminary data when a certain number of patients in the trial contracted COVID. At that point, the DSMB would analyze the findings and figure out how many of the cases were among patients who had received the vaccine, and how many occurred among the patients who had received the inert placebo. And the results would be reported to a small number of Pfizer colleagues who were unblinded to the data, while the rest of the large study remained blinded to everyone else. That morning the chair of the DSMB had a meeting with a small number of Pfizer scientists to tell them that the vaccine was highly effective and safe, and that the data could be compiled for submission to regulatory agencies. In all, ninety-four patients had contracted COVID in the study, and nearly all of them were on the placebo. In total, the vaccine was more than 90 percent effective at preventing the symptoms of COVID.

It was an emotional call. We had confidence that the vaccine was going to be effective, but I don't think any of us expected it to be this successful at preventing COVID. Albert shared a video with us, recording the moment when the results were first reported to the management team by Dr. Kathrin Jansen and Dr. Bill Gruber, Pfizer's senior vice president of vaccine clinical research and development. Everyone in the room stood up and screamed in joy. It was clear at that moment that the world had changed. It was the beginning of the end of a devastating global event. A vaccine capable of producing this level of benefit had the potential to end the pandemic. COVID was now a preventable disease. The key question was going to be how fast the company could ramp up the supply and distribute the vaccine quickly and equitably to the patients who needed it.

The next day, I was scheduled to appear on CNBC at 6 a.m. I knew the results, but I needed to wait for the company to release them later that morning. I was asked about the progress of the clinical trials, and I said simply that I would let Pfizer speak on updates about the vaccine.[33] It was a standard line of mine; I would routinely defer to the company to speak about its progress. So, there was nothing out of the ordinary in my reply. CNBC brought me back on the air after the results were formally released. Though I now had more freedom to speak, I remained cautious in my remarks, not wanting to overstate the impact of the interim results or draw a premature conclusion ahead of the FDA's careful analysis of the data. "This is a meaningful result, a very meaningful result, and exceeds the expectations based on the interim analysis," I said. "And if the data does hold up on the final read out from the full clinical trial, I think it's light at the end of the tunnel."[34]

It was indeed. Pfizer filed with the FDA on Friday, November 20, seeking authorization for emergency use of the vaccine.[35] It was the first company to seek permission to offer a COVID vaccine to US patients. It was just 248 days since the company had first set out with BioNTech to develop the vaccine. No vaccine had ever been developed that fast. The process involved 150 different clinical trial sites, 43,661 patient volunteers, the work of thousands of Pfizer colleagues, and the hopes of a weary nation: it was one of the largest, fastest, and most important scientific endeavors of its kind in modern history.[36]

Pfizer initially committed $1 billion to the project, a sum that would soon double as the company ramped up manufacturing sites across the US and Europe and started to produce the vaccine. Much of that money would pay for facilities specially designed to manufacture the mRNA vaccines. The specially formulated mRNA used in the vaccine was produced by batteries of highly specialized mixing machines that Pfizer had custom-made for the purpose at a cost of $200 million each. I remember the board meeting when we voted to buy more of them. We initially purchased five of the batteries of machines in all, at a cost of $1 billion—before we even knew if the vaccine was going to work. We would end up purchasing many more. While other companies took grants from the federal government to offset the cost of development and to pay for the investments they would need to expand their manufacturing facilities, Pfizer made a decision to forgo federal money to support the venture. Pfizer agreed in advance to sell an initial allotment of 100 million doses of vaccine to the federal government at an agreed-upon price, but only if the vaccine was successful and won FDA authorization. The US government would pay Pfizer only if the vaccine worked, and only for the doses they were buying.[37] Bourla wanted it that way. He was concerned that the strings attached with federal grant money could slow the process. He wanted his scientists unencumbered. He knew that speed to market would save lives.

The mRNA vaccines marked a historical turning point. COVID straddled a line between two states of scientific fitness, a point when the wet approach—producing flu vaccine in large vats that relied on eggs and cell cultures and live viruses—gave way to the intelligent design of synthetic biology: digitizing snippets of mRNA that could commandeer our cells and turn our bodies into breweries for viral proteins. For the first time, we had the ability to develop fully synthetic vaccines based entirely on information about a pathogen's genetic sequence. Under the new approach, maximizing potency turned on the ability to tweak these strands using computers in ways that would make the resulting proteins more immunogenic. A decade earlier, we might have struggled for a year to get SARS-CoV-2 to grow in cell cultures. Messenger RNA proved to be an agile approach, allowing us to develop highly effective vaccines directly from the virus's genetic composition.

Indeed, the two inactivated virus vaccines developed by the Chinese, and put into clinical trials long before the mRNA vaccines began human testing, wouldn't come to market until well after the mRNA vaccines were authorized in the US and Europe. And they wouldn't work nearly as well. Using inactivated virus proved to be less immunogenic than the newer mRNA technologies. Moreover, getting the inactivated virus to grow in cell cultures, and manufacturing it at large enough scale to supply the overwhelming demand, remained a bottleneck. The Chinese were able to pivot quickly to the inactivated-virus vaccines because the approach was well established, but it proved to be a less efficient and effective platform. Moreover, as SARS-CoV-2 mutated, swapping in a new variant and getting the modified virus to grow in cell cultures wasn't nearly as simple as swapping a new gene sequence directly into the vaccine.

The mRNA approach was basically plug-and-play. Substituting in a new, mutated strain was fast and straightforward.

As it turned out, when new strains emerged over the winter of 2020–21 in Brazil and South Africa and raised concerns that the existing mRNA vaccines might offer less robust protection against these new variants, Pfizer and Moderna were able to develop new constructs that targeted these mutant strains and get them into development in under a few months.[38] The speed and adaptability of the mRNA approach demonstrated its superior utility. The companies were able to generate boosters for new variants as soon as scientists isolated the new viral sequences, and have the modified vaccines ready if they were needed.

There are other approaches that allow us to develop countermeasures directly from sequence information, and we need to invest in these kinds of adaptable capabilities to prepare for future threats. These are the kinds of core competencies that will help us guard against unpredictable threats from nature and leave us better prepared for the next pandemic. The speed, agility, and effectiveness of the mRNA approach demonstrated the kinds of aptitudes for developing countermeasures that we need to be investing in. Among these promising approaches are efforts to develop small-molecule drugs that can be taken orally and serve as broad-acting inhibitors of viral polymerase and proteases, which are common machinery used by RNA viruses to replicate. In contrast to antibody drugs and vaccines, which are specific to each individual strain and target precise proteins on a virus, the protease inhibitors can act on all coronaviruses, along with other related viruses.

Polymerase inhibitors are also broad-based viral disruptors that have the potential to work against entire classes of viruses. One approach uses small interfering RNA, sometimes known as short interfering RNA, silencing RNA, or siRNA. This is a class of double-stranded RNA molecules that can be used to disrupt the ability of viruses to translate (or assemble) their RNA stands into proteins. The basic concept bears some similarities to the approach taken with the mRNA vaccines. Using siRNA, we can interfere with the expression of specific genes as a way to interrupt the principal activities of a virus. Also, like the mRNA vaccines, these drugs can be put into development based only on information about the virus's genetic sequence. If we can establish how siRNA can be used to disrupt certain innate aspects of viral replication within each of the key classes of viruses, from influenzas to coronaviruses, then these drugs can be modified as new strains and variants emerge. We should pursue these and similar capabilities against RNA respiratory viruses. We're going to need more of these kinds of flexible platforms.

The old approach to pandemic preparedness was to create a list of priority pathogens and cultivate specific countermeasures. COVID proved how hard it is to anticipate the next threat, which could be a novel class of viruses or a known virus that has mutated into a more dangerous form. To counter the unknowns, the most valuable capabilities are the ones that can be quickly adapted to novel threats, or countermeasures that can help defeat entire categories of pathogens—for example, a vaccine that will work against multiple strains of coronavirus.[39] These capabilities will allow us to field drugs and vaccines that can quickly counter an unforeseen danger shortly after it emerges, based on features—like its genetic sequence—that can be deduced and isolated.

The ability to deploy these nimble platforms turns on relatively new technology. Just ten or fifteen years ago, we didn't know how to implement a rapid response with antibody drugs targeted to viral proteins. Just five years ago, we couldn't develop vaccines through synthetic approaches that used genomic material as the starting point. Having flexible technology platforms paid off. The threats we would face, and how these capabilities would ultimately be used to reduce them, were hard to foresee and plan for.

Consider this: when mRNA was first embraced as a potential therapy, the initial vision was that it could be used to replace protein drugs. Instead of infusing proteins like insulin as a drug, the view was that you would deliver a sequence of mRNA that would code for the production of the same proteins. The body would become the manufacturing plant for the biological. But researchers quickly realized that they couldn't stimulate the production of enough protein from mRNA to use the platform as a protein replacement therapy. Their focus then shifted to using the technology to stimulate the production of T cells that could target cancer. It was even more recently that scientists realized that the platform could be used for development of vaccines that targeted viral proteins.

This brief history shows that innovation can't always be predicted. We don't know which platform will yield answers for future threats. As part of our national preparedness, it will be important to stockpile countermeasures to some of the known risks, but it's equally important to support the development of novel technology platforms that have broad applicability over a range of potential threats. We can't always see the future. We can't focus solely on the dangers that we can see today. And we can't guess which approaches to developing drugs and vaccines will yield the most effective solutions. The use of mRNA to customize synthetic vaccines showed the value of having agile competencies at the ready. These are the technologies we will need to reduce our vulnerability.

CHAPTER 18

A NEW DOCTRINE FOR NATIONAL SECURITY

In April 2000, the Clinton administration designated HIV/AIDS a risk to America's national security. Until that point, there was no similar precedent for treating a naturally occurring infectious disease as a strategic risk. Following the declaration, when federal health officials were pulled into an initial meeting with the White House national security staff, the medical leaders were so unversed in their new landscape that the meeting was delayed when the head of the Office of National AIDS Policy couldn't find the way to the White House Situation Room.[1]

The declaration was viewed as a historic affirmation of the growing nexus between infectious diseases and our national defense. In addition to the staggering humanitarian crisis in Africa, a recent intelligence estimate had concluded that HIV infections were also accelerating in Eastern Europe.[2] The expansion wasn't just a public health calamity. It was also a geopolitical threat, and it had the potential to fuel global instability. The analysis concluded that continued spread would "challenge democratic development and transitions and possibly contribute to humanitarian emergencies and military conflicts to which the United States may need to respond."[3]

The formal designation was a Rubicon—an indication of a growing movement coupling security and global public health. Another recent national intelligence estimate had established that global pathogens posed a "nontraditional threat" to American security and economic interests. Eight years earlier, in 1992, the National Academy of Science's Institute of Medicine had sparked the movement to link global public health to our national security in a widely cited report on emerging diseases that argued that Americans were no longer insulated from infections that were once rare or were a burden largely confined to remote parts of the world.[4] The report's prominent authors said that the US would need new capabilities to combat these threats.[5]

"Despite the appearance of security, however, there is only a thin veneer protecting humankind from potentially devastating infectious disease epidemics," the report noted. "There can be a delicate balance between maintaining control of a disease and the initiation of an epidemic. It's one thing to have this balance disrupted by essentially uncontrollable elements; it's quite another to have it go awry as a result of individual or organizational complacency." The key to identifying outbreaks of novel disease, and tracking the spread of known infections, the report went on, is better surveillance and rapid response to identified threats, placing these competencies "at the core of preventive medicine." It continued, "A well-designed and well-implemented infectious disease surveillance program can provide a means to detect unusual clusters of disease, document the geographic and demographic spread of an outbreak. . . . Unfortunately, there is insufficient awareness of and appreciation for the value of comprehensive surveillance."

Among the group's recommendations was the establishment of a consolidated system for reporting information about potentially concerning pathogens and outbreaks. This wasn't a new idea. The same recommendation had been made repeatedly to the CDC. In 1997, a follow-up report by the Institute of Medicine drew a similar conclusion, seeking to establish a new standard that linked the risk of biological events to US national security.[6]

Later, in 2006, Congress appropriated funds to improve the CDC's system for detecting and reporting on novel threats, directing the agency to "establish a near real-time electronic nationwide public health situational awareness capability" to "share data and information to enhance early detection of rapid response to, and management of, potentially catastrophic infectious disease outbreaks"; but the CDC never implemented the called-for changes.[7]

A respiratory pathogen poses especially grave risks. Because of its route of transmission, it has the potential to spread easily and extensively, and to instigate widespread fears along the way. It literally spreads through our breath, making it risky for people to breathe the air in a room with others. COVID showed how the ensuing anxiety can be especially crippling.

Along the path to COVID, there were plenty of warnings about how real the threat was from novel respiratory pathogens, beyond our experience with SARS-1 and MERS. There had been enough near misses to prove that coronaviruses were a present danger.

In January 2003, a novel coronavirus was isolated from a seven-month-old child suffering from bronchiolitis in Hong Kong.[8] Additional cases were soon found in other children.

In 2004, a novel coronavirus was found in a seventy-one-year-old man admitted to a hospital in Hong Kong with pneumonia three days after he had returned from a trip to Shenzhen, China.[9]

In 2013, twenty-one doctors and nurses in China's eastern Anhui Province were hospitalized after being diagnosed with severe viral pneumonia. All of them worked in the department of respiratory care at the general hospital of the Wanbei Coal-Electricity Group. The pathogen was never revealed, but a novel coronavirus was a rumored culprit. Chinese officials dismissed the disease as an ordinary pneumonia and said it was not a cause for public health concern.[10]

There was also that outbreak of a severe pneumonia in 2012,

believed to be caused by a novel coronavirus, among a group of miners in southern China. The mines were near a series of caves where bats had been found that carried viruses similar to SARS-1. There would be speculation among some scientists in 2020 that the 2012 strain bore similarities to SARS-CoV-2, but Chinese officials would continue to maintain that it was a fungus that had sickened the miners.[11] Chinese scientists would find eight other completely novel coronaviruses in bats they sampled from those same mine-shafts, but the full sequences of these additional strains would never be shared.[12] The six stricken miners were taken to Wuhan for treatment, and their blood work had been frozen and stored at the WIV.[13] It's possible that a predecessor to SARS-CoV-2 virus had made the jump from bats to humans before, testing our species repeatedly before it finally got a firm grip on us. It appears that none of these previous outbreaks resulted in major spread. But any one of them could have. Any one of these and other similar incidents could have been the spark that lit the fire of a new contagion.

In fact, coronaviruses as a class were added years earlier to the WHO's watch list as a pathogen with pandemic potential, after our brush with SARS-1 and MERS.[14] There was ample evidence that these viruses were evolving in ways that made them increasingly dangerous, but we didn't take this class of pathogens as seriously as we should have and didn't adequately prepare ourselves for this threat. In fact, the WHO never declared MERS a Public Health Event of International Concern (PHEIC) under the revised International Health Regulations; a decision that dismayed many at the CDC, who believed that the international body had set the bar too high for these declarations. Now, the death and devastation caused by COVID, and our failure to effectively identify, isolate, and reduce this risk, will result in "a shift in political preferences aimed at undoing the existing order," writes Nicholas Christakis in his history of the COVID pandemic, *Apollo's Arrow*.[15] The emphasis will be on building new institutions and capabilities to make sure such a system failure can't happen again.

Among the institutions we'll need to turn to are the agencies charged with guarding our national security. At one time, we viewed the public health mission more firmly through a lens of national security preparedness. In 1952, when the CDC created the Epidemic Intelligence Service, the inclusion of the word *intelligence* was deliberate. The agency's early leadership, Dr. Joseph Mountain and Dr. Alexander Langmuir, used the word because they wanted the CDC to be viewed as a national security asset. That was a 1950s mind-set, but this strategic orientation began to change and erode in the 1960s. Today the EIS is a globally recognized fellowship program renowned for its efforts to train experts to investigate and respond to outbreaks of disease around the world, but it has no formal relationship to our national security functions.

In recent decades the US government has often been reluctant to view the public health mission as a matter of national security, and even more reluctant to use the tools of national defense and intelligence gathering to identify and counter these threats. The Department of Defense initially resisted private calls, early in the COVID pandemic, to reprogram biodefense resources to focus on combatting COVID: some in the Pentagon argued that its fidelity was to counter deliberate biological threats and not a naturally occurring virus. The result was that nobody fully owned the mission of protecting the homeland from these dangers. Whatever parts of this obligation ostensibly fell to the CDC, either by design or by default, these commitments were not properly pursued. Going forward, we'll need to prescribe a much clearer role for how we also use the conventional tools of our national security infrastructure to help guard the nation against future pandemics.

In 2014, when the Obama administration's national security team worked on the global health security agenda, there was a deliberate decision to put the words *health* and *security* together, in that document, as part of an attempt to bring the intelligence and public health communities closer together.[16] However, this step engendered controversy, evidencing the challenges we would face in integrating national security agencies into public health efforts.

There were "epic arguments," according to one of the people involved, over whether the words *health* and *security* could be used together in a national security context. For many years, we'd treated pandemic planning as largely a public health function. The practical limits of this approach were revealed in our deficient response to COVID. Moreover, other countries would turn to their own intelligence services as part of biological threat planning and mitigation and we'd have no way to fully counter them. Sometimes other nations engaged their intelligence assets as a way to augment their public health response, and sometimes it was our foes trying to frustrate our own domestic efforts or siphon information off our strengths to buoy their own capabilities. What has become clear is that we need to leverage our intelligence assets—as much for advancing our public health goals as for guarding against adversaries who would try to exploit the chaos to weaken our response.

Among the activities we got wind of, during the summer of 2020, was that China was using its foreign intelligence services to collect trade secret information on US efforts to develop COVID vaccines. The *New York Times* reported, based on interviews with people tracking the espionage, that this spying was "one of the fastest peacetime mission shifts in recent times for the world's intelligence agencies, pitting them against one another in a new grand game of spy versus spy." There was evidence that Chinese intelligence hackers had managed to breach some sensitive networks. And in July, the Justice Department indicted several hackers who were working on behalf of China's Ministry of State Security, the country's spy service.[17]

Meanwhile, Russia's foreign intelligence service, the SVR, also targeted vaccine research networks in the US, Canada, and Britain. The Russian espionage was first detected by a British spy agency that was monitoring international fiber optic cables. American intelligence officials said that the Russians were seeking to steal research to accelerate development of their own vaccine, not to sabotage our efforts.[18] Then, once Russia had developed its own successful COVID vaccine, Sputnik V, its intelligence services mounted an elaborate campaign to undermine confidence in the COVID vaccine marketed by Pfizer.[19] It was believed that Russia waged this derogatory operation as part of an effort to gain commercial advantage over Western vaccines as Russia tried to compete with US drug makers for the market primacy of its own vaccine, especially in middle-income nations where Russia was using its vaccine for diplomacy. At one point, a head of state from an Eastern European nation called Pfizer CEO Albert Bourla and implored Bourla to try and encourage the Biden administration to allow US COVID vaccines to be shipped to Eastern Europe because those nations were being flooded with Russian products. Bourla's subsequent efforts helped instigate a broader action by the Biden administration to procure 500 million doses of Pfizer vaccine, at a not-for-profit price, and distribute it to low- and middle-income countries through COVAX. The actions of Russia, and the situation in Eastern Europe, had provided an added impetus to efforts by the US government to expand global distribution of the American products.[20]

Recognizing the threat, American security agencies had moved to protect the universities and private companies doing most of our vaccine research. Spy tactics that had once been reserved for the arms race were now playing out on the field of public health response—proof that the game had changed.[21] Pandemic preparedness had taken on a new strategic standing. In some respects, this was just spies being spies. However, the level of attention that foreign intelligence services focused on the American COVID vaccines foreshadows a broader shift in security priorities. In Beijing, Moscow, and other capital cities, public health was now clearly being viewed as a matter of national security. The US had to start playing this new game, too.

Across multiple administrations, the willingness to view pandemic preparedness through the lens of national security has waxed and waned, being downgraded only to be elevated after new threats emerged, and then marginalized again when the immediate risks seemed to recede. This cycle took place most visibly in relation to the National Security Council (NSC). At the beginning of the Bush administration in 2001, there had been a person charged with overseeing biothreats on the NSC, but the position was eliminated only to be reinstated after 9/11, with an emphasis on bioterrorism. In 2009, the incoming Obama administration disbanded the biothreats staff on the NSC, as the Bush administration initially had done, but then reinstituted the function after the H1N1 swine flu pandemic and the Ebola epidemic.[22] And yet once again, in 2018, this NSC function was disbanded by the Trump administration, not long before COVID struck. The NSC chief, John Bolton, was said to be very focused on the biothreats agenda, but the explicit function that had existed was consolidated under the work stream for weapons of mass destruction, reflecting—at least in part—an orientation that placed a lot of the focus on bioterrorism.

The volatile nature of that NSC role was a leitmotif for the shifting focus on these dangers and reflected an ambivalence toward elevating biological risks beyond agents of bioterrorism and engaging the nation's national security leadership to manage preparations against naturally occurring threats. The general refrain was that the pandemic piece of that portfolio was the CDC's domain. However, the CDC didn't know how to view these contingencies through the lens of national security. Most of the national security cooperation between the CDC, the public health community, and the intelligence community remained centered on bioterrorism, much like the focus of the NSC. And even these efforts never acquired a prominent place in the CDC's sprawling mission. The CDC's programs to mitigate the risk that smallpox could be weaponized by a rogue nation or terrorist group was one such example. The select agent program had existed at the CDC since the 1990s, but it was expanded after 9/11. Some in the CDC believed that the newly broadened program should have been transferred to the newly created Department of Homeland Security, but the academic community and organizations like the American Society for Microbiology strongly opposed such a move and prevailed, arguing that the program needed to be 'science based'—a principle that could be fulfilled only inside the CDC.

And yet the CDC wasn't the right agency to take ownership of anticipating and developing strategies to counter national security threats, because it possesses a retrospective mind-set. The CDC doesn't generally focus on actionable assessments of future risks; it produces academically oriented after-action reports in an effort to define new scientific principles. The intelligence community, by contrast, has a prospective mind-set; it's always scanning the horizon. Intelligence agencies have to make a call on future threats. They're willing to be wrong, but they're compelled by the nature of their work to make predictions. Even if the CDC were in the business of making assessments on future threats, it largely lacks the data science capabilities and the advanced analytics required to do this sort of forecasting of risk.

America's dismal experience with COVID leaves us little choice but to expand the tools we use to inform us of new risks. In bolstering our pandemic preparedness, our purpose shouldn't be merely to blunt the impact of the next pathogen that emerges, but to make sure that a calamity on the scale of COVID can never happen again, and the US can never be threatened in this way again. If our goal is to head off a potential pandemic entirely and prevent the next contagion from ever getting a grip on America the way COVID did, we'll have to involve our instruments of national security. We'll need to lean more heavily on nontraditional capabilities for gathering information on these threats, to better target our public health response, and improve our overall security. The COVID pandemic not only showed the lengths to which other countries would go to in order to gain an edge; it also revealed the gaps in the application of our own intelligence capabilities.

We had always relied on global cooperation to counter pandemic threats, but the SARS-1 outbreak in 2003 taught us that the global order could be frayed and feckless when it was most needed. Countries were reluctant to report these kinds of events and to share needed samples when new pathogens emerged. These behaviors were not exclusive to the Chinese government. We thought we had learned lessons after SARS-1 and had strengthened international conventions that required timely reporting. But the behaviors persisted. In 2005, Chinese authorities denied an H1N1 outbreak that Hong Kong scientists had identified.[23] In 2007, Indonesia refused to share specimens of H5N1 with the US and UK.[24] In 2015, authorities in Guinea took nearly three months to report to the WHO that Ebola was the cause of an infectious outbreak that was already spreading widely in that nation.[25] In 2019, the WHO issued an unusual notice censuring the government of Tanzania for not responding to requests for information about a possible Ebola outbreak there.[26] The list goes on. In times of biological crisis, nations have concealed their contagions. During COVID, the president of Tanzania would insist that his country didn't have any spread of SARS-CoV-2, a position that would be repudiated when he was hospitalized for the disease in March 2021 and died, reportedly of a heart ailment, just days later.[27]

Global cooperation around these kinds of threats dates back to a meeting in 1995 where public health authorities sought to update the International Health Regulations that were first adopted in 1969.[28] Taken together, these pacts governed how countries notified each other of emerging infectious threats. The revisions made in 1995 implemented a series of regional and global networks for monitoring infectious disease.[29] Then, in 2000, global public health authorities established the Global Outbreak Alert and Response Network, with the goal of creating regional hubs around the globe that would be a backstop for identifying emerging risks—another tripwire that wouldn't rely solely on national governments and the decisions of political leaders for informing the world of new outbreaks.[30] The idea was to create a "network of networks" for spotting new threats.

Sometimes the system worked. It was this new network that in February 2003 alerted world health authorities that a dangerous pneumonia was spreading in Guangdong Province in China.[31] The outbreak occurred among medical providers, which may be the only reason that international monitors spotted it. Clusters of unusual infections among healthcare workers tend to get noticed. The Chinese government officially notified the WHO of the outbreak only after its existence was revealed by the Response Network. But by then, there were already 305 known cases, including 105 infected healthcare workers, and five people who had died. However, the Chinese government still described the outbreak as "under control."[32] The illness was not being caused by a novel flu, as was first thought; it was a new coronavirus that had started to spread among humans. This was the start of what would become the SARS-1 pandemic. It didn't stay confined to China.

The infection had been spreading slowly in China since early November 2002. The first case was traced to people who worked in the food industry in the city of Foshan. The outbreak became amplified when it spread to healthcare workers, who contracted the virus from sick patients. The window to identify the virus and take steps to intervene to stop its spread had been open for more than a year, and we missed it. In response to SARS-1, in 2005, a substantial revision was made to the international regulations that administrated the reporting of outbreaks and novel pathogens. Once again, global health authorities came to believe that, through greater cooperation and more binding commitments, they would be able to prevent the kinds of obfuscations that had initially concealed SARS-1. However, countries remained caught between two competing priorities. On the one hand, nations had an overarching public health imperative to prevent infectious diseases from spreading and crossing their borders. However, they also had a political incentive to minimize the economic impacts when they were host to a new contagion and prevent restrictions on trade and travel that could imperil their economies. This tension would confound our response to COVID.

How the world reacted to the COVID pandemic may have promoted actions that will make future cooperation, and the early sharing of information about new outbreaks, even more difficult. At one time, isolating a nation with an epidemic was seen as far too damaging and ultimately self-defeating. Applying even a limited quarantine would weaken a country's ability to respond to an epidemic and might exacerbate spread. Moreover, by the time other nations contemplated travel restrictions, it was often much too late—the epidemic would already be well under way, with sparks of onward transmission already lit. This judgment was behind the reluctance to impose travel restrictions on West Africa during the Ebola epidemic in 2014. In these regards, the initial travel restrictions that the US and Europe imposed on China in response to the novel virus were unprecedented and deeply controversial at the time. Indeed, when the coronavirus first emerged in China in January, the WHO privately fought the Trump administration as the US contemplated its initial travel checkpoints. The opening actions weren't travel bans, either, but measures to subject passengers inbound from Wuhan to increased scrutiny at US airports. Part of the WHO's efforts were instigated by the Chinese government's intense lobbying to get the organization to intervene and try to stop the US restrictions from being implemented. However, a part of the resistance was also owed to a historical concern among public health experts that travel restrictions might undermine efforts to contain a virus rather than help achieve control. "It's part of the religion of global health: travel and trade restrictions are bad," Lawrence Gostin, a notable expert on global health law at Georgetown University Law Center and one of the original architects of the International Health Regulations would tell the *New York Times*. "I'm one of the congregants."[33]

In the case of the travel restrictions later put on Europe, there was a lot of tension in the White House as to whether to pursue them. Members of the president's economic team, including Treasury Secretary Steve Mnuchin, opposed the travel ban, worried about the economic fallout it might trigger.[34] It was not until Deborah Birx got a download of raw data from Italy and fed it into new models showing that the European epidemic was poised to explode and that cases from Europe were seeding a lot of the new spread in the US that officials were convinced to shut down travel from Europe. The announcement of a ban on travel from Europe came that same week.

COVID normalized the widespread use of these practices. When the next outbreak emerges, we can expect nations to act swiftly to impose strict travel controls. When the UK identified a more contagious variant of SARS-CoV-2, the first thing the French government did was close the Chunnel.[35] Even the US put new limits on travelers entering from Great Britain. While the evidence for the benefits of border closures and travel controls was weak, so was proof for the counterargument.[36] The case against travel restrictions is based on a small number of studies showing that closing borders will delay but not prevent spread and can do more harm than good by destabilizing regions where an outbreak is occurring, causing people to flee hot zones. However, there's no firm data to support the notion that travel bans are bad. After they were used in COVID, there was a belief that they helped some and little convincing data to say they didn't.

Given the intense isolation that nations are now likely to experience when they disclose that they're host to a menacing outbreak of a novel disease, and the economic repercussions they'll incur, we can expect countries to adjust their behaviors. They'll be even more reluctant to reveal the existence of a new pathogen or to share strains and sequence information. There's currently no requirement, under any international law or convention, that can compel nations to share samples of pathogens in a public health crisis.[37] To this day, China has never shared the initial strains of SARS-CoV-2, which are key if we wish to firmly establish the virus's origin. These source strains would also have helped support the early development of drugs and diagnostics. While it's true that researchers were able to use the sequence information China published, having an actual sample of the virus to validate the sequence is critical to effective development work. At one point, we tried to circumvent China's intransigence by arranging a transfer directly between the Wuhan Institute of Virology and the labs at the University of Texas Medical Branch, which had in place a long-standing collaboration.[38] At first the Wuhan lab agreed to share the samples, but officials in Beijing intervened and blocked the planned transfer. The inability to access early samples remains a major gap. When a WHO delegation visited China a year after the initial outbreak to identify the source of SARS-CoV-2 and investigate theories that the virus may have leaked out of a lab, the Chinese government still wouldn't allow the early specimens to be shared. In June 2021, information would emerge to show that Chinese researchers working at Wuhan University had deleted from NIH research servers the sequence information from dozens of samples collected in early December 2019. The scientists had initially submitted the data as part of a research project, and then inexplicably pulled it down. Perhaps the researchers had been pressured to withdraw the samples, or balked when the Chinese government changed the rules governing what sequence information scientists were permitted to share?[39]

The persistence of speculation that the pandemic might have been sparked by an accident in a Chinese lab, where it's rumored that some workers were inadvertently infected with a virus they were handling in the fall of 2019, and sick enough to seek hospital care, is due, in part, to the Chinese government's own obfuscations.[40] As one commentator noted, owing to the evasions, the withholding by China of pertinent data, and the efforts to frustrate and discomfort the WHO delegation that was sent to China with the mission of trying to investigate the source of the pandemic, there's likely never going to be a concrete fact pattern firmly establishing where the virus originated, only inferences and a fight over competing narratives.[41] And yet, independent of the Chinese government's shenanigans, there remains enough evidence on the other side of the ledger to nurture a theory that the initial epidemic may have been sparked by accidental infections that occurred as a result of the poor handling of a lab specimen inside China.

The proximity of the Wuhan Institute of Virology to the site of the initial outbreak fueled early theories that the WIV could have been the point of origin.[42] But it would take more than inference and coincidence to give these speculations traction, and other public facts would support these suppositions. Many of the speculations sprung from the kind of high risk studies that was known to take place at the WIV, and its singular role as a center for coronavirus research. If there were novel coronavirus specimens found by researchers in China, there was a good bet the WIV, given its prominent scientific role, would obtain access to them. It was well established that there was an outbreak of a novel coronavirus infection in pangolins in March 2019, when the Guangdong Wildlife Rescue Center, in Guangzhou, reported that it had taken custody of twenty-one pangolins that had been seized by customs police.[43] The animals were reported to be in poor health, with signs of respiratory distress. Samples were initially sent at the time to a government lab in Guangzhou and genomic evidence of coronaviruses were found.[44] It was established by research published in the journal *Nature* that SARS-CoV-2 bore close resemblance to this particular pangolin coronavirus, especially the critical part of the virus, the receptor binding domain, that attaches to the ACE2 receptors in our respiratory tract and gives the virus its pathogenicity in humans (based on these findings, pangolins were believed to be one possible intermediate host as the virus migrated from bats to humans). It's known that the WIV had worked with pangolin coronaviruses and was China's main research facility for work on coronaviruses following SARS-1. So if any lab in China was going to be called upon to investigate an outbreak of a novel coronavirus in pangolins or any other species, ostensibly it would have been the WIV, making it possible that the WIV was in possession of, and had worked with, novel coronavirus specimens that hadn't been previously shared or disclosed.[45]

The US State Department published a fact sheet on "Activity at the Wuhan Institute of Virology" on January 15, 2021, stating that "The WIV became a focal point for international coronavirus research after the 2003 SARS outbreak and has since studied animals including mice, bats, and pangolins."[46] Adding fuel to these speculations were some unusual genetic features that the SARS-CoV-2 strain harbored. These were features that weren't commonly found in naturally occurring variants of similar coronaviruses that had been—so far—cataloged by scientists; but, the theorists reasoned, they could be the kinds of adaptations that researchers would make in a laboratory as they tried to evaluate the characteristics of a virus. Among the most prominent of these findings was the fact that the SARS-CoV-2 virus has a specific addition to one of its key surface proteins (called a furin cleavage site)—an attribute that's absent in the strains of coronavirus most closely related to SARS-CoV-2 (called Sarbecoviruses), but which is found in other types of coronaviruses such as Merbecovirus (which causes MERS), Embecovirus (which causes the common cold), and Hibecovirus and Nobecovirus (which, so far, have only been found in bats).[47] Adding to the concerns is the fact that manipulation of the furin cleavage site is a common adaptation that's sometimes made to viruses in order to enable them to more easily grow in a cell culture in a lab, although the adaptation cuts the other way, too—sometimes the furin cleavage site isn't added or, when it's already present in a virus, it's removed in order to help control a virus's pathogenicity and improve lab safety as a virus is evaluated experimentally. It could be added, however, in circumstances where a researcher wants to make it easier to infect animals with a cultured virus. If a virus has been manipulated, whether by a seamless method where a characteristic is deliberately inserted into the virus's genetic sequence, or by serial passage through cell cultures, there's probably no way of knowing for certain in this case simply by evaluating the virus's sequence.[48]

These are all speculations, of course, and critics of these theories sometimes blame those who advance them for imprecision, but the concerns that a lab leak could have been the origin of COVID persist largely because the Chinese government has withheld information that could help put such theorizing to rest. For some observers, the question becomes: When do too many coincident facts become hard to overlook? Adding further weight to the theory that there was a connection between the lab and the epidemic were persistent concerns that the WIV had poor controls. When the lab was on the cusp of being cleared for operations in 2017, the journal *Nature* raised some of the first concerns, stating that "Some scientists outside China worry about pathogens escaping, and the addition of a biological dimension to geopolitical tensions between China and other nations."[49] France helped build the WIV and there were persistent concerns that it was improperly built. French scientists continued to work inside the facility, to assist their Chinese counterparts, but were forced out of the facility around 2017 when, it was believed, the Chinese started to make a pivot toward high-end research that the Chinese didn't want their French collaborators to see. "The Chinese increasingly cut them out and made the oversight impossible," one former US official told me, "so the French eventually withdrew those people." It was around 2017 that the WIV was also believed to have begun conducting classified research, including laboratory animal experiments, on behalf of the Chinese military.[50]

In an effort to dismiss the plausibility of a lab leak being the original source of SARS-CoV-2, the WHO team would cite the WIV's stringent controls in its higher-security (BSL-4) labs. The WIV's careful handling of specimens made an accident implausible, the WHO team argued. "It was a very well-run lab," said Dr. Peter Daszak, one of the WHO team members. "They did all the things you would expect them to do in terms of testing the staff, psychological evaluation before you're allowed to work in the biocontainment labs, all the things and checking the facility."[51] However, it was established that the WIV was doing its coronavirus research in its lower-security (BSL-2) labs, which didn't have the same rigorous controls.[52] The WHO investigators didn't request full access to the database of viruses held by the WIV, so they couldn't be certain what exactly was contained in the lab's inventory. Some of the WHO team members asserted that they already knew what viruses the WIV had worked on, because they had been collaborating with the lab, and so they said they were sure the WIV's inventory couldn't include any unrevealed strains that closely resembled SARS-CoV-2.[53] That unverified assertion shouldn't have sufficed for due diligence on this critical question.[54]

Yet the WIV's freezers held what is believed to be the most complete inventory of sampled bat viruses found in any research institution—22,000 samples of genomic material or viruses, including 15,000 samples derived from bats and covering more than 1,400 bat viruses.[55] The database also held "more than 100 unpublished sequences of bat coronaviruses that could significantly help the probe into the origins of the pandemic," according to the *Washington Post*.[56] Adding to the concerns was the WIV's purported dual use by civilian and military researchers. Matthew Pottinger, the former US deputy national security adviser, said that the US had strong reason to believe that the Chinese military was doing classified animal experiments in WIV's laboratories, going back to at least 2017. He told Margaret Brennan on *Face the Nation* that the WIV was infecting animals with coronaviruses as part of the lab's research.[57] These could include laboratory mice genetically engineered to carry the human version of the ACE2 protein found on the surface of cells that line the respiratory tract. The mice would thus mimic the human response to a coronavirus. It was well established that the WIV had been doing gain-of-function research, and some of this work involved the infection of these genetically modified mice with versions of coronaviruses that had been deliberately altered to bind more tightly to ACE2, to see how improved infectivity of the spike protein could yield clues on how to design better drugs and vaccines to counter it.[58] It was also asserted that the NIH had previously helped indirectly support some of this type of research with financial grants.[59] The process of infecting animals with coronaviruses raised the odds that the virus might have accidentally transferred to workers handling those animals.[60] Former CDC director Robert Redfield would be among the highest-profile public health officials to say publicly that he believed SARS-CoV-2 resulted from gain-of-function research. Following this line of thought, it was speculated that Chinese researchers may have deliberately altered the virus as a way to study it, but then the virus accidentally escaped from the WIV and first started to spread in September or October, not in December.[61]

Gain-of-function research had long alarmed other scientists. In 2014 a group of researchers calling themselves the Cambridge Working Group urged restraint on using genetic manipulations to create novel viruses. "Accident risks with newly created 'potential pandemic pathogens' raise grave new concerns," they wrote. "Laboratory creation of highly transmissible, novel strains of dangerous viruses, especially but not limited to influenza, poses substantially increased risks. An accidental infection in such a setting could trigger outbreaks that would be difficult or impossible to control."[62] A highly respected pair of researchers put the odds that such an effort (in this case involving a hypothetical strain of a novel influenza) could create a dangerous pathogen that would accidentally escape from a lab and trigger a pandemic at 1 in 10,000 to 1 in 100,000 per lab year.[63] This was a meaningful risk. The US government imposed a moratorium on gain-of-function studies in 2014. No such restrictions were imposed in China. The US moratorium was later lifted by the NIH in 2017 and gain of function experiments continued including one widely cited study that was done by US researchers in collaboration with the WIV.[64]

Wuhan is home to at least six other high-containment lab facilities, and published papers show that some of these labs were also very active in coronavirus research.[65] The Wuhan Center for Disease Control and Prevention, a satellite facility operated by the Chinese CDC, was similarly known to be doing research inside its lower-security BSL-2 lab with viruses that were derived from bats. It's located just a few blocks from the Wuhan food market that was originally implicated in the initial outbreak. The focus has been on the WIV. But, before COVID, research on coronaviruses wasn't typically done in the highest-security labs, so any one of these six other lab facilities in Wuhan could have been working with novel strains.

The theory of a lab accident involving a deadly coronavirus had precedent. The last known cases of SARS-1 were caused by lab accidents.

In 2004, there was an outbreak of SARS-1 in Beijing that infected nine people, killing one, that had originated from an accidental exposure in a lab. Two of the people infected had worked with the virus in the National Institute of Virology in Beijing, and their infections were linked to experiments using live and inactivated SARS-1 virus at the labs belonging to the Chinese CDC.[66] One of those infected researchers had her illness go unnoticed by officials at the Chinese CDC, even after she developed symptoms consistent with SARS-1, including fever. She presented on her own to the hospital where she infected a nurse before traveling to her home in Anhui by train.[67] The Chinese government initially tried to conceal the outbreak, along with its origin, before journalists uncovered the events. In all, SARS-1 had been the subject of six outbreaks since its last-known natural occurrence, each one the consequence of its escape from a laboratory: one time each in Singapore and Taiwan, and then four separate escapes from the same lab in Beijing.[68] Another instance where an experiment in China had gone awry, and triggered the global spread of a novel virus, had occurred in 1977 and involved a strain of H1N1 influenza. It's believed that the virus was carelessly released when several thousand Chinese military recruits were challenged with a live strain of the virus to test an experimental vaccine. This H1N1 strain (and its descendants) have circulated globally ever since.[69]

The lab leak theory has important strategic implications, and the possibility that SARS-CoV-2 accidentally escaped shouldn't be easily dismissed. The probability, even if it's judged to be remote, has important implications for how we prepare against future pandemic threats. There are hundreds of BSL-3 and BSL-4 labs around the world with active pathogen research, posing a potential health and security risk if they aren't appropriately staffed, managed, and resourced.

First, the risk reinforces how important it is for countries to share more information, not less. The Chinese government hurt itself and its standing by hiding pertinent evidence that could help the world firmly establish the origin of SARS-CoV-2. The genetic evidence derived from the SARS-CoV-2 genome supports the argument that the virus could have emanated from an entirely natural origin, jumping between animals and people before it finally broke out. Most of the notable features on the virus could be found in coronaviruses already identified in nature, including the furin cleavage site. So, in theory, all of its genetic diversity could have evolved through natural selection. The features didn't necessarily need to arise from a lab, or from any special circumstance.[70]

Take the issue of the furin cleavage site, which was held out by some as a feature that suggested a lab origin for the coronavirus. Other experts have also argued persuasively that the furin cleavage site is unlikely to have been engineered, precisely because it's an unusual occurrence of this particular genetic feature. If someone were going to engineer the cleavage site in a laboratory, as a way to change the characteristics of a coronavirus and facilitate research on a novel rendering, then they would have likely used a consensus sequence for the furin cleavage site that's already described in the scientific literature or seen in other beta coronaviruses (such as the furin site that's found in the MERS virus—which is also a virulence factor in that particular strain). Moreover, scientists have pointed out that there's a feature in the genetic sequence of SARS-CoV-2 that sits adjacent to the virus's furin cleavage site (called a stem loop structure) that can cause the enzymes that read and write on the sequence to "slip," creating the circumstances where errors can occur that lead to genetic insertions. Such insertions are a known "mistake" that happens in the presence of similar genetic sequences, and it adds to the natural diversity of viruses; and the presence of the stem loop structure could support a theory that the furin site arose spontaneously. There are also plenty of proxies in nature where this particular feature is naturally acquired by a virus as it undergoes widespread transmission. The acquisition of a furin cleavage site is sometimes seen in influenza during large outbreaks in chicken flocks. With flu, as with COVID, the addition of a furin cleavage site can expand the range of hosts the virus is able to infect, enhancing its spread and its pathogenicity.[71]

The debate over the furin cleavage site is a microcosm of the larger controversy over the virus's origin, and our inability to settle critical questions. If the virus had a natural origin, there's unlikely ever to be a definitive piece of evidence that firmly establishes its source unless we find its intermediate host among a population of animals—like civet cats in the case of SARS-1, or camels in the case of MERS. So far, despite an extensive effort, no such animal population has been identified for SARS-CoV-2. As the *Wall Street Journal*'s Holman Jenkins observed, the lab leak theory might appear, at first blush, less plausible than a natural origin, unless you were to assemble the world's largest repository of dangerous coronaviruses in a lab that's located in a densely populated city, experiment with them in a lower-security facility with weak biological controls, and start infecting transgenic mammals as a way to evaluate the pathogenicity of the viral collection in the human immune system, all of which the Chinese did.[72] Based on what's publicly known, Chinese researchers had fashioned the conditions that created these risks. Even if Chinese officials didn't accidentally let the virus escape from one of its labs, there was plenty of inferences and circumstantial evidence that would make it hard for them to fully disprove that possibility. As a consequence, when the WHO team finally visited Wuhan, Jenkins wryly noted, there was probably very little chance that the Chinese government would ever let the chips fall where they may.[73]

The WHO team, to some, fueled these speculations by seeming more intent on debunking the theory of a lab leak, rather than objectively examining the question. The Biden administration and European governments largely rejected the team's findings and called for a new and independent examination.[74] On May 26, 2021, President Biden ordered intelligence agencies to undertake a closer review of what he said were two equally plausible scenarios of the origins of SARS-CoV-2, instructing them to redouble efforts to gather new facts, and engage the scientific community in a broader examination of the available evidence, and report to the administration and Congress in 90 days. "Here is their current position," Biden said, in calling for the new 90-day analysis. "While two elements in the [intelligence community] lean toward the [zoonotic] scenario and one leans more toward the [lab leak scenario]—each with low or moderate confidence—the majority of elements do not believe there is sufficient information to assess one to be more likely than the other," Biden said.[75] The WHO team did visit the WIV, but not other labs in the city, and it wasn't able to properly investigate the WIV. Nor was a lab investigation included as part of its mandate.[76] The group sent by the WHO had to conduct their review as part of a joint process with their Chinese counterparts and were given a highly curated set of information to examine.[77] The WIV didn't even provide new and independent sequencing of the coronaviruses present in its freezers, which would have been a straightforward way to help put some of the speculations to rest. Several months later, a group of eminent virologists, including Dr. Ralph S. Baric (who had previously done collaborative research with the WIV) published a statement in the journal *Science* arguing that both a zoonotic source and a lab accident are "viable" explanations for the origin of SARS-CoV-2 and each deserves equal consideration and investigation. "We must take hypotheses about both natural and laboratory spillovers seriously until we have sufficient data," the virologists wrote. "A proper investigation should be transparent, objective, data-driven, inclusive of broad expertise, subject to independent oversight, and responsibly managed to minimize the impact of conflicts of interest."[78] It's also possible that the lab wasn't the initial source of the viral outbreak, but a place where its spread was intensified by an accidental leak that contributed to ongoing spread—that there were multiple points of origin from where SARS-CoV-2 got introduced into the local population, some occurring in the lab, and some occurring in nature, perhaps at the point where a sample that might have eventually made its way to the WIV had been first collected—perhaps from the sick pangolins or some other animal host.

Fully assessing the veracity of the lab leak theory, and not dismissing these potential risks prematurely, has important strategic implications for other reasons. Establishing the odds that a lab might have been a link in the initial outbreak isn't merely a curiosity, or a political intrigue. It's a critical question of public health preparedness. How we handicap the probability that a lab could have played a role in the initial contagion informs how we approach efforts to guard against pandemics in the future, including how we design and regulate high-risk labs that handle some of the world's most dangerous pathogens. We've traditionally focused those efforts on trying to reduce the risk of spillover events from zoonotic sources in nature as our primary means to shut off the potential origins of these kinds of catastrophic events. Going forward, to prevent the next pandemic, we have to focus equal attention on improving global lab safety.

Labs capable of handling the most dangerous pathogens, classified as BSL-4 labs, are being built all around the world, many in countries with a history of poor controls and oversight of research practices. There needs to be an international agency to oversee safety measures at these labs, and what kind of research they're conducting, because of the well-documented risk that pathogens can escape from research facilities. Such an agency could be modeled after the International Atomic Energy Agency, an independent, global organization that provides international safeguards against misuse of nuclear technology and nuclear materials, and reports to the United Nations General Assembly and Security Council. "If we scientists are not forced to confront the issues of laboratory safety and risky research in a serious and sustained manner, history suggests that we will not do so," wrote Dr. David Relman, an eminent Stanford scientist who was one of the authors of the editorial in *Nature* that called for a more thorough investigation of the lab leak hypothesis. "In 2012, controversy erupted when it transpired that two sets of researchers—at the University of Wisconsin at Madison and Erasmus Medical Center in Rotterdam, the Netherlands—were altering highly pathogenic avian influenza viruses to enhance their transmissibility among mammals (to understand their potential to cause a pandemic). The subsequent debate led to a three-year moratorium on the funding of experiments designed to enhance the transmissibility or disease-causing capabilities of influenza viruses or coronaviruses. And yet we still have only a framework for guiding funding decisions by the U.S. Department of Health and Human Services and not the funding by other public and private entities. What's more, there is still inadequate attention paid to the dangers created by the publication of the sequences of these enhanced pathogens, with which anyone around the world skilled in the art can synthesize these dangerous agents," Relman wrote.[79]

We'll also need target our surveillance to these potential hazards, not only labs that are engaging in unduly risky practices, but how synthetic biology could be used for rogue purposes. The mere possibility of a lab leak as a source of this pandemic could serve to reinforce for those with diabolical intent that an organism refined and released from a laboratory could be used as a trigger to a widespread contagion. Nobody believes COVID emanated from a deliberate act. But the theory that its origin may have been an accidental release from a lab could already serve as a sinister proof of principle for those with ill intent, even if a lab source is never fully proven. It could plant a diabolical idea that a pathogen released from a single site may be all it takes to trigger a global contagion. Magnifying these dangers, sometimes the highest-risk research gets exported to the countries with the weakest controls in place because those are the regions of the world where more dangerous experiments are still permitted.

Yet, the fact remains, the global response to COVID may have made the needed cooperation more unlikely. The travel bans imposed on China, and later Europe, conditioned the world on what to expect when a country is host to a new contagion: now, travel restrictions are the new normal. These political conditions make it less likely nations will share early information about future outbreaks, zoonotic spillover events, novel pathogens, or lab accidents. We'll see more erosion in public health collaboration as an outcome. Countries have even more reason now to hide the embers of the next epidemic.

This is all the more reason why the US needs to lean on the tools of the intelligence community as a way to guard against future threats. Historically, it was often hard to get our intelligence community engaged on these kinds of missions, which were not considered on par with other threats to national security. On the public health side of the ledger, there was equal reluctance to see the intel community take on a larger role in public health response. The work of the intelligence community was traditionally seen by the scientific community to run counter to the principles of science, which encourage open sharing of information for the broadest possible public good. The CDC and the public health community were often reluctant to receive classified information and submit to the constraints that came with the handling of such material. And the public health community was sometimes reluctant to share sensitive findings with national security agencies, even in circumstances where the information could help the US adjust its response to a novel threat. During the Ebola epidemic in West Africa in 2014, the CDC signed a bilateral agreement with the government of Liberia to share information, but the CDC subsequently refused to give the data it gathered to the US Department of Defense, arguing that its agreement with Liberia prohibited the CDC from sharing the information with "third parties." This was "unconscionable," one former senior US government official involved in the Ebola response said to me. "They wouldn't even share it across agencies of the U.S. government," he said. The US military was already on the ground in

Liberia, as part of a major mobilization to assist in the response to the Ebola outbreak, and the CDC wouldn't share information that the Pentagon believed could help adjust and guide its response.

Leveraging the eighteen different agencies making up the US intelligence community could provide important new capabilities for reducing the risk posed by naturally occurring pathogens that have pandemic potential. Many public health leaders will worry about a corrosive effect from engaging the tools of national security to achieve missions that were traditionally executed through scientific exchange and multilateral commitments and collaboration. They will fear that this new posture will poison the well of international cooperation. Inside the government, the pushback to a greater role for our spy agencies has often come from the operating components of HHS. Among other things, HHS officials have argued that closer collaboration with intelligence agencies might complicate the overseas activities of public health officials and cause them to be viewed with suspicion when they travel. They worry that everyone with a white coat will be presumed to be a spy. But many foreign adversaries already assume that any American official traveling abroad could be operating on behalf of US intelligence agencies. And the record shows that global commitments, multilateral agreements, and the work of our public health officials weren't enough to keep us safe. They weren't even enough to secure access to the source strains of SARS-CoV-2, or early information that the virus in China was spreading human-to-human, had infected healthcare workers, and was being transmitted through asymptomatic infections. Some major taboos need to be lifted. Some things need to change. The COVID pandemic was so devastating that it jeopardized all of our other national priorities when it came to health, security, and economic prosperity. It laid bare that multilateral commitments are, in the end, nonbinding.

We need to look across all of our capabilities to make sure that such an event cannot happen again. Other nations are looking to their national security instruments to identify and reduce pandemic risks. We will need to make the same adjustments. Yet, as we go down this same path, we need to make sure we engage our tools of national security in a way that preserves our multilateral institutions and the work of scientific agencies. We need to continue to invest in capacity building in nations with whom we collaborate. We need to make sure that scientific exchange doesn't become encumbered. Agencies like the NIH don't want to be burdened with having to handle reams of classified information. Researchers want to know the provenance of any samples they receive and the environment in which it was collected so they can properly assess its usefulness. The rules of engagement will need to be carefully worked out.

The flip side is that embedding public health information in our intelligence reporting could elevate the work of our scientific agencies, allowing this analysis to get in front of policymakers in a form and function where it's more actionable. For policymakers, making good use of intelligence requires repetition. You need to be able to read intelligence estimates over time to develop a cadence for the information and how threats evolve. When it comes to threats from emerging diseases, policymakers aren't reading academic papers. It's not how they digest information that's outside their domain. But they are reading the CIA's World Intelligence Review and other national security estimates. If more health reporting were regularly included in these assessments and treated on par with other threats, it would keep it in front of policymakers, so they would be in a more informed position to secure the nation against biological risks.[80]

The components of HHS have access to a lot of knowledge that fits into the national security dialogue, but most of these operating units don't view this information through that prism. It's not the world they live in. The national security agencies could help fit this reporting into a broader mosaic that would help identify risks. At the same time, intelligence agencies could better hone their collection and reporting through collaboration with health experts who could provide context to help guide intelligence collection. The National Security Agency scoops up an enormous amount of signals intelligence but may not know precisely what to look for without the help of health experts, or how to interpret the information they gather. If the NSA found itself in possession of laboratory data, for example, it might not have a clear sense for how to pull this information out of the signals intelligence or how to make an initial analysis to determine its importance; but collecting these data would fall feasibly into their existing mission given that most of this information is now shared digitally.

Leveraging our intelligence assets to inform us better on risks would also complement our diplomatic tools. More information could be shared with our diplomats to let them target their own efforts around the US public health mission. If intelligence estimates identified certain weaknesses in a foreign country's scientific institutions, our diplomats would be in a better position to know where to focus their efforts on capacity building. Perhaps it's a foreign lab with poor internal controls or certain regions where surveillance is especially weak. These sorts of missions, where diplomatic initiatives are being fed by intelligence agencies, has been a hallmark of arms control efforts, nuclear nonproliferation, and nuclear inspection activities and a similar approach can help improve our biosecurity and surveillance as well.

When it has been proposed to public health officials that they should work more closely with intelligence services, and they express reluctance, the unwillingness often turns on their contempt of an episode where the CIA was reported to have administered a fake hepatitis B vaccine campaign in an effort to collect DNA from Osama bin Laden's children as a way to target his location in Pakistan in 2011.[81] A local physician, Shakil Afridi, was reported to have organized a vaccine drive, first in a poorer neighborhood, to appear "more authentic." Public health leaders denounced the scheme and its potential impact on legitimate vaccination campaigns, including the ongoing effort to eradicate polio.[82] Far from enabling these kinds of potential incidents, if public health experts collaborated more closely with our intelligence community it might help forestall such an approach. In such a case, the public health perspective may have been woven into operational planning early and closed down the idea of using a vaccination campaign as cover for intelligence gathering because of the impact that the tactics would have on overarching public health goals. The public health imperative could prevail, and intelligence operators might find a different way to pursue a mission that didn't risk undermining a legitimate vaccination effort.[83] Had public health officials been at the table, they might have discouraged such an effort at the outset. The more engaged intelligence services are with public health experts, the more they'll be cognizant of the lines that shouldn't be crossed in trying to balance security interests with the fundamental goals of public health. In May 2014, the Obama White House announced that the CIA would no longer use vaccination programs as a cover for espionage.[84]

COVID showed the importance of timely, reliable reporting. Consider two hypothetical scenarios that might have unfolded in the early days of the outbreak, when the virus was still localized within China. In the first scenario, US intelligence officials are briefing the president sometime around December 20, 2019. They tell the president that intelligence agencies are tracking a mysterious outbreak of pneumonia in Wuhan. The intelligence officials are relying largely on open-source material—information they've gleaned from WeChat and other social media. They don't know the characteristics of the virus and haven't established whether it's spreading person-to-person. However, they've found some troubling postings on social media. Doctors in China are worried. A few local sources have told them that hospitals are seeing a higher number of flu cases than is normal for this time of year, and emergency rooms are full. Taken together, US agencies believe it could be an outbreak with a novel virus.

Is that enough to instigate any decisive action in the US? Probably not.

Now consider the second scenario. Intelligence officials tell the president that they were made aware of an outbreak in Wuhan based on the same open-source material. Digging deeper, they've been told by local assets, who work inside Chinese medical facilities, that ten healthcare workers have already become infected with the virus. This is a crucial piece of information that strongly suggests the pathogen is capable of spreading between people. Additionally, they've obtained samples of the virus and sequenced it. The genetic diversity between the different specimens suggests that the virus has been spreading for more than a month. Moreover, comparing the novel virus to a database of known pathogens, they've assessed that the virus bears disturbing similarities to SARS-1 and is likely deadly and highly contagious.

If US officials had had access to that second briefing rather than the first, it could have instigated actions that would have given us a monthlong head start on COVID. We squandered a lot of opportunities to get an edge on the epidemic, so it's not clear that, at least in this instance, more time would have made a significant difference in our preparations. But in the hands of federal officials who are astute to these threats, a month's head start can be enormous.

The other piece of this puzzle, in addition to the clandestine capabilities, is the sequencing data. The COVID pandemic was the first time that scientists used sequencing at a massive scale to evaluate the evolution of a virus's pathogenicity and trace its spread. In the future, this practice will be an essential tool for doing routine global surveillance.[85] Leveraging these data as a way to get a head start on gauging a pathogen's pandemic potential will require a more reliable means to gain access to samples, and then to link the unique characteristics of a virus's sequence with clinical information that can predict how these features correlate with its transmissibility and lethality.

With the current technology it's hard to deduce from just a sequence alone the behavior of a virus. Still, some things can be inferred. For example, we might be able to surmise from the RNA sequence, combined with a knowledge of the structural biology, that a novel pathogen has the potential to bind well to human tissue. That can be a key insight that tips us off to its ability to cause disease. In the case of SARS-CoV-2, there was early evidence that it bound well to the ACE2 receptor in the human respiratory tract. Its sequence also resembled SARS-1. Those two findings gave some key insights that the novel strain might pose a serious danger. These insights could also help with the early development of treatment options or the creation of ways to identify spread through the monitoring of symptoms. And as we expand the quality and quantity of sequence data that we have—correlating different genetic features with how viruses behave, how contagious they are, and how likely they are to cause disease—then the usefulness of sequence information as a predictive tool will greatly expand. Given the enormous potential of this approach, we need to build on these capabilities.

Achieving this goal will require a massive amount of data to better connect key genetic features found in sequencing data with their clinical significance. More information will be needed on how viruses behave in animals and humans.[86] This is the biggest obstacle right now to using sequencing data to deduce a pathogen's severity and infectivity. We don't know precisely how small changes in a virus's genetic sequence will translate into differences in how it behaves once it spreads among people. But as sequencing data get more robust, as we collect and store more information to correlate genetic changes with their clinical significance, and as the ability to use computational methods to make these predictions continues to improve, so will the practical application of these tools. We need to look toward this future and invest in these capabilities.

The first time that sequencing was used to help guide the investigation and response to an epidemic was with Ebola in 2014. Owing largely to improvements in the technology for sequencing, scientists at the Broad Institute were able to use sequencing information derived from Ebola samples collected in the field in West Africa to trace patterns of spread. This information, in turn, helped public health workers better understand how the virus was being passed from person to person, and gave them new ways to intervene to break off the chains of transmission.[87] But COVID was a dramatic step forward. It was the first time scientists used sequencing at a massive scale, in a near real-time fashion, to trace the spread of individual cases, characterize an epidemic, and monitor its expansion.[88] The biology and the informatics will improve to the point that we'll be able to make better predictions based on sequence information alone. Even partially predictive assessments are valuable if they can be made early, when new strains and variants first emerge.

There are other advantages offered by these capabilities.

When a new cluster of infections first emerges, we'll be able to deduce whether the infections are the result of transmission between people or repeated introductions from a single animal source to different people. The latter scenario was the first theory advanced by China to explain the spread of COVID. It was wrong. For weeks the prevailing belief was that the initial infections were each the result of people interacting with an infected animal in a local food market and not the result of person-to-person spread. As we've seen, in the right hands, even a few weeks' head start on recognizing that the virus was spreading between people can give us a crucial edge at containing it.

The ability to use sequencing to monitor for emerging pathogens, and evaluate their potential harm, is not a new idea. It was envisioned by the National Strategy for Pandemic Influenza released in May 2006, but no federal agency has ever fully implemented this vision as a strategic priority.[89] Back then, the technology didn't exist to make such a framework practical. The advent of fast, accurate, and less-costly sequencing, along with better experimental evidence, finally makes it more obtainable. In 2007, the cost of sequencing a single human genome was about $10 million, according to the National Human Genome Research Institute; today it's under $1,000, and most predict the cost will soon fall below $100.[90] For sequencing a virus's genome, the costs are typically about one-tenth the cost for sequencing a human genome. In recent years, public health agencies have been incorporating pathogen genome sequencing into their infectious disease surveillance with support from the Advanced Molecular Detection program, which Congress established at the CDC in 2014.[91] The FDA also relies on its GenomeTrakr network, the first distributed network of labs that use whole genome sequencing for pathogen identification in instances of outbreaks of foodborne illness. However, most of the prior uses for sequencing were retrospective. The tools were used to conduct deeper analysis of outbreaks that had already been identified or resolved. Such work can be valuable because it allows researchers to uncover outbreaks that they may not have previously recognized, which can yield important insights into the risks posed by pathogens and how they're transmitted. However, until COVID, sequencing was never deployed in a large scale to identify the origins of a burgeoning outbreak and help contain its present spread. Information on viral genomes was never widely exchanged outside the setting of influenza. During COVID, the framework that enabled viral sequence data to be shared around the world was a system originally built for flu surveillance, the Global Initiative on Sharing Avian Influenza Data (GISAID). We now know that such a collaborative network can be used to exchange information on other viruses.

At proper scale, sequencing can be used in a near real-time fashion to identify emerging threats and to predict their activity.[92] At Harvard, a project is under way to create a sentinel surveillance system that could detect novel strains of viruses before they evolve into the next pandemic. Started in response to COVID, the idea is to deploy sequencing equipment in hot zones to allow novel pathogens to be evaluated when they make their first jump into man. By uncovering these dangers early, and sharing the information quickly, the hope is that such a system can serve as a tripwire that will give us the capability to head off the next pandemic.[93] The CDC also committed to sharply expanding its use of sequencing for sentinel surveillance in the US. The agency is looking for ways to integrate its surveillance into municipal and national planning, such as through sequencing wastewater. The CDC launched a project called the National Wastewater Surveillance System (NWSS) that is monitoring for variants of SARS-CoV-2, but also starting to look for new viruses that may be circulating in the population.[94] Because we shed many viruses into our feces, city wastewater can serve as a "liquid biopsy" to help inform us of emerging outbreaks.

It now appears that COVID may have been spreading widely in China, and had broken out of the country to regions with close ties to Wuhan, much earlier than we first suspected.[95] In Italy, the first known COVID case was reported in the town of Codogno in Lombardy on February 21, 2020, but since then, a few studies have suggested it may have been spreading earlier, including one study that found a positive sewage sample in northern Italy in mid-December 2019 and another that detected the novel coronavirus in a banked patient sample that was taken in the same region in early December. It was from the throat culture of a child first suspected of having measles.[96]

In France, another banked throat swab, this one from a patient with pneumonia who was admitted to a hospital on December 27, 2019, was later found to have SARS-CoV-2 RNA in it.[97]

In Brazil, testing of sewage found SARS-CoV-2-positive results in samples collected on November 27, 2019, much earlier than the first reported case in the Americas.[98]

If these and similar findings are true, the window to detect the virus's initial jump into people may have been open much longer, along with the opportunity to intervene to avert a global pandemic. Better surveillance, using sequencing tools, might have uncovered its spread earlier. In the future, using these approaches, we can build a system that alerts us sooner to when a new virus has gained the capacity for human transmission, so we can start expanding our surveillance, expanding our testing, and developing therapies and vaccines, much earlier.

The ability to turn raw sequence data into actionable, digital information is a complex and computationally intensive endeavor. It will turn on the availability of a well-curated and up-to-date database of sequences that can serve as a reference for comparing novel strains to known viruses, along with information on how genomic features correlate with the clinical behavior of different pathogens. COVID proved the importance of having these proficiencies. In addition to being able to identify new viruses, we'll need these capabilities to monitor known viruses and identify when they mutate and undergo a dangerous evolutionary change. We've already seen the appearance of variants that are more contagious and less susceptible to our antibody drugs and vaccines. The UK had in place a systematized approach for sequencing samples and was able to detect these variants in late 2020. The United States had no similar framework, and so we became heavily seeded by the time we started to look for and find these new strains.[99] It was an echo of what had happened in February and March 2020, when we lacked the PCR testing to detect the first wave of infection until it was too late.

These efforts should be expanded to focus on all RNA respiratory pathogens. As a category of pathogens, these viruses share a plethora of the basic features that give them pandemic potential.[100] RNA viruses mutate relatively quickly, and many, like influenza, are able to undergo a process known as antigenic drift, by which the virus is able to alter the surface antigens that are the targets of our antibodies—thus evading our existing immunity. Some viruses, like measles, cannot change their genomic sequence in ways that substantially alter enough of their surface proteins, so measles remains susceptible to our vaccines or the immunity that we get from prior infection. However, for viruses like influenza, as their surface proteins undergo change, the virus is able to dodge the protective antibodies that we've developed from past infection or vaccination.[101] This is how flu slips past our immunity every season and why we need to constantly update our vaccines. Being transmitted through respiratory secretions means that these viruses can spread widely. By having a global enterprise for sampling, sequencing, and describing the clinical features of RNA respiratory pathogens and the disease they cause, we can detect troubling variants when there's still time to target them with new antibody drugs or vaccines.

These same capabilities can help us identify the next novel virus—the new syndrome X. We need to encourage systematic sampling of viral specimens, and reporting to public databases. This needs to become a routine part of disease surveillance. Viruses don't have privacy rights, they aren't entitled to withhold their genetic information from public view, and they aren't the property of the country in which they first arise. Pathogens that have the potential to spread globally belong to the world and should be shared by scientists everywhere. The rapid distribution and sequencing of viral samples is going to become an increasingly important part of effective surveillance.[102]

Disease surveillance should be a global effort enabled through the voluntary cooperation of researchers in different countries and supported by our tools of national security. It can be layered onto the global network that already exists for flu surveillance. Effective monitoring for new threats is an exercise in viral sleuthing. We'll need the ability to identify clusters of unusual infection and get access to source strains. Where we can, we should try to build these capabilities on top of cooperative models and multilateral efforts already under way.

A good example of such an existing effort is the PREDICT program that was previously maintained by the US Agency for International Development (USAID). The goal was to seek out and catalog animal viruses that had the potential to threaten people. The program was focused on zoonotic risks—viruses collected from animals that had not yet made the jump to humans. The PREDICT program spent more than $200 million to train about five thousand scientists dispersed across the world, with many working in regions where new infections were most likely to emerge, including Africa and parts of Asia. The effort aimed to join the researchers together in a global hunt for new zoonotic diseases.[103] It helped build new capacities for monitoring threats in resource constrained nations.

Like a lot of our pandemic planning, the creation of this program was triggered by fears surrounding the emergence of the H5N1 bird flu in 2005.[104] Over its first ten years, PREDICT funded about sixty labs that became part of an ambitious global alliance to find viruses in animals that had the potential to leap into humans. Through this effort, a network of scientists collected and examined more than 140,000 biological samples from animals, identifying about 1,000 viruses that were never before cataloged. Among them was a new strain of Ebola.[105] In 2016, PREDICT helped find a virus in China that was closely related to the 2002 SARS-1 virus, but distinct enough that experimental vaccines developed against SARS-1 didn't seem to work against this novel virus.[106] Despite the usefulness of this global effort as a potential tripwire, US funding for PREDICT was allowed to expire in 2019, just weeks before COVID emerged, but was subsequently restored in the spring of 2020.[107]

The global cooperation that supported PREDICT is a model for how we might conduct disease surveillance at proper scale, and layer sequencing onto these efforts. To line up that global mission we first need to get our own house in order and align US domestic capabilities behind a more coordinated, coherent framework that elevates our public health response as a matter of our national security. Right now, pieces of the necessary capabilities—the task of monitoring the antigenic drift of known viruses and the task of identifying new ones and the mission to respond to these threats—are scattered across different agencies. The work of viral sleuthing for "spillover events" from zoonotic sources falls under USAID, which oversees PREDICT.[108] The CDC is charged with investigating outbreaks. NIH does research into emerging pathogens and the development of therapeutic countermeasures, but it shares these responsibilities with the FDA, BARDA, and the Defense Advanced Research Projects Agency (DARPA), and coordination among all of these agencies is far from seamless. Each agency pursues its own mission, agenda, and countermeasures. This entire enterprise needs to be expanded and better synchronized.

Viral detection is a matter of national security. So are the means to develop diagnostic tests to uncover new pathogens, infer their pathogenicity, and trace their origins. COVID proved that all of these functions are beyond the reach of the CDC alone, the agency that was historically charged with many of these obligations. We need the equivalent of a Joint Special Operations Command for biothreats—a coordinating entity that can bring together the capabilities of different agencies that contribute to this mission, that has an operational focus, that has a national security mind-set, and has the ability to work across the classified and unclassified world. In many respects, our lack of preparedness for a threat like COVID is not a lack of technology to deal with these threats, it's the absence of policy coordination and an integrated approach between the different components that had relevant capabilities and expertise to inform and advance our response.

COVID raised the stakes for making sure that we get these capabilities right. The pandemic proved that a way to create asymmetric harm to America is through the spread of a respiratory pathogen. While COVID likely emerged from a natural origin, and if it was the product of a lab, its release was not a deliberate act, some adversaries may be watching its disproportionate effect on the US and thinking that they can weather an engineered respiratory pathogen better than America can. It was once thought that a terrorist or nation-state would be unlikely to weaponize a pathogen that could easily blow back on them. COVID may have changed some of that calculus. Western democracies proved uniquely incapable of implementing respiratory precautions and defending from this threat. The US was one of the worst hit among nations. Terrorists along with more traditional nation-state adversaries may now see that respiratory pathogens are a poor man's nuclear weapon. Viruses are far more accessible than enriched uranium, and easier to engineer—but still capable of unleashing mass destruction. We need to remain focused on the risk from select agents and deliberate acts of bioterrorism, but given the poor response the US mounted to COVID and the outsize impact it had on our economy, we need to revise our planning and add respiratory viruses to the list of pathogens that rogue nations and terrorist groups could seek to use against us.[109]

When it comes to new threats arising from nature, the pathogens that demand our greatest focus are respiratory RNA viruses. Not just influenza, not just coronaviruses, but the entire category of respiratory pathogens that derive their genetic instruction set from RNA. The fact that these viruses replicate through RNA means that the pathogens can undergo rapid mutation when they face selective pressure and can adapt to threaten us in unforeseen ways or evade our drugs and vaccines. Something that spreads through inhalation is also hard to isolate and contain.[110]

Our planning must prepare against the threat from any one of the respiratory RNA viruses. We need to involve our national security agencies to monitor for these risks, alongside our public health agencies and our multilateral efforts to improve global surveillance. Intelligence activities are no substitute for multilateral conventions and capacity building in nations, but the two missions—the public health engagement guided by diplomacy, and the security efforts led by intelligence agencies—need to find more seamless and effective ways to share resources.

We also need to make our tools for responding to these threats more resilient. We need diagnostic panels that can test for a broad range of respiratory pathogens at the same time, and they should be deployed to sentinel surveillance sites that are paid to screen for unusual respiratory illnesses. We'll need to use sequencing to monitor for new viruses and evaluate known pathogens for evolutionary features that may make them more dangerous. We also need to invest in technologies for fashioning drugs and vaccines that can defeat entire classes of respiratory viruses. These approaches must allow us to adapt our countermeasures as new strains emerge. And we need to develop drugs that can work to defeat entire classes of viruses, like coronaviruses or influenzas, by disrupting some of the core features they use to replicate and spread.

Any respiratory RNA virus can emerge to create the next pandemic. One will. COVID won't be the last time that such a virus tries to overwhelm us.

CONCLUSION

Nobody knew how the coronavirus got into the White House. After all, every guest who arrived for the Rose Garden ceremony on September 26, 2020, to formally nominate Judge Amy Coney Barrett to the Supreme Court had to test negative before entering the compound.[1]

What "testing negative" meant had changed over time. The White House testing protocol had been a constant work in progress. At the beginning of the pandemic, the White House was testing only those staff members who were thought to have been exposed to the virus or were coming off quarantine. When the staff drove through the White House gates, they would get swabbed from their cars and undergo other screening for fever and symptoms. As the pandemic worsened and the risk increased, testing was broadened. Everyone who would be in the West Wing and might come into contact with the president was getting tested on a regular schedule. Then, after the president's valet developed COVID in May, the White House moved to daily testing that was done in the medical clinic in the Eisenhower Executive Office Building, the ornate building that's adjacent to the White House and located inside the White House compound.[2]

The Secret Service oversaw the logistics of the testing operation. There would be a line outside the medical clinic as people waited to get screened. Once a staff member tested negative, he or she could be cleared to come into work. When visitors came in to see the president, they too would be tested out of that clinic. However, you could still get inside the White House compound without being tested, so long as you didn't work in the West Wing. And people who worked in West Wing jobs where they didn't directly come into contact with the president were usually not tested unless they developed symptoms or had an exposure. The virus could find its way past the White House gates and into the compound from someone who worked there, or by someone who was visiting. Eventually, it did.

By the time of the Barrett event, the White House was relying on testing alone to keep COVID out. Inside the White House, there were no masks or distancing or other meaningful measures to stop the virus. For months, some White House staff had condoned an attitude of emphatic nonchalance about the pandemic, often mocking personal precautions as marks of weakness, taunting colleagues who wore masks, and criticizing public health measures as overly nervous.[3] Part of that boldness sprung not from outright defiance but misplaced confidence in their testing protocol. However, the tests they were using were imprecise, especially when they were being used to screen an asymptomatic population. On most days, that testing relied on a combination of the rapid molecular test from Abbott called the ID Now, and a rapid antigen test that the company had also developed called BinaxNow. The Secret Service were also being trained to use a PCR test, the Cepheid GeneXpert. It was a potpourri of testing. Different people would get tested with different systems with no discernible pattern as to who had received which test and why. For a while, the White House had relied heavily on the Abbott ID Now. Based partly on advice they were getting from the Department of Health and Human Services (which favored the use of antigen tests for asymptomatic and more routine screening), the White House was starting to shift more of the testing to the BinaxNow.[4]

It was a leaky protocol. The tests were not fit to the purpose for which they were being used. Without masks and other mitigation inside the compound, the White House needed an airtight testing system. They didn't have one. Not with the tests they were using. Data from the CDC would later show that the antigen test that the White House deployed could miss as many as two-thirds of asymptomatic infections.[5] Yet, the testing done at the White House gate was the compound's only layer of defense from the coronavirus. And on this day, that screening system may have let the virus through. Once inside, the pathogen could have hitched a ride on someone's breath, traveling to the West Wing, where it mingled in the air, perhaps at an indoor reception for the prospective justice and, from there, it may have found its way to the president.

A chain of transmission would likely be lit that spread to more than thirty people, including two US senators and former governor Chris Christie, who was in close proximity to Trump a few days later, during preparation for a presidential debate with then-former Vice President Joe Biden.[6] Given the timing of his illness, Trump would have been infected at the Barrett event or sometime that same week.

We'll never know for sure the full scope of the spread that was set in motion by the White House cluster.[7] The White House didn't make any visible effort to use contact tracing to isolate the source of the outbreak or to identify those who became infected through successive chains of transmission. High-level officials who attended the event told me they were never called by White House contact tracers.[8] A group of CDC epidemiologists were already detailed to the White House—they had been there for months to implement the White House testing protocols and assist with contact tracing when cases emerged inside the compound. However, these CDC staff were never fully engaged in trying to get to the bottom of the chain of transmission that was ignited that day.

It seemed clear that someone may have walked SARS-CoV-2 into the most secure space in the world and brought a deadly infection within breathing distance of the nation's leadership. It was a dangerous and deeply regrettable event that shouldn't have been allowed to occur. The virus shouldn't have been able to get that close to the president and put his life at risk. However, it was a little more than a month before the presidential election. By all public appearances, some in the White House didn't want to know where the virus came from or where it went. Reporters mused that Bloomberg's White House correspondent, Jennifer Jacobs, became the chief contact tracer through her successive articles and tweets identifying and linking the cases that emerged as part of the cluster.[9] Even if the White House engaged no outward signs that it was conducting robust contact tracing, everyone else wanted to know: Who was patient zero?

After all, the outbreak put the president's life in serious danger.

It turns out, two reporters for the *New York Times*, Michael Shear and Al Drago, were among those infected in the outbreak. So the newspaper set out to do some contact tracing of its own. With two of its own staff members carrying the strain that sickened people that day, the *Times* had what it needed to begin its sleuthing.

The newspaper turned to Trevor Bedford, the Seattle researcher who was refining the use of genomic sequencing as a way to track the spread of COVID. Bedford was pioneering a new scientific endeavor, genomic epidemiology, that marries the traditional tools of epidemiology with genomic information derived by sequencing the viral samples taken from thousands of different patients. Using the genetic fingerprints he uncovered, Bedford was able to construct a more reliable mosaic for how the virus was spreading around the globe. Given the timing of their infections, the two journalists believed they were exposed to the virus at the same time. By looking at the sequence of the two strains that they carried, the hope was that Bedford could help pinpoint the origin of this particular occurrence of the virus.

Shear and Drago were probably infected while covering events surrounding the Barrett ceremony at the White House. As the *Times* reported, the two journalists had repeated contact with White House officials over the period when the virus was believed to be coursing through the West Wing. However, neither man spent time near the other in the weeks before their positive tests. So it was thought that each had to have been infected by some unspecified third party. Shear had traveled with the president on Air Force One on September 26, the same day as the Barrett event, and he had interacted with Trump and members of the White House staff. Drago covered the same event from the Rose Garden.[10] Based on the timing of when their symptoms started, the two reporters could have been sickened by the same person who had also infected the president.

Bedford sequenced the virus isolated from both journalists to establish its origin. By comparing the two samples and identifying the subtle mutations that viruses acquire as they undergo replication and spread, Bedford was able to trace their viral lineage. It turned out that these were not ordinary strains. The two cases were indeed linked to a common source, in what appeared to be a single introduction by someone carrying a seldom seen strain of SARS-CoV-2.

By the end of 2020, there were about 250,000 SARS-CoV-2 genomes available in public databases. These were samples of the virus that had been sequenced by labs around the world and uploaded to a public repository of genomic data to help other researchers track the virus's spread and monitor how it was evolving. By comparing the sequence of the virus taken from the two *New York Times* reporters against the samples in this database, Bedford was able to conclude that the White House infections fell into a family of strains that first emerged in the US in April or May 2020 and were initially detected in Virginia and Michigan.[11] The viral ancestors of these two particular samples had spread to the US from Europe and were circulating widely in the early summer. But then the trail went cold. Scientists didn't see another instance of this particular variant until it reemerged in the White House, meaning it was probably spreading quietly, at a relatively low level, before it found its way to 1600 Pennsylvania Avenue.[12]

Genomic epidemiology, or "gen epi" as it's called in the field, has become a powerful tool, and Trevor Bedford helped advance the modern application of this approach as part of the global response to COVID. Bedford's work with the *New York Times* was a vivid, albeit small application of these methods for tracing patterns of transmission. When done at a larger scale, these could be powerful tools for evaluating the evolution of an epidemic and uncovering how a virus spreads across a population. As the repository of genomic information expands and our capacity to store and mine this information improves, along with the ability to sequence genetic material faster and more cheaply, it will accelerate our ability to read across viral genomes to identify novel pathogens and detect when a known virus has taken a dangerous evolutionary turn.

"Historians of science sometimes talk about new paradigms, or new modes of thought, that change our collective thinking about what is true or possible. But paradigms often evolve not just when new ideas displace existing ones, but when new tools allow us to do things—or to see things—that would have been impossible to consider earlier," observed science writer Jon Gertner. The advent of commercial genome sequencing, he says, is one such tool. Its application to the COVID pandemic proved its modern utility to transform whole fields of science.[13]

As computational power and storage capabilities improve, so will the precision of the sequencing techniques. Critics point out the shortcomings in using large-scale sequencing as a way to identify emerging pathogens, chief among them the limitations on what can be deduced from sequence data alone. However, as it becomes easier, less costly, and more routine to sequence respiratory samples; as we build a bigger global repository of this information and expand open-source platforms that allow these data to be easily shared; and as genetic features are correlated with more and better clinical and experimental evidence to give us added insight into how aspects of a virus's sequence change the way it behaves, the usefulness of these approaches will continue to quickly expand. A committee of the National Academy of Medicine, writing on steps to modernize our pandemic response, reflected on the application of gen epi to COVID, observing that "advances in the speed, granularity, affordability, and portability of genomic sequencing technologies have created transformative potential for widespread rapid genomic surveillance during infectious disease outbreaks, particularly when data from genomic sequencing are integrated with and analyzed alongside patient-based clinical and population-based epidemiological data."[14]

The public repository of sequences had made it possible to use the genetic features of the strain found at the White House, to track its migration around the globe and into the West Wing. With the virus that infected the president lifted off the two journalists and placed in those same public databases, scientists can now trace infections that have descended from the White House cluster.[15] If other people got sick as a result of the outbreak, and if scientists happen to sequence those subsequent infections, then they may be able to identify the future travels of the strain.

This ability to track the spread of a virus using sequencing information, to deduce insights about its risk based on its genetic features, and to use these data to derive fully synthetic drugs and vaccines on an expedited timetable—all these marked a new inception in the use of science to uncover biological risk and fashion ways to counter it. We've crossed a threshold between two states of science, entering a post-pandemic paradigm where genomic information will drive more of our public health response.

The emergence of SARS-CoV-2 and the pandemic that it caused marked a historical turning point in many aspects of our society and culture, and the application of genomics was one visible crossroad. The scientific advances we achieved in combating COVID gave us one of our few bright spots in this tragic event. The pandemic defined other historical inflection points, in other areas of our life, in addition to its impact on our use of science to advance public health. Among them was a turning point for how we must view our public health preparedness as a matter of national security and a crossroad for how we must address protracted social ills that leave us more vulnerable to health threats. COVID exposed wounds of inequity that plague our nation, and one hopes it will start a cycle of profound change for how we address them.

On the technology: With COVID, America embraced a genomically driven public health response, leaning heavily on gene-based approaches to track and trace the spread of SARS-CoV-2, target our mitigation, and develop synthetically derived drugs and vaccines directly from the sequence of the virus's genetic code. Bedford's integration of sequencing with traditional epidemiology helped redefine how outbreaks are traced, allowing us to more accurately reveal patterns of spread so we can better inform our tactics for breaking chains of transmission. It allowed us to identify and track new variants in the winter that posed heightened risks, and then develop vaccines and antibody drugs that target them.[16]

The speed by which companies like Pfizer, Moderna, Regeneron, Johnson & Johnson, and Eli Lilly were able to develop countermeasures to the SARS-CoV-2 threat owed in large part to this same genomic transformation. This scientific aptitude was one of the inspirations in the US response. We developed a safe and effective vaccine against a completely novel target in less than a year—a record time. Before COVID, the fastest that we had ever developed a vaccine was the original preparations against mumps, a process that took about four years.[17] The new speed turned largely on our ability to deploy novel approaches for developing vaccines, where the initial constructs were derived entirely from the virus's genome. This enabled us to use pieces of synthetic genetic material that could be made quickly available as the vaccine starting point. The application of these tools transformed the way we develop vaccines against a novel viral target. After COVID, these approaches will be routine.

On our public health system: COVID revealed dangerous gaps in the US public health preparedness, medical infrastructure, and healthcare system. We lacked the public health capacity and resiliency we thought we had. In the most advanced healthcare system in the world, we ran out of medical masks. We had to retrofit anesthesia machines and turn them into respirators. We didn't have enough swabs to collect samples from patients' noses.

Our system was set up well to handle singular, technology-intensive, and complex problems like developing a novel vaccine or antibody drugs. We do this better than anyone. But it faltered when we were faced with more mundane problems like manufacturing those vaccines in bulk, deploying testing centers, or making nose swabs to collect respiratory samples. When we finally developed safe and effective therapeutics and vaccines that could treat or prevent infection, we couldn't manufacture enough of them in time to supply the nation for the winter surge. We had to set up elaborate rationing schemes. Then, we were unable to establish an efficient distribution plan. Antibody drugs went unused because we couldn't deliver them. It took more than a month after the authorization of the first vaccine to begin vaccinating the 1.34 million residents of US nursing homes, where the most COVID deaths were occurring.[18] It took even longer to set up mass delivery sites to the general public. The vaccine was our only backstop against a relentless surge of infection in winter 2021, and we failed to amplify its timely use.

The CDC couldn't deploy a test to screen for the virus, allowing the nation to become heavily seeded with infection before it was detected. This was a historic failure that we would never overcome. By the fall, testing still couldn't keep up with demand. With millions of infections occurring in the US, the CDC didn't systematically collect and report information on the clinical outcomes. It didn't deploy sequencing as a tool to detect and evaluate new variants in time to uncover their spread, and it didn't use the tracking and tracing of sick patients to firmly establish the social compartments where spread was most likely to occur, or to identify the circumstances that were contributing to transmission. Public health authorities overestimated the role of fomites because collectively US agencies underestimated the impact of asymptomatic spread. The virus proved how underfunded our public health system really was.

On our social ills: COVID had a particularly devastating impact on communities of color in the United States. Infectious diseases often hit especially hard those communities that already lack access to good medical care and are overlooked or face bias. Black, Hispanic, and indigenous Americans bore a disproportionately heavy burden from the virus, as it crept through wounds of inequity and racism that have long plagued our society and our medical system. This increased vulnerability sprung from social factors that were made more glaring by this public health crisis. When we look at the unequal impact COVID had on communities of color, the epidemic revealed broader injustices we have to resolve.

This is one piece of a much larger problem of inequity and intolerance. Often Black and Hispanic communities had higher rates of COVID disease because of low income and related factors: overcrowded housing, a reliance on public transportation, the need to work in lower-wage but essential jobs interacting with the public. Many people couldn't do their work over Zoom or socially distance on their job site. They couldn't easily isolate themselves at home when they became infected, because they lived in crowded multigenerational homes where, if one member of a family got sick, others were made excessively vulnerable. Too many Americans didn't have the social equity at work to demand proper protective equipment and safer working conditions.

In many cases, Black and Hispanic Americans also faced delays in diagnosis of COVID, difficulties in getting access to testing, and more barriers in acquiring care. And when people from communities of color got sick, they also suffered higher rates of serious disease and death. Years of inequity in healthcare—poor access to medical services, distrust of the healthcare system, and sometimes discrimination in how care is delivered—as well as higher rates of chronic diseases like diabetes and hypertension, which also owed in part to the inequities in income, environment, and access to medical care, were all aggravated by COVID's advance. The virus struck hardest those who were most vulnerable. Some were vulnerable by age. Others were made vulnerable by a system that left them behind. Stopping the next pandemic will depend on making sure we address the reasons why some communities were more likely to experience suffering and death from COVID-19. The social unrest of the summer of 2020 was a defining moment that brought these inequities into sharp focus. The hope is that it will be a turning point in how we address them.

Our political union also turned out to be socially fragile. We fought over the collective action we needed to take to reduce spread, perpetuated by an often-false dichotomy between the exercise of liberty and the application of public health. Masks became one unfortunate flashpoint. However, some of that anger and opposition was born of understandable frustrations. The mitigation measures we took had unduly hard impacts on certain segments of society: small business owners who ran restaurants, parents who couldn't work unless their kids could go to school, kids who depended on schools for meals. Just as the virus would disproportionately sicken some segments of society, the steps taken to control infection would also excessively burden certain people. Yet here again, we failed to take special measures to help those most affected. Debate and division over some of these actions were understandable, but over others, like the wearing of masks, such rancor was inexcusable. We couldn't even agree on the easy stuff.

The pandemic also exposed weaknesses in the international conventions and scientific collaboration that were supposed to help safeguard us from a global pandemic. COVID crushed the global order that governed public health cooperation by revealing that the established conventions were unenforceable in a moment of crisis. Many thought that China had been chastened after SARS-1.[19] In response, nations signed agreements and created institutions, with a promise to share information about new contagions so countries would be alerted earlier to novel threats. When COVID appeared, those conventions crumpled. Whether early notice would have made a difference in the global response to COVID is debatable. In the US, we made plenty of our own mistakes, downplaying the risk of the threat for too long and failing to get testing in place despite ample warning. However, the proof that even after SARS-1, countries like China would still sidestep many of these international commitments, underscored a simple orthodoxy. In a moment of crisis, every country will follow its own self-interest.

We had already gotten a glimpse of this reality in 2009, during the H1N1 swine flu pandemic. Friendly nations like Canada and Australia refused to ship to the US vaccine that had been purchased by our government and manufactured at our behest but made at facilities in those two nations. Only after Canada had satisfied the needs of its own citizens did it allow American products to flow to the US for American patients.[20] With COVID, the same behaviors would reoccur. Italy exercised its export control regulations under European Union rules to block the shipment of 250,000 doses of the AstraZeneca vaccine that was bound for Australia, on the grounds that Australia had controlled its epidemic and was in less immediate need of the doses.[21] It was the first time that these EU regulations had ever been invoked.[22] A few weeks later, the EU implemented new emergency legislation that gave it broad powers to curb exports of COVID vaccines that were manufactured in facilities located in member nations. The rules hit Britain especially hard, since the country was far ahead of other European nations in vaccinating its population at the time and was dependent on shipments from other European countries to keep up its vaccination campaign.[23] The European Union was putting its interests first. The same tenets influenced US policy. In May 2021, even as the US developed a surplus of COVID vaccine—stockpiling as many as 100 million vaccine doses by the beginning of the month—the Biden administration wouldn't allow those doses to be shipped to

Brazil, which was experiencing an uncontrolled epidemic. We were vaccinating healthy and low-risk sixteen-year-olds in the US while hospitals in India and Brazil were running out of oxygen for patients hospitalized with the disease.

For those who still cling to the gauzy hope that nations will join hands to better identify and coordinate around global risks, the gloomy truth was revealed by the collective international response to COVID—the application of trade and travel barriers as a way to isolate the virus and the nations that hosted it; the nationalization of production facilities that made critical medical products; the deliberate withholding of drugs and equipment needed for the global response.

If anything, COVID normalized the breakdowns in a global order that it was presumed, perhaps naively, would protect us, just as COVID pierced our own perception of domestic resiliency, cooperation, and fortitude. By implementing global travel bans, we made it clear that any country that's host to a novel pathogen in the future will quickly find itself isolated. Our hasty decision to pull out of the WHO, and then rejoin it just as quickly, with no preconditions for its reform, hardened these lessons. No matter what steps we take to rebuild international alliances, the overarching lesson will be the one proven by the crisis. In the future, when even friendly nations become host to an outbreak of a new and worrisome respiratory pathogen, we shouldn't expect them to raise their hand so easily.

These harsh, practical lessons won't be quickly unwound by sanguine rhetoric about international cooperation and gratifying agreements signed in COVID's aftermath. Already, the Biden administration is focused on strengthening the International Health Regulations that govern global cooperation around outbreaks as a safeguard against future pandemics. It's a sound effort, and we need to improve our multilateral relationships and the capacity for sharing information and supporting nations that are host to outbreaks, but we shouldn't rely on these efforts alone.

The virus made clear that we'll need to fundamentally alter the way we approach all of these risks. If we don't, our society will remain excessively vulnerable. For starters, we'll have to lean much more on our intelligence agencies, and in a different fashion. International agreements alone haven't provided us with the information we need about emerging threats. There's little reason to believe they'll perform much better in the future. The devastation caused by the pandemic proved that these risks, and our preparedness for them, is a matter of national security on par with other threats. We're going to have to build the capacity to seek out the information we need to protect ourselves. Sometimes that will demand that we avail ourselves of the tools and tradecraft of our clandestine services. The challenge will be to maintain collaboration and multilateral efforts even as we turn more heavily toward intelligence services to guard against the risk of new contagions. We manage to walk this line in other nonmilitary areas of national security. We'll have to make a similar posture work when it comes to global public health.

Going forward, extending an approach to public health preparedness that views these risks through a lens of national security protection also means building more resilience into our public health infrastructure. It means expanding our healthcare system's capacity to deal with a crisis, particularly our ability to conduct massive diagnostic testing and to implement large-scale manufacturing of biologicals. It also means reexamining the global supply chain that has left us vulnerable to shortages when every country that we depend on was suddenly hoarding the same equipment. It means we need a no-regrets ethos when it comes to pandemic preparedness. There must be no repentances or recriminations if we overinvest in preparedness when a new pathogen emerges only to see the virus fizzle out rather than explode into the next pandemic.[24]

Finally, it means bringing back to the US more of the manufacturing of critical healthcare components and finished goods. To do this, we may have to pay higher prices for these parts and services to make sure that we maintain a reliable domestic capacity.

There's a reason why much of this manufacturing left the US in the first place. It was cheaper to produce overseas. But we can no longer afford the luxury of valuing low price for these goods above all else.

America was long perceived to be the country best prepared for a public health crisis of this magnitude, with our extensive healthcare system and our ability to quickly field advanced countermeasures. However, we ended up being one of the nations brought lowest by the virus, with continuous waves of infection that stalled our economy and claimed more than 600,000 of our fellow citizens' lives.

Learning from what went wrong, we have a chance to build a safer future. COVID was the worst pandemic in modern times. It won't be the last. Weak leadership exacerbated the pandemic's toll, but even with a stronger and more coordinated federal response, we were poorly prepared for this threat. We didn't anticipate that a novel coronavirus would cause a pandemic, but now, looking back, we should have been more attuned to these risks. SARS-1 and MERS were a brutal warning. Looking even further back, it's now believed that past pandemics dating back many years, before we had the capability to pinpoint a causative virus, where we had once presumed that the culprit was a novel strain of influenza, may have instead been the first entry of a novel coronavirus into human circulation.[25] More recently, we've feared that a deadly pandemic involving a novel strain of bird flu would sweep the globe. We've known that we were long overdue for such an event. COVID was a devastating global calamity, but it could have been a lot worse. Odds are the next pandemic will be. And odds are that it will be caused by something that we don't fully expect.

We had plenty of warning that we were unprepared for an event like COVID, despite the evident risks. Zika, Ebola, SARS-1, MERS, the swine flu, and countless other contagions all showed that the US was exposed in critical ways and lacked some key capabilities for an effective response. In many respects, we prepared for the wrong pathogen.

Now, our vulnerability is unmistakable.

In February 2020, just as COVID was spreading to Europe, there was a large outbreak of H5N1 bird flu in chickens located on farms just to the south of Wuhan. The pandemic event we long feared was always expected to be caused by a bird flu.[26] The next pandemic very well could be, perhaps a new variant of H5N1 or a novel form of the even more fearsome H7N9 strain.

With COVID, we turned a chapter in history, and now we are more aware of our strengths and frailties for countering these threats, and more mindful of the carnage that such events will inflict on our health, economy, and culture. "Life can only be understood backward; but it must be lived forwards," said the Danish philosopher Søren Kierkegaard. In the end we succeeded on the scientific part of the response but failed badly on its human aspects. COVID crushed us as a result, and it left our society permanently altered. What we learn from it, and how we change, will determine if we are better prepared for the next pandemic, or whether we remain just as vulnerable.

ACKNOWLEDGMENTS

I want to thank all the scientists, policy and public health experts, and government professionals who spent time with me, reviewed portions of this manuscript, and provided critical insights and feedback. Many of my former colleagues from the FDA, the Department of Health and Human Services, and the White House helped inform these pages. I also learned a great deal from the scientists and executives with whom I've had the privilege to work, at various life science companies. In many cases, I've preserved the discretion of individuals who have helped me develop the ideas and content on these pages, but I received a great deal of feedback and guidance as I researched this book and I'm grateful for all of it.

I was supported in this endeavor by the team at the American Enterprise Institute, led by its president, Robert Doar. I was greatly aided by the work of a talented researcher and colleague, Abigail Keller. I was also helped immensely by an outstanding assistant, Lisa Castro.

I had the good fortune to hone many of the threads of argument that I lay out in this book during television appearances on CNBC, MSNBC, NBC, and CBS News's *Face the Nation*. I am grateful to Margaret Brennan, Mika Brzezinski, John Dickerson, Sara Eisen, Wilfred Frost, Joe Kernen, Norah O'Donnell, Becky Quick, Stephanie Ruhle, Joe Scarborough, Shepard Smith, Andrew Ross Sorkin, Jake Tapper, Meg Tirrell, Chuck Todd, and Scott Wapner, among many others, as well as their executive teams and producers, including Kim Angle, Erin Barry, Shani Benezra, Michael Biette, Donna Burton, Jacqueline Corba, David Gernon, Mary Hager, Carol Joynt, Terry King, Max Meyers, Jesse Rodriguez, Andy Rothman, and Toby Taylor.

Paul Gigot, James Taranto, and Kate Bachelder gave me the opportunity to share many of the perspectives embedded in this book on the editorial pages of the *Wall Street Journal*, as did Mark Lasswell at the *Washington Post* and Dawn Kopecki at CNBC. I also owe deep thanks to my literary agent, Rafe Sagalyn at ICM, who first suggested that I write this book during the summer of 2020 and guided me through the process, and to my editor at HarperCollins, Eric Nelson, who helped shape the manuscript and provided invaluable feedback that refined the content and prose.

I've also benefited greatly from the editorial input of Paul Golob, who cast a skillful eye over the manuscript and shared his sharp insight and well-considered suggestions. I especially want to thank my former colleagues at the FDA who provided many important insights and inspired me with their dedication to our shared public health interests.

Many people reviewed portions of this book, some who chose to remain anonymous, but among them were Jerome Adams, Scott Becker, Trevor Bedford, Luciana Borio, David Boyer, Andrew Bremberg, Carlos del Rio, Shami Feinglass, Richard Hatchett, Leslie Kiernan, Yuval Levin, John Martin, Chris Mason, Mark McClellan, Stephen Ostroff, Kavita Patel, Caitlin Rivers, Veronique Rodman, Lowell Schiller, Joshua Sharfstein, Lauren Silvis, Michael Strain, Marc Thiessen, and Steve Usdin.

I was also inspired by the work of many great reporters, commentators, and thought leaders who chronicled this crisis and provided critical insights on the threats and how we could address them. The real-time exchange of information, often through social media, helped shape the nation's response and improve our collective outcomes. Where political leadership fell short, the informal networks forged between people with shared public health goals helped to identify what was going wrong in real time, and to light the way on how we could reduce death and disease and ease suffering. Our global response benefited from the close engagement of a broad community of experts.

Some of the first detailed reporting of the outbreak in Wuhan was from Helen Branswell of STAT News. She deserves special mention for alerting America with her insightful dispatches throughout this crisis. Among many others whose outstanding ideas, reporting, and commentary shaped and informed my own writing, and whose work is sometimes quoted in these pages, are: Yasmeen Abutaleb, Amesh Adalja, Lisa Alexander, Kristian Anderson, Stephanie Armour, Drew Armstrong, Frances Arnold, Samira Asma, Jeannie Baumann, Emily Baumgaertner, Guy Benson, Carl Bergstrom, Isaac Bogoch, Alexandre Bolze, Tom Bossert, Zachary Brennan, Robert Califf, Adam Cancryn, Michael Cannon, Tim Carney, Muge Cevik, Caroline Chen, Esther Choo, Nicholas Christakis, Jon Cohen, Jonathan Cohn, Rory Cooper, Michelle Fay Cortez, Stacey Cunningham, Josh Dawsey, Natalie Dean, Dan Diamond, Andrew Dunn, Katherine Eban, Lev Facher, Jeremy Farrar, Jeremy Faust, Phillip Febbo, Nicholas Florko, Tom Frieden, Atul Gawande, Walid Gellad, Julie Gerberding, Gregg Gonsalves, Lawrence Gostin, Celine Gounder, Denise Grady, Nathan Grubaugh, Youyang Gu, Sanjay Gupta, Maggie Haberman, Peggy Hamburg, Chris Hamby, Chris Hayes, Sue Desmond-Hellmann, Matt Herper, Emma Hodcroft, Eddie Holmes, Jared S. Hopkins, Peter Hotez, Thomas Inglesby, Jeneen Interlandi, Akiko Iwasaki, Ashish Jha, Sheila Kaplan, Sarah Karlin-Smith, Dhruv Khullar, Sarah Kliff, Jeremy Konyndyk, Florian Krammer, Josh Kraushaar, Krutika Kuppalli, Adam Kucharski, Kai Kupferschmidt, Martin Landry, Robert Langreth, Michael O. Leavitt, Larry Levitt, Benhur Lee, David Lim, Berkeley Lovelace Jr., Rich Lowry, Ian Mackay, Debora Mackenzie, Syra Madad, Marty Makary, Apoorva Mandavilli, Chris Mason, Megan McArdle, Laurie McGinley, Donald G. McNeil Jr., Carter Mecher, Rich Mendez, Michael Mina, Chris Mooney, Bob More, Farzad Mostashari, Siddhartha Mukherjee, Trevor Mundel, Christopher J. L. Murray, David Nakamura, Robert Nelsen, Jayne O'Donnell, Paul Offit, Saad Omer, Emily Oster, Michael Osterholm, Caitlin Owens, Sarah Owermohle, Ramesh Ponnuru, Saskia Popescu, Vinay Prasad, Tatiana Prowell, Carl Quintanilla, Megan Ranney, Angela Rasmussen, Christopher Rowland, Philip Rucker, Will Saletan, David Sanger, Nicole Saphier, Sally Satel, Ben Shapiro, David Shaywitz, Michael D. Shear, Marc Siegel, Nate Silver, Andy Slavitt, Tara Smith, Craig Spencer, Jennifer Steinhauer, Sheryl Gay Stolberg, Lena H. Sun, Teena Thacker, Katie Thomas, Derek Thompson, Eric Topol, Zeynep Tufekci, Bob Wachter, Noah Weiland, Nathaniel Weixel, Leana Wen, Lawrence Wright, Ed Yong, Eunice Yoon, Alison Young, and Carl Zimmer, among many others.

This list is by no means exhaustive, and I know I'm leaving out some important names of people who've informed me and helped me shape my own efforts to inform others about these events. I spoke with many leaders in the life science industry, and I'm grateful to all of the scientists and management teams that conferred with me during the crisis. I'm fortunate to work for great firms engaged in advancing innovation in life science, all of whom were deeply committed to the public health response to COVID; my colleagues at Pfizer, Illumina, Tempus Labs, Aetion, National Resilience, New Enterprise Associates, and the Mount Sinai Health System supported me through this effort, shared key insights, and gave me important feedback along the way. They encouraged me to engage broadly in the public dialogue around our COVID response and never once flinched through my many op-ed articles and television appearances. I'm fortunate to work with people and firms that are deeply committed to the nation's public health.

The stories of harrow and valor shared by my New York–based medical colleagues, who worked on the front lines of the brutal first wave of the American epidemic, stirred me to lay out some of the ways we can work collectively to make sure a contagion of this scale never happens again.

My greatest thanks go to my wife, Allyson, and my three wonderful daughters. They wanted to know why Daddy was always working. I promised to make it up to them.

NOTES

Introduction

1. Emma Goldberg, "Early Graduation Could Send Medical Students to Virus Front Lines," *New York Times*, March 26, 2020; and Emma Goldberg, *Life on the Line: Young Doctors Come of Age in a Pandemic* (New York: Harper Collins, 2021).
2. Dhruv Khullar, "'Adrenaline, Duty, and Fear': Inside a New York Hospital Taking on the Coronavirus," *New Yorker,* April 3, 2020; and Kasey Grewe, "Headlines Don't Capture the Horror We Saw," *Atlantic*, December 6, 2020.
3. Noah Higgins-Dunn, Berkeley Lovelace Jr., and Will Feuer, "FEMA Sends Refrigerator Trucks to NYC to Serve as Temporary Mortuaries for Coronavirus Victims," CNBC, March 30, 2020; and Michael Rothfeld et al., "13 Deaths in a Day: An 'Apocalyptic' Coronavirus Surge at an N.Y.C. Hospital," *New York Times*, April 14, 2020.
4. Alan Feuer and William K. Rashbaum, "'We Ran Out of Space': Bodies Pile Up as N.Y. Struggles to Bury Its Dead," *New York Times,* April 30, 2020.
5. Carl Campanile and Kate Sheehy, "NY Issues Do-Not-Resuscitate Guideline for Cardiac Patients amid Coronavirus," *New York Post,* April 21, 2020.
6. Catherine Graham, "Johns Hopkins Engineers Develop 3-D Printed Ventilator Splitters," Johns Hopkins University HUB, April 2, 2020, and Rothfeld et al., "13 Deaths in a Day."
7. Michelle A. Jorden et al., "Evidence for Limited Early Spread of COVID-19 within the United States, January–February 2020," *Morbidity and Mortality Weekly Report* 69, no. 22 (June 5, 2020): 680–4.
8. Gardiner Harris, "Bush Announces Plan to Prepare for Flu Epidemic," *New York Times*, November 2, 2005.
9. Monali C. Rahalkar and Rahul A. Bahulikar, "Lethal Pneumonia Cases in Mojiang Miners (2012) and the Mineshaft Could Provide Important Clues to the Origin of SARS-CoV-2," *Frontiers in Public Health* 8, no. 581569 (2020).
10. Editorial Board, "We're Still Missing the Origin Story of This Pandemic: China Is Sitting on the Answers," *Washington Post,* February 5, 2021; Jane Qiu, "How China's 'Bat Woman' Hunted Down Viruses from SARS to the New Coronavirus," *Scientific American,* June 1, 2020; and Peng Zhou et al., "Addendum: A Pneumonia Outbreak Associated with a New Coronavirus of Probable Bat Origin," *Nature* 588, no. E6 (2020).

11. It was reported that the doctor who cared for the stricken miners sent patient samples to the Wuhan Institute of Virology, a lab that's a focal point of coronavirus research in China. There, according to the thesis and subsequent reporting in the *New York Post* and an analysis posted online, scientists found that the source of infection may be a SARS-like coronavirus from a Chinese rufous horseshoe bat. See Isabel Vincent, "COVID-19 First Appeared in a Group of Chinese Miners in 2012, Scientists Say," *New York Post*, August 15, 2020; and Li Xu, "The Analysis of 6 Patients with Severe Pneumonia Caused by Unknown Viruses," master's thesis, First Clinical Medical College of Kunming Medical University, Kunming Medical University, Kunming, 2013.
12. Xing-Yi Ge et al., "Coexistence of Multiple Coronaviruses in Several Bat Colonies in an Abandoned Mineshaft," *Virologica Sinica* 31, no. 1 (2016): 31–40; Maciej F. Boni et al., "Evolutionary Origins of the SARS-CoV-2 Sarbecovirus Lineage Responsible for the COVID-19 Pandemic," *Nature Microbiology* 5, (2020): 1408–17; Jon Cohen, "Trump 'Owes Us an Apology.' Chinese Scientist at the Center of COVID-19 Origin Theories Speaks Out," *Science*, June 24, 2021; and Zeynep Tufekci, "Where Did the Coronavirus Come From? What We Already Know Is Troubling.," *New York Times*, June 25, 2021.
13. Xing-Yi Ge et al., "Isolation and Characterization of a Bat SARS-Like Coronavirus That Uses the ACE2 Receptor," *Nature* 503 (2013): 535–8.
14. Xing-Lou Yang et al., "Isolation and Characterization of a Novel Bat Coronavirus Closely Related to the Direct Progenitor of Severe Acute Respiratory Syndrome Coronavirus," *Journal of Virology* 90, no. 6 (2016).
15. Hongying Li et al., "Human-Animal Interactions and Bat Coronavirus Spillover Potential among Rural Residents in Southern China," *Biosafety and Health* 1, no. 2 (2019): 84–90.
16. Ning Wang et al., "Serological Evidence of Bat SARS-Related Coronavirus Infection in Humans, China," *Virologica Sinica* 33, no. 1 (2018): 104–7.
17. Christophe Fraser et al., "Factors That Make an Infectious Disease Outbreak Controllable," *Proceedings of the National Academy of Sciences* 101, no. 16 (2004): 6146–51.
18. Scott Gottlieb, "The Casualties That Never Arrived," *British Medical Journal* 323, no. 7314 (2001): 654.
19. Keith Bradsher and Liz Alderman, "The World Needs Masks. China Makes Them, but Has Been Hoarding Them.," New York Times, April 2, 2020; and Ed Silverman, "India Bans Exports of Drug Touted by Trump as Potential Covid-19 Treatment," STAT, March 25, 2020.
20. Dake Kang, Maria Cheng, and Sam McNeil, "China Clamps Down in Hidden Hunt for Coronavirus Origins," Associated Press, December 30, 2020.
21. Craig Mauger and Beth LeBlanc, "Trump Tweets 'Liberate' Michigan, Two Other States with Dem Governors," *Detroit News*, April 17, 2020; and Noah Higgins-Dunn and Lora Kolodny, "Trump Says He 'Totally Disagrees' with Georgia Gov. Kemp's Decision to Reopen Businesses in the Middle of Coronavirus Pandemic," CNBC, April 23, 2020.
22. Amy Davidson Sorkin, "At the White House, Trump Takes Off His Mask and Sends a Dangerous Message," *New Yorker*, October 6, 2020.

23. Scott Gottlieb, "Moderation on Masks Might Make More Get a Shot," *Wall Street Journal*, May 16, 2021.
24. Higgins-Dunn and Kolodny, "Trump Says He 'Totally Disagrees' with Georgia Gov. Kemp's Decision to Reopen Businesses in the Middle of Coronavirus Pandemic."
25. Daniel Stadlbauer et al., "Repeated Cross-Sectional Sero-Monitoring of SARS-CoV-2 in New York City," *Nature* 590 (2021): 146–50.
26. Rothfeld et al., "13 Deaths in a Day."
27. Scott Gottlieb and Yuval Levin, "The Trump Coronavirus Spread," *Wall Street Journal*, October 4, 2020.

Chapter 1: America the Vulnerable

1. SARS-1 was never officially declared a pandemic by the World Health Organization.

   James W. LeDuc and M. Anita Barry, "SARS, the First Pandemic of the 21st Century," *Emerging Infectious Diseases* 10, no. 11 (2004).
2. Lisa Schniming, "Wuhan nCoV Outbreak Quadruples, Spreads within China," CIDRAP News, January 19, 2020.
3. Walter Sim, "Japan Confirms First Case of Infection from Wuhan Coronavirus; Vietnam Quarantines Two Tourists," *Straits Times*, January 17, 2020.
4. World Health Organization (@WHO), "Preliminary Investigations Conducted by the Chinese Authorities Have Found No Clear Evidence of Human-to-Human Transmission of the Novel #coronavirus (2019-nCoV) identified in #Wuhan, #China," Twitter, January 14, 2020, 6:18 a.m.
5. Elliot Lefkowitz et al., "Virus Taxonomy: The Database of the International Committee on Taxonomy of Viruses (ICTV)," *Nucleic Acids Research* 46, no. D1 (2018).
6. US Centers for Disease Control and Prevention, "AFM Cases and Outbreaks," February 19, 2021.
7. Rebecca Ballhaus and Stephanie Armour, "Health Chief's Early Missteps Set Back Coronavirus Response," *Wall Street Journal*, April 22, 2020; and Yasmeen Abutaleb et al., "The U.S. Was Beset by Denial and Dysfunction as the Coronavirus Raged," *Washington Post*, April 4, 2020.
8. Javier Hernández, "Deadly Mystery Virus Reported in 2 New Chinese Cities and South Korea," *New York Times*, January 21, 2020.
9. Wright, "The Plague Year."
10. Alexandra Ma, "A Video of Medics in Hazmat Suits Scanning Plane Passengers for China's Mysterious Wuhan Coronavirus Is Stoking Fears of a Pandemic," *Business Insider*, January 21, 2020.
11. The video was retweeted January 19 by David Paulk, a journalist at Sixth Tone, a news site based in Shanghai. See David Paulk(@davidpaulk), "Passengers Screened for Coronavirus Symptoms on a Domestic Flight out of Wuhan," Twitter, January 19, 2020, 11:24 p.m..
12. Gao Yu et al., "In Depth: How Early Coronavirus Signs Were Spotted and Throttled," *Caixin*, March 3, 2020.
13. Lisa Schniming, "WHO Eyes Possible Sustained nCoV Spread in China," CIDRAP News, January 21, 2020.

14. Chaolin Huang et al., "Clinical Features of Patients Infected with 2019 Novel Coronavirus in Wuhan, China," *Lancet* 395, no. 10223 (2020): 497–506.
15. Based on email communications reviewed by the author.
16. Matthew Belvedere, "Trump Says He Trusts China's Xi on Coronavirus and the US Has It 'Totally Under Control,'" CNBC, January 22, 2020.
17. Dakin Andone, "US Travel Restrictions Go into Effect to Combat Coronavirus Spread," CNN, February 3, 2020.
18. World Health Organization, "Coronavirus," February 12, 2020; and World Health Organization, "Updated WHO Recommendations for International Traffic in Relation to COVID-19 Outbreak," February 29, 2020.
19. Sarah Owermohle and Adam Cancryn, "Top Health Officials Warn Senators More Wuhan Virus Cases Likely," *Politico*, January 24, 2020.
20. James Areddy and Liza Lin, "U.S. Working to Evacuate American Citizens From Epidemic-Stricken Chinese City," *Wall Street Journal*, January 25, 2020.
21. Areddy and Lin, "U.S. Working to Evacuate American Citizens From Epidemic-Stricken Chinese City."
22. European Centre for Disease Prevention and Control, "Outbreak of Acute Respiratory Syndrome Associated with a Novel Coronavirus, China: First Local Transmission in the EU/EEA—Third Update," January 31, 2020.
23. Scott Gottlieb, "What Must Be Done to Head Off the Coronavirus Threat," *Washington Post*, January 23, 2020.
24. Scott Gottlieb, "We Need to Prepare for US Outbreak of Wuhan Coronavirus," CNBC, January 27, 2020.
25. Luciana Borio and Scott Gottlieb, "Act Now to Prevent an American Epidemic," *Wall Street Journal*, January 28, 2020.

Chapter 2: Confusion and Subterfuge

1. Amanda Woods, "Shrimp Vendor at Wuhan Market May Be Coronavirus 'Patient Zero,'" *New York Post*, March 27, 2020.
2. Jeremy Page, Wenxin Fan, and Natasha Khan, "How It All Started: China's Early Coronavirus Missteps," *Wall Street Journal*, March 6, 2020.
3. *Paper*, "Looking for the First Infected Person in South China Seafood Market," March 25, 2020.
4. *China Daily*, "The Central Hospital of Wuhan," July 26, 2019.
5. Page, Fan, and Khan, "How It All Started."
6. Woods, "Shrimp Vendor at Wuhan Market May Be Coronavirus 'Patient Zero.'"
7. Page, Fan, and Khan, "How It All Started."
8. Novel Coronavirus Pneumonia Emergency Response Epidemiology Team, "The Epidemiological Characteristics of an Outbreak of 2019 Novel Coronavirus Diseases (COVID-19)—China, 2020," *China CDC Weekly* 2, no. 8 (2020): 113–22.
9. Page, Fan, and Khan, "How It All Started."
10. Gao Yu et al., "In Depth: How Early Coronavirus Signs Were Spotted and Throttled," *Caixin*, March 3, 2020.
11. Ibid.

12. Amos Zeeberg, "Piecing Together the Next Pandemic," *New York Times*, February 16, 2021; and Ellen C. Carbo et al., "Coronavirus Discovery by Metagenomic Sequencing: A Tool for Pandemic Preparedness," *Journal of Clinical Virology* 131, (2020): 104594.
13. Michael T. Osterholm and Mark Olshaker, *Deadliest Enemy* (New York: Little, Brown Spark, 2017).
14. Ibid; and Laura-Isobel McCall, Jair Siqueira-Neto, and James McKerrow, "Location, Location, Location: Five Facts About Tissue Tropism and Pathogenesis," *PLOS Pathogology* 12, no. 5 (2016): e1005519.
15. Ulrich Spengler and Jacob Nattermann, "Immunopathogenesis in Hepatitis C Virus Cirrhosis," *Clinical Science* 112, no. 3 (2007): 141–55.
16. Jennifer Tisoncik et al., "Into the Eye of the Cytokine Storm," *Microbiology and Molecular Biology Review* 76, no. 1 (2012): 16–32; and Alison George, "Cytokine Storm: An Overreaction to the Body's Immune System," *New Scientist*.
17. Yu et al., "In Depth: How Early Coronavirus Signs Were Spotted and Throttled."
18. Ibid.
19. Ibid; and Jon Cohen, "Mining Coronavirus Genomes for Clues to the Outbreak's Origins," *Science*, January 31, 2020.
20. Edward Wong, Julian E. Barnes, and Zolan Kanno-Youngs, "Local Officials in China Hid Coronavirus Dangers From Beijing, U.S. Agencies Find," *New York Times*, September 17, 2020.
21. Yu et al., "In Depth: How Early Coronavirus Signs Were Spotted and Throttled."
22. Kristin Huang, "Coronavirus: Wuhan Doctor Says Officials Muzzled Her for Sharing Report on WeChat," *South China Morning Post*, March 11, 2020.
23. Reporters Without Borders, "Whistleblowing Doctor Missing After Criticizing Beijing's Coronavirus Censorship," April 14, 2020.
24. Jianxing Tan, "新冠肺炎"吹哨人"李文亮: 真相最重要" ["Covid-19 "Whistleblower" Li Wenliang: the Truth is Most Important"], *Caixin*, January 31, 2020.
25. Lin Zehong, "'Whistleblower' of Wuhan Pneumonia: I Knew It Could Be 'Person-to-Person' Three Weeks Ago," *United Daily News*, January 31, 2020.
26. Michelle Chan, "Beijing Urges Calm over Mysterious Virus Ahead of Lunar New Year," *Nikkei Asia*, January 8, 2020.
27. Chan, "Beijing Urges Calm over Mysterious Virus Ahead of Lunar New Year."
28. Jeff Kao and Mia Shuang Li, "How China Built a Twitter Propaganda Machine Then Let It Loose on Coronavirus," ProPublica, March 26, 2020.
29. Masashi Crete-Nishihata et al., "Censored Contagion II: A Timeline of Information Control on Chinese Social Media during COVID-19," *Citizen Lab*, August 25, 2020.
30. On December 30, the health commission sent an "urgent notice" to all hospitals about the existence of "pneumonia of unclear cause"—but omitted any mention of SARS or a coronavirus. The health commission also ordered all health departments to immediately compile and report information about known cases.

    See Gerry Shih, Emily Rauhala, and Lena H. Sun, "Early Missteps and

State Secrecy in China Probably Allowed the Coronavirus to Spread Farther and Faster," *Washington Post*, February 1, 2020.

31. Drew Hinshaw, Jeremy Page, and Betsy McKay, "Possible Early Covid-19 Cases in China Emerge during WHO Mission," *Wall Street Journal*, February 10, 2021.
32. World Health Organization, "Strengthening Health Security by Implementing the International Health Regulations (2005): States Parties to the International Health Regulations (2005)," June 15, 2007.
33. Jane McMullen, "Covid-19: Five Days That Shaped the Outbreak," BBC, January 26, 2021.
34. US Centers for Disease Control and Prevention, "International Health Regulations (IHR)," August 19, 2019.
35. US Centers for Disease Control and Prevention, "International Health Regulations (IHR)," February 19, 2021; and Lawrence Gostin, "The International Health Regulations and Beyond," *Lancet Infectious Disease* 4, no. 10 (2004): 606–7.
36. Yu et al., "In Depth: How Early Coronavirus Signs Were Spotted and Throttled."
37. Du Juan, "Wuhan Wet Market Closes Amid Pneumonia Outbreak," *China Daily*, January 1, 2020.
38. National Health Commission of the People's Republic of China, "Life and Death at Wuhan's Jinyintan Hospital," April 2, 2020.
39. David Hui et al., "The Continuing 2019-nCoV Epidemic Threat of Novel Coronaviruses to Global Health—the Latest 2019 Novel Coronavirus Outbreak in Wuhan, China," *International Journal of Infectious Diseases* 91 (2020): 264–6.
40. By May, the head of China's CDC would reveal that testing of samples from the market had failed to show links between animals being sold in the market and the SARS-CoV-2, and that officials had ruled out the market as the origin of the virus. Dr. Gao Fu, director of the Chinese CDC, would say at the time that "It now turns out that the market is one of the victims," and was probably just an early site of one or more superspreading events.

    See James Areddy, "China Rules Out Animal Market and Lab as Coronavirus Origin," *Wall Street Journal*, May 26, 2020.
41. Jon Gertner, "Unlocking the Covid Code," *New York Times Magazine*, March 25, 2021.
42. Eddie Holmes (@edwardcholmes), "All, an initial genome sequence of the coronavirus associated with the Wuhan outbreak is now available at http://Virological.org here," Twitter, January 10, 2020, 8:08 p.m.
43. Zhuang Pinghui, "Chinese Laboratory That First Shared Coronavirus Genome with World Ordered to Close for 'Rectification,' Hindering Its Covid-19 Research," *Southern China Morning Post*, February 28, 2020, and Victoria Gill, "Coronavirus: Virus Provides Leaps in Scientific Understanding," BBC, January 10, 2021.
44. Lisa Schniming, "China Releases Genetic Data on New Coronavirus, Now Deadly," CIDRAP News, January 11, 2020, and Edward Holmes, "Novel 2019 Coronavirus Genome," *Virological*, January 10, 2020.
45. Catharine Paules, Hilary Marston, and Anthony Fauci, "Coronavirus

Infections—More Than Just the Common Cold," *JAMA* 323, no. 8 (2020): 707–8.

46. World Health Organization (@WHO), "BREAKING: WHO has received the genetic sequences for the novel #coronavirus (2019-nCoV) from the Chinese authorities. We expect them to be made publicly available as soon as possible," Twitter, January 11, 2020, 4:23 p.m.
47. Associated Press, "China Delayed Releasing Coronavirus Info, Frustrating WHO," June 2, 2020.
48. Selam Gebrekidan et al., "In Hunt for Virus Source, W.H.O. Let China Take Charge," *New York Times*, February 9, 2021.
49. World Health Organization, "World Health Organization Declares Coronavirus a Global Health Emergency," January 30, 2020.
50. Heritage Institute, "Virtual Event: One Year Later: Lessons from the Early COVID-19 Response," January 14, 2021.
51. Cohen, "Mining Coronavirus Genomes for Clues to the Outbreak's Origins."
52. World Health Organization, "Pandemic Influenza Preparedness Framework," May 24, 2011.
53. Emily Baumgaertner, "China Has Withheld Samples of a Dangerous Flu Virus," *New York Times*, August 27, 2018.
54. Heritage Institute, "Virtual Event: One Year Later: Lessons from the Early COVID-19 Response."
55. Shane Harris et al., "U.S. Intelligence Reports from January and February Warned About a Likely Pandemic," *Washington Post*, March 20, 2020.
56. Nick Paton Walsh, "The Wuhan Files: Leaked Documents Reveal China's Mishandling of the Early Stages of Covid-19," CNN, December 1, 2020, and Mi Liu et al., "Influenza Activity During the Outbreak of Coronavirus Disease 2019 in Chinese Mainland," *Biosafety and Health* 2, no. 4 (2020): 206–9.
57. Sanjay Gupta and Robert Kadlec, "CNN Special Report, COVID WAR: The Pandemic Doctors Speak Out," CNN, March 28, 2021.
58. Drew Hinshaw, Jeremy Page, and Betsy McKay, "Possible Early Covid-19 Cases in China Emerge During WHO Mission," *Wall Street Journal*, February 10, 2021.
59. Nick Paton Walsh, "The Wuhan Files."
60. Jonathan Pekar et al., "Timing the SARS-CoV-2 Index Case in Hubei Province," *Science* 372, no. 6540 (2021): 412–7.
61. Josephine Ma, "Coronavirus: China's First Confirmed Covid-19 Case Traced Back to November 17," *South China Morning Post*, March 13, 2020.
62. Jeremy Page and Drew Hinshaw, "China Refuses to Give WHO Raw Data on Early Covid-19 Cases," *Wall Street Journal*, February 12, 2021.
63. Charles Hutzler, "Mistaken Identity of Germ Culprit Cost the Chinese Time and Prestige," *Wall Street Journal*, June 4, 2003.
64. Debora Mackenzie, *The Pandemic That Never Should Have Happened, and-How to Stop the Next One* (New York: Hatchette Book Group, 2020).
65. Swissinfo, "A Swiss Lab Made the First Synthetic Clone of SARS-CoV-2," March 3, 2020.
66. *Renwu Magazine*, "The One Who Provided the Whistle," March 13, 2020.

67. Amy Qin and Javier C. Hernandez, "A Year After Wuhan, China Tells a Tale of Triumph (and No Mistakes)," *New York Times*, January 14, 2021.
68. Nicholas A. Christakis, *Apollo's Arrow* (New York: Little, Brown Spark, 2020); and Keng Jin Lee, "Coronavirus Kills Chinese Whistleblower Ophthalmologist," *American Academy of Ophthalmology*, February 10, 2020.
69. *People's Daily*, "Why Wuhan Central Hospital Has the Highest Rate of Medical Staff Infected with #COVID19," Facebook, March 11, 2020.
70. World Health Organization, "Emergencies: International Health Regulations and Emergency Committees," December 19, 2019.
71. Inside Justice, "Swine Flu: Legal Obligations and Consequences When the World Health Organization Declares a 'Public Health Emergency of International Concern,'" April 29, 2009.
72. Xinhua News Agency, "Xi Jinping Made Important Instructions on the Pneumonia Epidemic Caused by the New Coronavirus, Emphasizing That the Safety and Health of the People Should Be Put First, Resolutely Curbing the Spread of the Epidemic, Li Keqiang Issued Instructions," January 20, 2020, and Ban Gong Ting, "The Mayor of Shanghai: Effective Measures Have Been Taken to Prevent and Treat Pneumonia Caused by the New Coronavirus, and the Confirmed Cases Will Be Announced in Time," *Shanghai Observer*, January 20, 2020.
73. Se Young Lee and Lusha Zhang, "China Virus Spreads to U.S., Curbing Travel Plans and Spooking Markets," Reuters, January 20, 2020.
74. Associated Press, "How China Blocked WHO and Chinese Scientists Early in Coronavirus Outbreak," NBC News, June 2, 2020.
75. Tedros Adhanom Ghebreyesus, "WHO Director-General's Opening Remarks at the Media Briefing on COVID-19," March 11, 2020, https://www.who.int/director-general/speeches/detail/who-director-general-s-opening-remarks-at-the-media-briefing-on-covid-19---11-march-2020.
76. Shengje Lai et al., "Effect of Non-Pharmaceutical Interventions to Contain COVID-19 in China," *Nature* 585, no. 7825 (2020): 410–3.
77. Huaiyu Tian et al., "An Investigation of Transmission Control Measures during the First 50 Days of the COVID-19 Epidemic in China," *Science*, May 8, 2020, and Matteo Chinazzi et al., "The Effect of Travel Restrictions on the Spread of the 2019 Novel Coronavirus (COVID-19) Outbreak," *Science*, April 24, 2020.

Chapter 3: Pandemics as National Security Threats

1. The CDC had shifted a majority of its staff out of China earlier in the Trump administration because they had been working HIV/AIDS, and US support to China over decades on issues related to HIV had been judged successful, and so the CDC was gradually moving those personnel to other regions. In 2019 the US had fourteen CDC staff in China (three Americans and eleven local staff). At the start of the Trump administration, the US had eight Americans in China (and additional local staff). Most of the five Americans who were relocated were focused on HIV.
2. James Bandler et al., "Inside the Fall of the CDC," ProPublica, October 15, 2020.

3. Chaolin Huang et al., "Clinical Features of Patients Infected with 2019 Novel Coronavirus in Wuhan, China," *Lancet* 395, no. 10223 (2020): 497–506.
4. Abutaleb et al., "The U.S. Was Beset by Denial and Dysfunction as the Coronavirus Raged."
5. Michael Shear et al., "The Lost Month: How a Failure to Test Blinded the U.S. to Covid-19," *New York Times*, April 1, 2020.
6. Bandler et al., "Inside the Fall of the CDC."
7. Ibid.
8. Jennifer Hansler, Curt Merrill, and Isaac Yee, "The Many Times Trump Has Praised China's Handling of the Coronavirus Pandemic," CNN, May 19, 2020. That same day, I published an op-ed for CNBC stating that "the epidemic in China is likely much broader than official statistics currently suggest. A lot of mild cases probably remain unrecognized. Even a lot of severe cases are unreported since diagnostic tests were only recently deployed to the front lines of China's healthcare system. . . . Global spread appears inevitable. So too are the emergence of outbreaks in the U.S., even if a widespread American epidemic can still be averted." By this point, it was clear that China had lost control of its outbreak.

   See Gottlieb, "We Need to Prepare for US Outbreak of Wuhan Coronavirus."
9. Associated Press, "Timeline: China's COVID-19 Outbreak and Lockdown of Wuhan," January 22, 2021.
10. Harris et al., "U.S. Intelligence Reports from January and February Warned about a Likely Pandemic."
11. Heritage Institute, "Virtual Event: One Year Later: Lessons from the Early COVID-19 Response."
12. Brianna Ehley and Alice Miranda Ollstein, "Trump Announces U.S. Withdrawal from the World Health Organization," *Politico*, May 29, 2020.
13. Susan Jakes, "Beijing Hoodwinks WHO Inspectors," *Time*, April 18, 2003.
14. Wright, "The Plague Year."
15. Mackenzie, *The Pandemic That Never Should Have Happened*.
16. Nicholas King, "The Scale Politics of Emerging Diseases," *Osiris* 19, (2004): 62–76.
17. Will Feuer, "World Health Organization to Send Delegation to China to Help Combat Coronavirus Outbreak," CNBC, January 28, 2020.

Chapter 4: The Outbreak We Didn't Want to See

1. Michelle L. Holshue et al., "First Case of 2019 Novel Coronavirus in the United States," *New England Journal of Medicine* 382 (2020): 929–36.
2. Peter Robison, Dina Bass, and Robert Langreth, "Seattle's Patient Zero Spread Coronavirus Despite Ebola-Style Lockdown," Bloomberg Businessweek, March 10, 2020.
3. Holshue et al., "First Case of 2019 Novel Coronavirus in the United States."
4. Robert Costa and Philip Rucker, "Woodward Book: Trump Says He Knew Coronavirus Was 'Deadly' and Worse than the Flu While Intentionally Misleading Americans," *Washington Post*, September 9, 2020.

5. Wright, "The Plague Year."
6. University of Washington Medicine, "Paul S. Pottinger, MD, DTMH, FACP, FIDSA," https://www.uwmedicine.org/bios/paul-pottinger; and David Nakamura, Carol Leonnig, and Ellen Nakashima, "Matthew Pottinger Faced Communist China's Intimidation as a Reporter: He's Now at the White House Shaping Trump's Hard Line Policy Toward Beijing," *Washington Post*, April 29, 2020.
7. Matt Pottinger, "Return of SARS Sparks Concerns about Lab Safety," *Wall Street Journal*, April 26, 2004.
8. Costa and Rucker, "Woodward Book."
9. Wright, "The Plague Year."
10. Lisa Schenker and Jessica Villagomez, "First US Person-to-Person Case of Coronavirus Reported in Chicago. 'We Believe People in Illinois Are at Low Risk.,'" *Chicago Tribune*, January 30, 2020.
11. Isaac Ghinai et al., "First Known Person-to-Person Transmission of Severe Acute Respiratory Syndrome Coronavirus 2 (SARS-CoV-2) in the USA," *Lancet* 395, no. 10230 (2020): 1137–44.
12. Denise Grady, "Chicago Woman Is Second Patient in U.S. with Wuhan Coronavirus," *New York Times*, February 24, 2020.

    CDC's Dr. Nancy Messonnier, director of CDC's National Center for Immunization and Respiratory Diseases, stated on a press call on January 24, 2020: "To date, we have 63 of what we are calling patients under investigation or PUIs from 22 states. So far, only two have been confirmed positive and 11 tested negative. We anticipate by next week we'll begin regular reporting of case information on our website." See US Centers for Disease Control and Prevention, "Transcript of 2019 Novel Coronavirus Response," January 24, 2020.
13. Elvia Malagon, Lauren Zumbach, and Dawn Rhodes, "Chicago Woman Who Traveled to China Diagnosed with Coronavirus, Health Officials Say," *Chicago Tribune*, January 25, 2020.
14. Rob Barry, Joel Eastwood, and Paul Overberg, "Coronavirus Hit the U.S. Long Before We Knew," *Wall Street Journal*, October 8, 2020.
15. Washington et al, "Genomic Epidemiology Identifies Emergence and Rapid Transmission of SARS-CoV-2 B.1.1.7 in the United States," medRxiv, February 7, 2021, www.medrxiv.org/content/10.1101/2021.02.06.21251159v1.
16. Trevor Bedford, "Cryptic Transmission of Novel Coronavirus Revealed by Genomic Epidemiology," Bedford Lab, March 2, 2020, https://bedford.io/blog/ncov-cryptic-transmission/; Michael Worobey et al., "The Emergence of SARS-CoV-2 in Europe and North America," *Science* 370, no. 6516 (2020): 564–70, and Trevor Bedford et al., "Cryptic Transmission of SARS-CoV-2 in Washington State," *Science* 370, no. 6516 (2020): 571–5.
17. Sheri Fink and Mike Baker, "'It's Just Everywhere Already': How Delays in Testing Set Back the U.S. Coronavirus Response," *New York Times*, March 10, 2020.
18. The primers are for cDNA conversion and DNA amplification. The probes are for DNA detection (the cycle number is the Ct). The primers correspond to genomic segments that are found at the end of the RNA sequence that you're analyzing, and the probe usually corresponds to a segment found

somewhere in the middle of the specific RNA sequence that you are trying to detect with a diagnostic test.

19. US Centers for Disease Control and Prevention, "When You Can Be around Others after You Had or Likely Had COVID-19," February 11, 2021, https://www.cdc.gov/coronavirus/2019-ncov/if-you-are-sick/end-home-isolation.html.
20. Advisory Board, "Coronavirus Tests Are Extremely Sensitive. (That Could Be a Problem, Experts Say)," September 1, 2020, https://www.advisory.com/daily-briefing/2020/09/01/covid-tests.
21. Apoorva Mandavilli, "Your Coronavirus Test Is Positive: Maybe It Shouldn't Be.," *New York Times*, January 19, 2021, Michael Mina, Roy Parker, and Daniel Larremore, "Rethinking Covid-19 Test Sensitivity—a Strategy for Containment," *New England Journal of Medicine* 383, no. 120 (2020): e120, and Michael Tom and Michael Mina, "To Interpret the SARS-CoV-2 Test, Consider the Cycle Threshold Value," *Clinical Infectious Diseases* 71, no. 16 (2020): 2252–4.
22. US Centers for Disease Control and Prevention, "Interim Guidance on Duration of Isolation and Precautions for Adults with COVID-19," February 13, 2021, https://www.cdc.gov/coronavirus/2019-ncov/hcp/duration-isolation.html.
23. Shatavia S. Morrison, "COVID-19 Genomic Epidemiology Toolkit: Module 1.2," US Centers for Disease Control and Prevention, March 12, 2021, https://www.cdc.gov/amd/pdf/slidesets/ToolkitModule_1.2.pdf.
24. Amelia Swift, Roberta Heale, and Alison Twycross, "What Are Sensitivity and Specificity?," *Evidence-Based Nursing* 23 (2020): 2–4.
25. Diego Lozano and Miguel Cantero, "Difference between Analytical Sensitivity and Detection Limit," *American Journal of Clinical Pathology* 107, no. 5 (1997): 619.
26. Amy Maxmen, "The Race to Unravel the Biggest Coronavirus Outbreak in the United States," *Nature*, March 7, 2020.
27. Fink and Baker, "'It's Just Everywhere Already.'"
28. Matt Markovich, "Seattle Flu Study Researchers Defy Federal, State Guidelines to 'Save Lives,'" KOMO News, March 11, 2020.
29. "Additional Cases of COVID-19 in Washington State," Washington State Department of Health press release, February 28, 2020, https://www.doh.wa.gov/Newsroom/Articles/ID/1103/Additional-Cases-of-COVID-19-in-Washington-State.
30. Chu et al., "Early Detection of Covid-19 Through a Citywide Pandemic Surveillance Platform."
31. Maxmen, "The Race to Unravel the Biggest Coronavirus Outbreak in the United States."
32. Fink and Baker, "'It's Just Everywhere Already.'"
33. Chu et al., "Early Detection of Covid-19 Through a Citywide Pandemic Surveillance Platform."
34. Ibid.
35. Eric Lipton et al., "The C.D.C. Waited 'Its Entire Existence for This Moment.' What Went Wrong?," *New York Times*, June 3, 2020.

36. Benedict Carey and James Glanz, "Hidden Outbreaks Spread Through U.S. Cities Far Earlier Than Americans Knew, Estimates Say," *New York Times*, July 6, 2020.
37. World Health Organization, "Report of the WHO-China Joint Mission on Coronavirus Disease 2019 (COVID-19)," February 28, 2020.
38. Alex Azar, "The President's FY 2021 Budget," testimony before the Committee on Finance, US Senate, February 13, 2020, and US Centers for Disease Control and Prevention, "Transcript for CDC Media Telebriefing: Update on COVID-19," February 14, 2020, https://www.cdc.gov/media/releases/2020/t0214-covid-19-update.html.html.
39. The author reviewed Draft of the CDC COVID-19 Surveillance Enhancement Plan dated February 14, 2020.
40. Lawrence Wright, *The Plague Year* (New York: Knopf, 2021), page 111.
41. Among the pan-respiratory panels that are typically ordered in these circumstances are tests manufactured by Luminex and BioFire. Among the viruses that the BioFire panel can screen for using a single test are Adenovirus, Coronavirus HKU1, Coronavirus NL63, Coronavirus 229E, Coronavirus OC43, Severe Acute Respiratory Syndrome Coronavirus 2 (SARS-CoV-2), Human Metapneumovirus, Human Rhinovirus/Enterovirus, Influenza A, Influenza A/H1, Influenza A/H3, Influenza A/H1-2009, Influenza B, Parainfluenza Virus 1, Parainfluenza Virus 2, Parainfluenza Virus 3, Parainfluenza Virus 4, and Respiratory Syncytial Virus.

    See Biofire Diagnostics, "Potential Reimbursement CPT Codes," 2019, https://docs.biofiredx.com/wp-content/uploads/2019/03/FLM1-MKT-0119-FilmArray-CPT-Code-Brochure-Insert-RGB.pdf.
42. Jason Horowitz, "Behind the Curve: The Lost Days That Made Bergamo a Coronavirus Tragedy," *New York Times*, February 2, 2021.
43. US Food and Drug Administration, "FDA Statement on the FDA's Ongoing Investigation into Valsartan and ARB Class Impurities and the Agency's Steps to Address the Root Causes of the Safety Issues," January 25, 2019, https://www.fda.gov/news-events/press-announcements/fda-statement-fdas-ongoing-investigation-valsartan-and-arb-class-impurities-and-agencys-steps; and US Food and Drug Administration, "FDA Statement on FDA's Ongoing Investigation into Valsartan Impurities and Recalls and an Update on FDA's Current Findings," August 30, 2019, https://www.fda.gov/news-events/press-announcements/fda-statement-fdas-ongoing-investigation-valsartan-impurities-and-recalls-and-update-fdas-current.

Chapter 5: Looking for Spread in the Wrong Places

1. US Centers for Disease Control and Prevention, "U.S. Influenza Surveillance System: Purpose and Methods," October 6, 2020, https://www.cdc.gov/flu/weekly/overview.htm.
2. The CDC's current best estimate on asymptomatic spread is 40 percent, with a lower bound of 10 percent and an upper bound of 70 percent. See US Centers for Disease Control and Prevention, "COVID-19 Pandemic Planning Scenarios," September 10, 2020, https://www.cdc.gov/coronavirus/2019-ncov/hcp/planning-scenarios.html.

Other studies similarly show that asymptomatic persons are believed to account for approximately 40–45 percent of SARS-CoV-2 infections, and they can transmit the virus to others for an extended period, perhaps longer than fourteen days. See Daniel Oran and Eric Topol, "Prevalence of Asymptomatic SARS-CoV-2 Infection: A Narrative Review," *Annals of Internal Medicine* 173, no. 5 (2020): 362–7.

Notably, in one studied conducted in Vo, Italy, 42.5 percent (95 percent CI: 31.5–54.6 percent) of the confirmed SARS-CoV-2 infections detected across the two surveys were asymptomatic. See Enrico Lavazzo et al., "Suppression of a SARS-CoV-2 Outbreak in the Italian Municipality of Vo," *Nature* 584, (2020): 425–29; and Roberto Pastor-Barriuso et al., "Infection Fatality Risk for SARS-CoV-2 in Community Dwelling Population of Spain: Nationwide Seroepidemiological Study," *British Medical Journal* 371 (2020): m4509.

3. US National Institutes of Health, "Covid Treatment Guidelines: Overview of COVID-19," December 17, 2020, https://www.covid19treatmentguidelines.nih.gov/overview/.
4. Jason Leopold, "NIH FOIA Anthony Fauci Emails," Buzzfeed, June 1, 2021. NBC News (@NBCNews), "WATCH: 'The best estimates now of the overall mortality rate for COVID-19 is somewhere between 0.1% and 1%,' Adm. Brett Giroir, assistant secretary for health at HHS, says. 'That's lower than you heard probably in many reports . . . it's not likely in the range of 2-3%,'" Twitter, March 6, 2020, 7:53 a.m.
5. US Centers for Disease Control and Prevention, "FluView Weekly Surveillance Reports," February 15, 2020, https://www.cdc.gov/flu/weekly/pastreports.htm.
6. US Centers for Disease Control and Prevention, "CDC Guidance for Expanded Screening Testing to Reduce Silent Spread of SARS-CoV-2," January 21, 2021, https://www.cdc.gov/coronavirus/2019-ncov/php/testing/expanded-screening-testing.html.
7. Sharon Begley, "Trump Said More Covid-19 Testing 'Creates More Cases': We Did the Math," STAT, July 20, 2020.
8. CSPAN, "Coronavirus News Conference," January 28, 2020, https://www.c-span.org/video/?468647-1/hhs-secretary-azar-update-us-response-coronavirus.
9. Jasper Fuk-Woo Chan et al., "A Familial Cluster of Pneumonia Associated with the 2019 Novel Coronavirus Indicating Person to Person Transmission: A Study of a Family Cluster," *Lancet* 395, no. 10223 (2020): 514–23.
10. Andy Slavitt, *Preventable: The Inside Story of How Leadership Failures, Politics, and Selfishness Doomed the U .S. Coronavirus Response* (New York: St. Martin's Press, 2021), pages 104–5.
11. Luciana Borio and Scott Gottlieb, "Why Does the U.S. Have So Few Confirmed Coronavirus Cases?," *Wall Street Journal*, February 20, 2020.
12. Meg Tirrell and Wilfred Frost, "Coronavirus Cases in US and Europe Confirmed," CNBC, February 21, 2020.
13. Jayne O'Donnell, "Top Disease Official: Risk of Coronavirus in USA is 'Minuscule'; Skip Mask and Wash Hands," *USA Today*, February 19, 2020.

14. Rob Barry, Joel Eastwood, and Paul Overberg, "Coronavirus Hit the U.S. Long Before We Knew," *Wall Street Journal*, October 8, 2020.
15. Scott Gottlieb (@ScottGottliebMD), "U.S. officials express a lot of confidence that if #COVID19 was spreading we'd see it in our epi surveillance. We should be hopeful. But providers and front line personnel should still have caution, awareness that community spread could become apparent in U.S. Be safe. Be alert.," Twitter, February 22, 2020, 10:25 a.m.
16. US Centers for Disease Control and Prevention, "Transcript for the CDC Telehbriefing Update on COVID-19," February 26, 2020, https://www.cdc.gov/media/releases/2020/t0225-cdc-telebriefing-covid-19.html; and Yasmeen Abutaleb and Damian Paletta, *Nightmare Scenario: Inside the Trump administration's Response to the Pandemic That Changed History* (New York: HarperCollins, 2021).
17. Brett Murphy and Letitia Stein, "Coronavirus Response Delayed Despite Health Officials' Private Alarm," *USA Today*, January 26, 2021. On February 29, CDC director Robert Redfield said at a White House press briefing: "The American public needs to go on with their normal lives. Okay? We're continuing to aggressively investigate these new community links. . . . But at this stage, again, the risk is low."

    See C-SPAN, "President Trump with Coronavirus Task Force Briefing," February 29, 2020, https://www.c-span.org/video/?469892-1/president-trump-holds-news-conference-amid-us-death-coronavirus.
18. Carter Mecher, Senior Medical Advisor for the Department of Veterans Affairs, in discussion with the author, August 16, 2020.
19. New York City Department of Health and Mental Hygiene, "Influenza Surveillance Report: Week Ending March 7, 2020 (Week 10)," March 7, 2020, https://www1.nyc.gov/assets/doh/downloads/pdf/hcp/weekly-surveillance03072020.pdf.
20. Carl Zimmer, "Most New York Coronavirus Cases Came From Europe, Genomes Show," *New York Times*, April 8, 2020.
21. Steve Eder et al., "430,000 People Have Traveled from China to U.S. Since Coronavirus Surfaced," *New York Times*, April 15, 2020.
22. Ibid.
23. Andrew Chatzky, "China's Belt and Road Gets a Win in Italy," Council on Foreign Relations, March 27, 2019, https://www.cfr.org/in-brief/chinas-belt-and-road-gets-win-italy.
24. Silvia Aloisi, "Insight: Italy's Chinese Garment Workshops Boom as Workers Suffer," Reuters, December 29, 2013.
25. Bandler et al., "Inside the Fall of the CDC."
26. Greg Miller, Josh Dawsey, and Aaron Davis, "One Final Viral Infusion: Trump's Move to Block Travel from Europe Triggered Chaos and a Surge of Passengers from the Outbreak's Center," *Washington Post*, May 23, 2020; BBC, "Coronavirus: Trump Suspends Travel from Europe to US," March 12; and Philip Rucker and Ann Gearan, "Besieged Trump Announces Europe Travel Ban in Effort to Stem Coronavirus Pandemic," *Washington Post*, March 11, 2020.
27. Wright, *The Plague Year*, page 91.

28. L. Hufnagel, D. Brockmann, and T. Geisel, "Forecast and Control of Epidemics in a Globalized World," *Proceedings of the National Academy of Sciences* 101, no. 42 (2004): 15124–9.
29. Neil Ferguson et al., "Strategies for Mitigating an Influenza Pandemic," *Nature* 442 (2006): 448–52.
30. Gina Samaan et al., "Border Screening for SARS in Australia: What Has Been Learnt?," *Medical Journal of Australia* 180, no. 5 (2004): 220–3.
31. David M. Bell et al., "Public Health Interventions and SARS Spread, 2003," *Emerging Infectious Diseases* 10, no. 11 (2004): 1900–6.
32. Patricia C. Priest et al., "Thermal Image Scanning for Influenza Border Screening: Results of an Airport Screening Study," *Proceedings of the National Academy of Sciences* (2011).
33. David Bell, "Public Health Interventions and SARS Spread, 2003," *Emerging Infectious Diseases* 10, no. 11 (2004): 1900–6; and Christakis, *Apollo's Arrow.*
34. Associated Press, "US Will End Current Health Screening of Some Travelers," September 10, 2020.
35. Ian Duncan, "More than 1,000 TSA Employees Have Tested Positive for Coronavirus," *Washington Post*, July 9, 2020.
36. Sandi Doughton, "New Analysis May Rewrite the History of Washington State's Coronavirus Outbreak," *Seattle Times*, May 26, 2020.

Chapter 6: The Zika Misadventure

1. CDC grows the virus, then sends to NIH's partners BEI and UTMB for further growth, processing, and controlled distribution.
2. Shawn Boburg et al., "Inside the Coronavirus Testing Failure: Alarm and Dismay among the Scientists Who Sought to Help," *Washington Post*, April 3, 2020.
3. Michelle Jorden et al., "Evidence for Limited Early Spread of COVID-19 within the United States, January–February 2020," *Morbidity and Mortality Weekly Report* 69, no. 22 (2020): 680–4.
4. US Centers for Disease Control and Prevention, "2009 H1N1 Pandemic Timeline," May 8, 2019, https://www.cdc.gov/flu/pandemic-resources/2009-pandemic-timeline.html.
5. US Centers for Disease Control and Prevention, "The 2009 H1N1 Pandemic: Summary Highlights, April 2009-April 2010," June 16, 2010, https://www.cdc.gov/h1n1flu/cdcresponse.htm.
6. David Willman, "The CDC's Failed Race Against Covid-19: A Threat Underestimated and a Test Overcomplicated," *Washington Post*, December 26, 2020.
7. Ibid.
8. In the US, only laboratories certified under the Clinical Laboratory Improvement Amendments to perform high complexity testing can develop and run their own tests.
9. Willman, "The CDC's Failed Race Against Covid-19: A Threat Underestimated and a Test Overcomplicated."
10. "Facilitating the Use of Medical Countermeasures in an Emergency: Project

BioShield establishes the Emergency Use Authorization (EUA) to provide access to the best available medical countermeasures following a Declaration of Emergency by the Secretary of Health and Human Services. The Declaration could be based on either the Secretary's determination of a public health emergency with the significant potential to affect national security, or on a heightened risk of a CBRN attack on the public or U.S. military forces (as determined by the Secretary of Homeland Security or the Secretary of Defense, respectively)."

See US Food and Drug Administration, "MCM-Related Counterterrorism Legislation: Project BioShield Act of 2004," January 5, 2021, https://www.fda.gov/emergency-preparedness-and-response/mcm-legal-regulatory-and-policy-framework/mcm-related-counterterrorism-legislation#bioshield.

11. US Food and Drug Administration, "Determination of a Public Health Emergency and Declaration that Circumstances Exist Justifying Authorizations Pursuant to Section 564(b) of the Federal Food, Drug, and Cosmetic Act, 21 U.S.C. § 360bbb-3," February 4, 2020, https://www.fda.gov/media/135010/download.
12. The national PHE declaration was on January 31. The section 564 determination was on the same day as the EUA. They are two separate declarations; the 564 determination follows the PHE declaration. As stated by HHS, "Please note: a determination under section 319 of the Public Health Service Act that a public health emergency exists, such as the one issued on January 31, 2020, does not enable FDA to issue EUAs. On February 4, 2020, the HHS Secretary determined that there is a public health emergency that has a significant potential to affect national security or the health and security of United States citizens living abroad, and that involves the virus that causes COVID-19."

    See US Food and Drug Administration, "Emergency Use Authorization," April 2, 2021, https://www.fda.gov/emergency-preparedness-and-response/mcm-legal-regulatory-and-policy-framework/emergency-use-authorization.
13. Rev.com, "CDC Official Testimony Transcript to Senate on Coronavirus," March 3, 2020, https://www.rev.com/blog/transcripts/cdc-officials-testimony-transcript-to-senate-on-coronavirus; and US Centers for Disease Control and Prevention, "Previous Testing Data Aug 27, 2020," August 27, 2020, https://www.cdc.gov/coronavirus/2019-ncov/cases-updates/previous-testing-in-us.html.

    In response to a question from ABC News anchor George Stephanopoulos, HHS Secretary Alex Azar would state, "Well, we've already tested over 3,600 people here in the United States. I'm not sure what he meant when he said there's no lab kit, because we, with historic speed, the CDC developed a lab test."

    See ABC, "'This Week' Transcript 3-1-20: Joe Biden, Sen. Bernie Sanders, Alex Azar," March 1, 2020, https://abcnews.go.com/Politics/week-transcript-20-joe-biden-sen-bernie-sanders/story?id=69320081; US Centers for Disease Control and Prevention, "Previous Testing Data Aug 27, 2020"; and Knvul Sheikh, "U.S. Plans 'Radical Expansion' of Coronavirus Testing," *New York Times*, March 3, 2020.

14. Anne Schuchat, "An Emerging Disease Threat: How the U.S. Is Responding to COVID-19, the Novel Coronavirus," testimony before the Health, Education, Labor, and Pensions Committee, US Senate, March 3, 2020.
15. US Centers for Disease Control and Prevention, "Previous Testing Data Aug 27, 2020."
16. Mackenzie, *The Pandemic That Never Should Have Happened.*
17. World Health Organization, "The History of Zika Virus," February 7, 2016, https://www.who.int/news-room/feature-stories/detail/the-history-of-zika-virus.
18. Nuno Rodrigues Faria et al., "Zika Virus in the Americas: Early Epidemiological and Genetic Findings," *Science* 352, no. 6283 (2016): 345–9.
19. US Government Accountability Office, "Emerging Infectious Diseases: Actions Needed to Address the Challenges of Responding to Zika Virus Disease Outbreaks," May 2017, https://www.gao.gov/assets/690/684835.pdf.
20. Paul S. Mead, Susan L. Hills, John T. Brooks, "Zika Virus as a Sexually Transmitted Pathogen," *Current Opinion in Infectious Diseases* 31, no. 1 (2018): 39–44.
21. David William, "Lessons Unlearned: Four Years before the CDC Fumbled Coronavirus Testing, the Agency Made Some of the Same Mistakes with Zika," *Washington Post*, July 4, 2020.
22. US Centers for Disease Control and Prevention, "Trioplex Real-Time RT-PCR Assay: Instructions for Use," April 6, 2017, https://www.cdc.gov/zika/pdfs/trioplex-real-time-rt-pcr-assay-instructions-for-use.pdf.
23. Alison Young, "CDC Lab Shipped Virus without Following Key Safety Steps, *USA Today*, October 1, 2016.
24. Lena H. Sun, "CDC Whistleblower Claims Agency Has Been Using Wrong Zika Test," *Washington Post*, September 28, 2016.
25. US Centers for Disease Control and Prevention, "Transcript for CDC Telebriefing: Updates on Zika Response Efforts," March 10, 2016, https://www.cdc.gov/media/releases/2016/t0310-zika.html; and Thomas Frieden, "CDC 24/7: On the Front Lines of America's Health Defense," testimony before the House Appropriations Subcommittee on Labor, Health and Human Services, Education and Related Agencies, US House of Representatives, March 23, 2016.
26. Sun, "CDC Whistleblower Claims Agency Has Been Using Wrong Zika Test."
27. Robert S. Lanciotti, "Response to Investigative Team Explanation," *Washington Post*, September 15, 2016, https://osc.gov/Documents/Public%20Files/FY16/DI-16-3709/DI-16-3709%20Whistleblower%20Comments.pdf; and Sun, "CDC Whistleblower Claims Agency Has Been Using Wrong Zika Test."
28. Robert Lanciotti et al., "Rapid Detection of West Nile Virus from Human Clinical Specimens, Field-Collected Mosquitoes, and Avian Samples by a TaqMan Reverse Transcriptase-PCR Assay," *Journal of Clinical Micobiology* 38, no. 11 (2000): 4066–71.
28. US Government Accountability Office, "Emerging Infectious Diseases."
30. US Food and Drug Administration, "Fact Sheet for Healthcare Providers:

Interpreting Trioplex Real-Time RTPCR Assay (Trioplex rRT-PCR) Results," March 1, 2017.

31. Susan E. Sharp, Charles E. Hill, and Alexandra Valsamakis, "ASM, AMP, and PASCV Express Concerns on New Molecular Test for Zika," American Society for Microbiology, October 17, 2016, https://asm.org/Articles/Policy/ASM,-AMP-AND-PASCV-EXPRESS-CONCERNS-ON-NEW-MOLECUL.
32. The supplement authorization that was granted, (1) clarified the volume of lysis buffer preferred for use with the authorized extraction instrument, (2) allowed the addition of a Singleplex reaction option for the Trioplex test, and (3) clarified the expected positive control values/ranges in the Trioplex test.
33. William, "Lessons Unlearned: Four Years Before the CDC Fumbled Coronavirus Testing, the Agency Made Some of the Same Mistakes with Zika."
34. Sun, "CDC Whistleblower Claims Agency Has Been Using Wrong Zika Test."
35. Ibid; and Lanciotti, "Response to Investigative Team Explanation."
36. Sun, "CDC Whistleblower Claims Agency Has Been Using Wrong Zika Test."
37. US Government Accountability Office, "Emerging Infectious Diseases."
38. Ibid.
39. Sun, "CDC Whistleblower Claims Agency Has Been Using Wrong Zika Test"; and Lanciotti, "Response to Investigative Team Explanation."

Chapter 7: The CDC Fails

1. US Food and Drug Administration, "Accelerated Emergency Use Authorization (EUA) Summary ORIG3N 2019 Novel Coronavirus (COVID-19) Test (ORIG3N, INC.)," April 4, 2020, https://www.fda.gov/media/136873/download; and Ruth McBride, Marjorie van Zyl, and Burtram Fielding, "The Coronavirus Nucleocapsid Is a Multifunctional Protein," *Viruses* 6, no. 8 (2014): 2991–3018.
2. Willman, "The CDC's Failed Race Against Covid-19."
3. David Willman, "The CDC's Failed Race Against Covid-19: A Threat Underestimated and a Test Overcomplicated," *Washington Post*, December 26, 2020.
4. Harriet Ryan, Paige St. John, and Xinlu Liang, "The Surprising Story of the Salesman Who Became L.A.'s First Known COVID-19 Patient," *Los Angeles Times*, August 21, 2020; Scott Becker, executive director of the Association of Public Health Laboratories, in conversation with the author; and author reviewed the notes from Scott Becker, conference call, January 28, 2020.
5. Dina Temple-Raston, "CDC Report: Officials Knew Coronavirus Test Was Flawed but Released It Anyway," National Public Radio, November 6, 2020; and Bandler et al., "Inside the Fall of the CDC."
6. Stephanie Armour et al., "What Derailed America's Covid Testing: Three Lost Weeks," *Wall Street Journal*, August 18, 2020; and Boburg et al., "Inside the Coronavirus Testing Failure."
7. World Health Organization, "WHO Director-General's Opening Remarks at the Media Briefing on 2019 Novel Coronavirus," February 6, 2020,

https://www.who.int/director-general/speeches/detail/who-director-general-s-opening-remarks-at-the-media-briefing-on-2019-novel-coronavirus.

The decision to purchase and distribute the test was made by WHO's procurement center and it was believed to have been done independent of any formal scientific review by the WHO. The German manufacturer did not initially submit data to the WHO to receive an emergency use listing. The WHO procurement center opted to purchase the test without these steps, which they can do in the circumstance of a public health emergency.

8. Wright, *The Plague Year,* page 60.
9. Wright, "The Plague Year."
10. Willman, "The CDC's Failed Race Against Covid-19."
11. Matthew M. Hernandez et al., "Molecular Evidence of SARS-CoV-2 in New York before the First Pandemic Wave," *Nature Communications* 12, (2021).
12. David Lim, "Problems with CDC Coronavirus Test Delay Expanded U.S. Screening," *Politico,* February 20, 2020.
13. US Centers for Disease Control and Prevention, "Transcript for CDC Telebriefing: Update on COVID-19," Feburary 21, 2020, https://www.cdc.gov/media/releases/2020/t0221-cdc-telebriefing-covid-19.html; and US Centers for Disease Control and Prevention, "Transcript for CDC Telebriefing: Update on COVID-19."
14. Jesse McKinley, Luis Ferré-Sadurní, and Christina Goldbaum, "Cuomo Pledges $40 Million to Combat Coronavirus," *New York Times,* February 28, 2020.
15. The primers and probe kits are manufactured under ISO 13485:2016 conditions. Each lot of IDT primer and probe kits undergoes quality control testing by the CDC. Once that qualification is received, those lots can be used as a key component of the CDC Emergency Use Authorization protocol noted above. Company documents.
16. FEMA Advisory, "Coronavirus (COVID-19) Pandemic: International Reagent Resource," April 13, 2020, https://www.ntciss.org/wp-content/uploads/2020/04/4-FEMA_Advisory_COVID19_FactSheet_IRR_20200413.pdf.
17. Staff reporter, "Integrated DNA Technologies Authorized under CDC EUA to Provide SARS-CoV-2 Test Kits," Genome Web, March 3, 2020.
18. David Willman, "CDC Coronavirus Test Kits Were Likely Contaminated, Federal Review Confirms," *Washington Post,* June 20, 2020.
19. Alexander Borst et al., "False-Positive Results and Contamination in Nucleic Acid Amplification Assays: Suggestions for a Prevent and Destroy Strategy," *European Journal of Clinical Microbiology & Infectious Diseases* 23, no. 4 (2004): 289–99.
20. Greg Smith et al., "A Simple Method for Preparing Synthetic Controls for Conventional and Real-Time PCR for the Identification of Endemic and Exotic Disease Agents," *Journal of Virological Methods* 135, no. 2 (2006): 229–34.
21. Bandler et al., "Inside the Fall of the CDC."
22. Willman, "The CDC's Failed Race Against Covid-19."
23. David Willman, "Contamination at CDC Lab Delayed Rollout of Coronavirus Tests," *Washington Post,* April 18, 2020; and *Washington Post,* "Sum-

mary of the Findings of the Immediate Office of the General Counsel's Investigation Regarding CDC's Production of COVID-19 Test Kits," June 20, 2020.

24. Stephanie Armour et al., "What Derailed America's Covid Testing: Three Lost Weeks," *Wall Street Journal*, August 18, 2020.
25. Self-dimerization occurs when some portion of short stretch of genomic material like RNA or DNA (called an oligonucleotide) is complementary to itself, resulting in a piece of genomic material that can hybridize to another stretch of genomic material of the exact same sequence.
26. *Washington Post*, "Summary of the Findings of the Immediate Office of the General Counsel's Investigation Regarding CDC's Production of COVID-19 Test Kits."
27. Ibid; and Willman, "CDC Coronavirus Test Kits Were Likely Contaminated, Federal Review Confirms."
28. The CDC ultimately made the virus available to test manufacturers through BEI Resources, a quasi-governmental entity established by the National Institutes for Allergy and Infectious Diseases to provide access to priority pathogens that require special handling. Under the existing system, in the setting of an emerging pathogen, early clinical specimens from the CDC are sent to NIAID, whose two partners, the BEI Resources Repository and the University of Texas Medical Branch, grow the virus and are able to make live virus, viral RNA, and inactivated virus available with close oversight from NIAID. The NIAID ultimately approves where the virus and viral material can be sent. It can be a slow process. The alternative to using live or inactivated virus was to use synthetically derived genomic material. However, in some cases, using synthetically derived material rather than real virus can lead to an overestimation of a test's performance, so the preference is to use real virus that's grown, typically inactivated, and then shared with labs for the express purpose of helping them validate the performance of their tests. Initially, BEI told the government it would take two to three months to grow and inactivate the virus. But the University of Texas said they could have a limited supply of extracted RNA available by February 14. FDA's career staff identified 10 priority test developers with cleared platforms that were already installed widely in the US who could make immediate use of the samples to help advance tests they were already working on. It was not until February 20 that NIH contacted these ten developers, along with one identified by BARDA, to inform them of the availability of the samples.
29. From May 1, when the CDC initially fielded the H1N1 test through September 1, 2009, the CDC shipped only about 1,000 kits to 120 domestic and 250 international public health laboratories. The clinical market needed millions of more tests than they got from commercial manufacturers. The CDC only supplied a small fraction of the market and only for select public health labs. See US Centers for Disease Control and Prevention, "The 2009 H1N1 Pandemic: Summary Highlights, April 2009–April 2010," June 16, 2020, https://www.cdc.gov/h1n1flu/cdcresponse.htm.
30. "Secretary Azar Declares Public Health Emergency for United States for 2019 Novel Coronavirus," US Department of Health and Human Services

press release, January 31, 2020, https://www.hhs.gov/about/news/2020/01/31/secretary-azar-declares-public-health-emergency-us-2019-novel-coronavirus.html; and US Food and Drug Administration, "Determination of a Public Health Emergency and Declaration that Circumstances Exist Justifying Authorizations Pursuant to Section 564(b) of the Federal Food, Drug, and Cosmetic Act, 21 U.S.C. § 360bbb-3," February 4, 2020, https://www.fda.gov/media/135010/download.

31. Scott Gottlieb (@ScottGottliebMD), "THREAD: A brief review of Catch-22 when it comes to #Coronavirus and need for more diagnostic screening capability. Hospitals can roll out RT-PCR based test CDC developed. They all have Roche systems to run these tests. The technology is fairly straightforward. Here's the rub 1/9," Twitter, February 2, 2020, 1:49 p.m.
32. On February 2, in the long thread on Twitter that starts in the previous note, I continued, "Hospitals could advance these tests as Laboratory Developed Tests or LDTs, meaning they develop them in house. But since HHS declared a public health emergency related to #Coronavirus, hospitals are now expected to get FDA permission to use their own test." But there was a straightforward way that the problem could be addressed, I said. "#FDA and #CDC can allow more labs to run the RT-PCR tests starting with public health agencies. Big medical centers can also be authorized to run tests under EUA," and "For now, they're not permitted to run the tests, even though many labs can do so reliably." See Scott Gottlieb (@ScottGottliebMD), Twitter, February 2, 2020, 1:49 p.m.
33. In a political act that reflected the consternation of HHS political officials, who saw the FDA regulatory requirements as part of the reason why fewer tests were available, this policy was revoked in a statement issued by HHS in August. US Department of Health and Human Services, "Rescission of Guidances and Other Informal Issuances Concerning Premarket Review of Laboratory Developed Tests," August 19, 2020, https://www.hhs.gov/coronavirus/testing/recission-guidances-informal-issuances-premarket-review-lab-tests/index.html.
34. Kathleen McLaughlin, "HHS Relaxed Oversight of Problematic Covid-19 Tests Despite Being Told of Accuracy Concerns," STAT, November 2, 2020.
35. The letter read: "On behalf of our nation's state and local public health laboratories (PHLs), we are writing to urge you to consider enforcement discretion to allow this select group of governmental laboratories the ability to create and implement a laboratory developed test (LDT) for the detection of SARS-CoV-2 (COVID-19)." See *Washington Post*, "Letter from APHL to FDA Commissioner Stephen Hahn," February 24, 2020, https://www.washingtonpost.com/context/letter-from-aphl-to-fda-commissioner-stephen-hahn/547133a7-6e13-4995-b14d-c3e0748e90b8/; and Boburg et al., "Inside the Coronavirus Testing Failure."
36. Scott Gottlieb (@ScottGottliebMD), "THREAD ON DIAGNOSTICS: In recent public health emergencies there's been a stepwise approach to ramp up U.S. testing capacity. CDC develops test early in outbreak, then pursuant to Emergency Use Authorization (EUA) from FDA CDC makes test available to network of public health labs," Twitter, February 24, 2020, 9:09 a.m.

37. "Coronavirus (COVID-19) Update: FDA Issues New Policy to Help Expedite Availability of Diagnostics," US Food and Drug Administration press release, February 29, 2020, https://www.fda.gov/news-events/press-announcements/coronavirus-covid-19-update-fda-issues-new-policy-help-expedite-availability-diagnostics.
38. Robinson Meyer and Alexis Madrigal, "The Dangerous Delays in U.S. Coronavirus Testing Haven't Stopped," *Atlantic*, March 9, 2020.
39. Mayor Bill DeBlasio (@NYCMayor), "One of the greatest challenges in fighting COVID-19 is the lack of fast federal action to increase testing capacity. We need the CDC to send more kits. We need the FDA to fast track approval for testing methods developed by private institutions. And we need it NOW.," Twitter, March 5, 2020, 6:54 p.m.
40. Mayor Bill DeBlasio (@NYCMayor), "We need the FDA to approve faster, more efficient testing. The faster the results, the faster we can limit the spread.," Twitter, March 6, 2020, 9:35 p.m.
41. Mayor Bill DeBlasio (@NYCMayor), "To the federal government: we need you to do the simplest thing in the world and have the FDA approve automated COVID-19 tests. We're bringing back hundreds of results. We could be bringing thousands. The FDA could do it with the stoke of a pen TODAY.," Twitter, March 9, 2020, 6:34 p.m.
42. Angelica LaVito, "Facebook, Google and Other Tech Giants Meet with Federal Regulators to Tackle Online Opioid Sales," CNBC, June 27, 2018.
43. Ashley May, "Romaine Lettuce Linked to E. Coli Scare Likely Came from California: FDA Commissioner," *USA Today*, November 23, 2018.
44. Ibid.
45. WholeFoods Magazine Staff, "Lettuce Producers Agree on Transparency," WholeFoods Magazine, November 26, 2018, https://wholefoodsmagazine.com/news/main-news/lettuce-producers-agree-on-transparency/.
46. Andrew Cuomo, *American Crisis* (New York: Crown Books, 2020), page 63.
47. New York submitted their EUA request for their Wadsworth Lab test on February 28, and it was authorized on February 29. The FDA had convinced the CDC to give the New York State Department of Health a right of reference to the CDC's data on the agency's test design. This would allow New York State to generate just a limited amount of data with their new test to gain authorization to use it with patients, basically proving that it matched the performance data from the CDC's own test in what is called a bridging study. With this data, the state would be able to seek authorization to do their own testing using New York State's laboratory developed test.
48. "Coronavirus (COVID-19) Update: FDA Provides More Regulatory Relief During Outbreak, Continues to Help Expedite Availability of Diagnostics," US Food and Drug Administration press release, March 16, 2020, https://www.fda.gov/news-events/press-announcements/coronavirus-covid-19-update-fda-provides-more-regulatory-relief-during-outbreak-continues-help.
49. Caroline Chen, Marshall Allen, and Lexi Churchill, "Internal Emails Show How Chaos at the CDC Slowed the Early Response to Coronavirus," ProPublica, March 26, 2020.
50. Ibid.

51. Gerry Shih et al., "As Deadly Coronavirus Spreads, U.S. to Expand Screening of Passengers from China at 20 Airports," *Washington Post*, January 27, 2020.
52. Meyer and Madrigal, "The Dangerous Delays in U.S. Coronavirus Testing Haven't Stopped."
53. Brett Murphy and Letitia Stein, "Coronavirus Response Delayed Despite Health Officials' Private Alarm," *USA Today*, January 26, 2021.
54. Tim Dickinson, "The Four Men Responsible for America's COVID-19 Test Disaster," *Rolling Stone*, June 2020.
55. Murphy and Stein, "Coronavirus Response Delayed Despite Health Officials' Private Alarm."
56. This policy would be changed in a footnote included in the CDC's policy statement on March 8, 2020: "For healthcare personnel, testing may be considered if there has been exposure to a person with suspected COVID-19 without laboratory confirmation. Because of their often extensive and close contact with vulnerable patients in healthcare settings, even mild signs and symptoms (e.g., sore throat) of COVID-19 should be evaluated among potentially exposed healthcare personnel." See US Centers for Disease Control and Prevention, "Updated Guidance on Evaluating and Testing Persons for Coronavirus Disease 2019 (COVID-19)," March 8, 2020.
57. The CDC ultimately made the virus available to test manufacturers through BEI Resources.

    The University of Texas said they could have a limited supply of extracted RNA available by February 14.
58. US Government Accountability Office, "Zika Supplemental Funding: Status of HHS Agencies' Obligations, Disbursements, and the Activities Funded," May 2018, https://www.gao.gov/assets/gao-18-389.pdf.
59. Stephen Hahn, "Remarks by FDA Commissioner Stephen Hahn to the American Clinical Laboratory Association," March 4, 2020, https://www.fda.gov/news-events/speeches-fda-officials/remarks-fda-commissioner-stephen-hahn-american-clinical-laboratory-association-03042020.
60. Jayne Byakika-Tusiime et al., "Steady State Bioequivalence of Generic and Innovator Formulations of Stavudine, Lamivudine, and Nevirapine in HIV-Infected Ugandan Adults," *PLOS One* 3, no. 12 (2008); World Health Organization, "WHO Statement on Removal of Two AIDS Medicines from List of Prequalified Products," June 17, 2004, https://www.who.int/mediacentre/news/statements/2004/statement_aidsprequal/en/; and Lancet Editorial Board, "The Important World of Drug Prequalification," *Lancet* 364, no. 9448 (2004).
61. C-SPAN, "Vice President Pence Meets with Diagnostic Lab CEOs," March 4, 2020, https://www.c-span.org/video/?470004-1/vice-president-pence-meets-diagnostic-lab-ceos.
62. Scott Gottlieb (@ScottGottliebMD), "1/n On #coronavirus testing: The nations big clinical labs meet in Washington today for a convention and their CEOs are in town meeting with federal elected leaders. This is their moment. They ought to step up. They may be judged by what they now do.," Twitter, March 4, 2020, 6:31 a.m.

Chapter 8: Not Enough Tests and Not Enough Labs

1. Recent publications have suggested there may still be viral replication, just at a very low level. RNA is labile, making low level replication one explanation for the continued presence of low levels of detectable viral RNA.
2. Apoorva Mandavilli, "Your Coronavirus Test Is Positive. Maybe It Shouldn't Be.," *New York Times*, January 19, 2021.

   The science around using the quantity of RNA found in a sample as a way to determine continued infectivity is inexact. The same cycle threshold between two tests can reflect different amounts of RNA and even the same cycle threshold for the same test performed in two different labs can reflect different amounts of RNA. Also, there's some data showing that even at high cycle thresholds, some individuals are still likely infectious.
3. "Coronavirus (COVID-19) Update: FDA Authorizes First Antigen Test to Help in the Rapid Detection of the Virus that Causes COVID-19 in Patients," US Food and Drug Administration press release, May 9, 2020, https://www.fda.gov/news-events/press-announcements/coronavirus-covid-19-update-fda-authorizes-first-antigen-test-help-rapid-detection-virus-causes.
4. In the case of the Abbott test, the antibodies on the paper surface were designed to bind to the coronavirus's nucleocapsid, which is the protein shell of a virus, enclosing its genetic material. The protein is typically highly unique to each particular strain of a virus.

   See US Food and Drug Administration, "EUA Authorized Serology Test Performance," accessed February 26, 2021, https://www.fda.gov/medical-devices/coronavirus-disease-2019-covid-19-emergency-use-authorizations-medical-devices/eua-authorized-serology-test-performance.
5. "Abbott's Fast, $5, 15-Minute, Easy-to-Use COVID-19 Antigen Test Receives FDA Emergency Use Authorization; Mobile App Displays Test Results to Help Our Return to Daily Life; Ramping Production to 50 Million Tests a Month," Abbott press release, August 26, 2020.
6. Initially, Abbott was able to manufacture 50 million tests a week from production sites it built in advance of its FDA approval; but the federal government cornered almost the entire initial supply, purchasing 150 million of the tests to direct them to nursing homes. See Michelle Fay Cortez, Cristin Flanagan, and Jordan Fabian, "U.S. Buys Almost All Abbott's $5 Rapid Tests Made This Year," Bloomberg, August 26, 2020.
7. "Abbott Begins Shipping BinaxNOW COVID-19 AG Self Test to Retailers Today," Abbott press release, April 19, 2021.
8. As the *Washington Post* reported, "Health officials in several states say they have been allowed no say in where the new tests are being sent and sometimes don't know which nursing homes will receive them until the night before a shipment arrives. That has left some facilities ill-trained in how to use the tests and what to do with results. And it may be contributing to false-positive test results—when people are identified as being infected but aren't." See William Wan and Lena H. Sun, "Trump Administration's New Rapid Coronavirus Tests Plagued by Confusion and a Lack of Planning," *Washington Post*, September 29, 2020.
9. Christopher Weaver, Anna Wilde Mathews, and Tom McGinty, "Many

Nursing Homes Shun Free Covid-19 Testing Equipment," *Wall Street Journal*, November 7, 2020.

10. US Centers for Disease Control and Prevention, "Interim Guidance for Antigen Testing for SARS-CoV-2," December 16, 2020, https://www.cdc.gov/coronavirus/2019-ncov/lab/resources/antigen-tests-guidelines.html.
11. FDA did make clear through a variety of means, including an FAQ, that the test can be ordered by a provider off-label for screening but given the lower sensitivity, negative results should be considered presumptive.
12. Daniel Larremore et al., "Test Sensitivity Is Secondary to Frequency and Turnaround Time for COVID-19 Surveillance," medRxiv, September 8, 2020, https://www.medrxiv.org/content/10.1101/2020.06.22.20136309v3; and Michelle Fay Cortez and Emma Court, "Abbott Test May Curb Virus Even Missing Some Cases, CDC Says," Bloomberg, January 19, 2021.
13. Lisa Song, "Rapid Testing Is Less Accurate Than the Government Wants to Admit," ProPublica, November 16, 2020.
14. Wan and Sun, "Trump Administration's New Rapid Coronavirus Tests Plagued by Confusion and a Lack of Planning."
15. "Of 3,725 antigen tests performed at 12 facilities, 60 came back positive, and on follow up testing with PCT, they identified 39 as false positives, or 60%." See Ken Ritter and Sam Metz, "Nevada Reverses Ban on Rapid Tests After Federal Pushback," Associated Press, October 9, 2020.
16. Nevada Department of Health and Human Services, "Discontinue the Use of Antigen Testing in Skilled Nursing Facilities Until Further Notice," Technical Bulletin, October 2, 2020, http://dpbh.nv.gov/uploadedFiles/dpbhnvgov/content/Resources/Directive%20to%20Discontinue%20Use%20of%20Antigen%20POC_10.02.2020_ADA_Compliant.pdf.
17. Alexis Madrigal and Robinson Meyer, "Why Trump's Rapid-Testing Plan Worries Scientists," *Atlantic*, October 9, 2020.
18. Alex Spanko, "As Federal Government Rolls Out Point-of-Care Units, Not All States Allow Nursing Homes to Use Them," Skilled Nursing News, July 31, 2020.
19. Ritter and Metz, "Nevada Reverses Ban on Rapid Tests after Federal Pushback."
20. US Food and Drug Administration, "Coronavirus (COVID-19) Update: FDA Takes Steps to Streamline Path for COVID-19 Screening Tools, Provides Information to Help Groups Establishing Testing Programs," March 16, 2021, https://www.fda.gov/news-events/press-announcements/coronavirus-covid-19-update-fda-takes-steps-streamline-path-covid-19-screening-tools-provides.
21. "Trump Interrupts Health Officials: 'Anybody That Wants a Test Can Get a Test,'" *Washington Post*, March 7, 2020.
22. Matthew Perrone and Mike Stobbe, "US Labs Await Virus-Testing Kits Promised by Administration," Associated Press, March 5, 2020.
23. "With current estimates (and this could change), 2.1 million tests would roughly translate to 850,000 Americans being able to undergo testing. "IDT and other manufacturers believe they can scale up production so that by the end of next week an additional 4 million tests could be shipped. This

does not include the ramp up expected by large commercial or academic labs." See US Food and Drug Administration, "Coronavirus (COVID-19) Update: White House Press Briefing by FDA Commissioner Stephen M. Hahn, M.D.," March 7, 2020, https://www.fda.gov/news-events/speeches-fda-officials/coronavirus-covid-19-update-white-house-press-briefing-fda-commissioner-stephen-m-hahn-md-03072020.

24. Covid-19 Test Capacity (@COVID2019tests), Twitter, https://twitter.com/COVID2019tests; and Scott Gottlieb (@ScottGottliebMD), "Our last update on U.S. #COVID19 testing capacity. As high throughput systems come online this week, testing capacity will substantially increase nationwide. Any limitations will be the testing supply chain (reagents, swabs, sites, etc)—not platforms for conducting testing," Twitter, March 16, 2020, 1:51 p.m.
25. Robinson Meyer and Alexis C. Madrigal, "The Dangerous Delays in U.S. Coronavirus Testing Haven't Stopped," *Atlantic*, March 9, 2020; and Chandelis Duster and Jacqueline Howard, "Health and Human Services Chief Says 'We Don't Know' How Many Americans Have Been Tested for Coronavirus," CNN, March 10, 2020.
26. Stephen Hahn, "Food and Drug Administration Budget Request for FY2021," testimony before the Subcommittee on Agriculture, Rural Development, Food and Drug Administration, and Related Agencies, Committee on Appropriations, US House of Representatives, March 11, 2020.
27. Sanjay Gupta and Brett Giroir, "CNN Special Report, COVID WAR: The Pandemic Doctors Speak Out," CNN, March 28, 2021.
28. "Roche's Cobas SARS-CoV-2 Test to Detect Novel Coronavirus Receives FDA Emergency Use Authorization and Is Available in Markets Accepting the CE Mark," Roche press release, March 13, 2020.
29. Puneet Souda, Westley Dupray, and Scott Mafale, "Life Science Tools and Diagnostics: Update #3: The Current State of COVID-19 Testing in the U.S.," SVB Leerink, March 30, 2020; and David Lim, "Latest Coronavirus Testing Glitch: Not Enough Cotton Swabs," *Politico*, March 16, 2020.
30. "Inslee Statement on Statewide Shutdown of Restaurants, Bars and Limits on Size of Gatherings Expanded," Washington Governor press release, March 15, 2020, https://www.governor.wa.gov/news-media/inslee-statement-statewide-shutdown-restaurants-bars-and-limits-size-gatherings-expanded.
31. Weizhen Tan and Riya Bhattacharjee, "California Governor Issues Statewide Order to 'Stay at Home' as Coronavirus Cases Soar," CNBC, March 20, 2020.

Chapter 9: Shortage after Shortage

1. WECT Staff, "CDC Recommends Healthcare Providers Assume Those with Mild Symptoms to Be Positive with COVID-19," WCET NBC News 6, March 23, 2020.
2. Emma Court, Kristen Brown, and Neil Weinberg, "U.S. Finally Ramps Up Virus Testing, but Demand Still Outpaces Supply," Bloomberg, March 27, 2020.
3. Katherine Eban, "How Jared Kushner's Secret Testing Plan 'Went Poof into Thin Air,'" *Vanity Fair*, July 30, 2020.
4. Lauren Webber and Christina Jewett, "Testing Swabs Run In Short Supply

as Makers Try to Speed Up Production," National Public Radio, March 18, 2020.

5. Lim, "Latest Coronavirus Testing Glitch: Not Enough Cotton Swabs."
6. Charles Eichacker, "With $75.5M from the Feds, this Guilford Swab Maker's Expansion Came Together in Weeks," *Bangor Daily News*, June 16, 2020; and Maureen Milliken, "Puritan, Hardwood Suit Is Tale of Family Business Going in Opposite Directions," Mainebiz, Feburary 25, 2020.
7. Olivia Carville, "America's Covid Swab Supply Depends on Two Cousins Who Hate Each Other," Bloomberg, March 18, 2021.
8. Charles Eichacker, "How a Family-Owned Company in Guilford Came to Host a Visit from the President," *Bangor Daily News*, June 16, 2020.
9. Company documents.
10. Lewis Kamb, "Fed Shipment of Q-Tip-Style Coronavirus Swabs Puzzles Washington State Officials, Latest Wrinkle in Supply Woes," *Seattle Times*, May 16, 2020.
11. The preferred materials for nasopharyngeal swabs are flock-tipped with nylon or spun with polyester. The preferred material for nasal swabs is spun with a foam tip. The polyurethane material in the foam-tipped swab is best for absorption in the nose and transferring to nitrocellulose test paper of lateral flow tests. But special—and in each case, different—machinery is required to manufacture the two kinds of swabs.
12. Carville, "America's Covid Swab Supply Depends on Two Cousins Who Hate Each Other."
13. "DOD Details $75 Million Defense Production Act Title 3 Puritan Contract," US Department of Defense press release, April 29, 2020, https://www.defense.gov/Newsroom/Releases/Release/Article/2170355/dod-details-75-million-defense-production-act-title-3-puritan-contract/.
14. Scott Gottlieb, "Are We Prepared? Protecting the US from Global Pandemics," statement before the Committee on Homeland Security and Governmental Affairs, US Senate, February 12, 2020.
15. Katherine J. Wu, "'It's Like Groundhog Day': Coronavirus Testing Labs Again Lack Key Supplies," *New York Times*, August 15, 2020.
16. US Food and Drug Administration, "Coronavirus (COVID-19) Update: December 28, 2020," December 28, 2020, https://www.fda.gov/news-events/press-announcements/coronavirus-covid-19-update-december-28-2020.
17. Staff Reporter, "COVID-19 Testing Scales Up with Automation but Supply Chain Disruptions Persist," *Modern Healthcare*, July 7, 2020.

    One company, Hamilton, made many of the pipettes and also manufactured the tips. They maintained the intellectual property over the devices, and the service agreements that they had in place often required that the Hamilton pipettes be used with Hamilton tips. This was done to maintain oversight of the quality of the testing systems. Hamilton worked hard to stand up a new manufacturing line and started to allow labs to use other vendors for the pipette tips without violating their service agreements. But when Hamilton allowed other tips to be used, they needed to be validated to work with the systems. Most alternate vendor tips didn't meet the established criteria and could not be used as substitutions for Hamilton tips. It proved once again how hard it is to improvise in a crisis. These supply issues needed to be worked out in advance. We hadn't prepared for what would be needed in a crisis.

18. Caroline Chen (@CarolineYLChen), "You know what's infuriating? That Utah's public health lab could do more genomic sequencing to look out for variants . . . if only it had more pipette tips.," Twitter, February 7, 2021, 9:43 a.m.
19. Kate Sheridan, "How Blackouts, Fires, and a Pandemic are Driving Shortages of Pipette Tips— and Hobbling Science," STAT, April 28, 2021.
20. US Food and Drug Administration, "Medical Device Shortages During the COVID-19 Public Health Emergency," accessed February 26, 2021, https://www.fda.gov/medical-devices/coronavirus-covid-19-and-medical-devices/medical-device-shortages-during-covid-19-public-health-emergency.
21. Farhad Manjoo, "How the World's Richest Country Ran Out of a 75-Cent Face Mask," *New York Times*, March 25, 2020.
22. Amanda Watts and Alison Main, "The US Has a Stockpile of Masks, Health Secretary Says," CNN, February 26, 2020.
23. Lawrence Wright, "The Plague Year," *New Yorker*, December 28, 2020.
24. David Sanger, Zolan Kanno-Youngs, and Nicholas Kulish, "A Ventilator Stockpile, with One Hitch: Thousands Do Not Work," *New York Times*, April 20, 2020; and Rick Sobey, "Donald Trump: U.S. Has 9,000 Ventilators in Federal Stockpile," *Boston Herald*, April 6, 2020.
25. Sanger, Kanno-Youngs, and Kulish, "A Ventilator Stockpile, with One Hitch: Thousands Do Not Work."
26. Kim Chandler, "Some States Receive Masks with Dry Rot, Broken Ventilators," Associated Press, April 4, 2020.
27. Lena H. Sun, "Inside the Secret U.S. Stockpile Meant to Save Us All in a Bioterror Attack," *Washington Post*, April 24, 2018.
28. Jon Swaine, Robert O'Harrow Jr., and Aaron C. Davis, "Before Pandemic, Trump's Stockpile Chief Put Focus on Biodefense. An Old Client Benefited," *Washington Post*, May 4, 2020.
29. Scott Gottlieb, "Beat Tomorrow's Pandemic Today," *Wall Street Journal*, May 23, 2021.
30. Interagency Task Force in Fulfillment of Executive Order 13806, "Assessing and Strengthening the Manufacturing and Defense Industrial Base and Supply Chain Resiliency of the United States," US Department of Defense, September 2018, https://media.defense.gov/2018/Oct/05/2002048904/-1/-1/1/ASSESSING-AND-STRENGTHENING-THE-MANUFACTURING-AND%20DEFENSE-INDUSTRIAL-BASE-AND-SUPPLY-CHAIN-RESILIENCY.PDF.
31. Andrew Jacobs, "Health Care Workers Still Face Daunting Shortages of Masks and Other P.P.E.," *New York Times*, December 20, 2020.
32. The market value of N95 respirators amounted to more than USD $802 million in 2019. See M. Garside, "Market Value of N95 Masks Worldwide in 2019 and 2027," April 29, 2021, Statista, https://www.statista.com/statistics/1148474/global-n95-mask-market-size/.

    The global N95 respirator market size was valued at USD 1.10 billion in 2019. See Grandview Research, "N95 Mask Market Size, Share & Trends

Analysis Report by Product (with Exhalation Valve, without Exhalation Valve), by Distribution Channel (Online, Offline), by End Use, by Region, and Segment Forecasts, 2020—2027," November 2020, https://www.grandviewresearch.com/industry-analysis/n95-mask-market.

33. Health Industry Distributors Association, "2020 Market Report: Personal Protective Equipment," December 2020, https://www.cardinalhealth.com/content/dam/corp/web/documents/Report/cardinal-health-2020-PPE-market-report.pdf.
34. C. Todd Lopez, "Domestic N95 Mask Production Expected to Exceed 1 Billion in 2021," June 10, 2020, DoD News, https://www.defense.gov/Explore/News/Article/Article/2215532/domestic-n95-mask-production-expected-to-exceed-1-billion-in-2021/.
35. Health Industry Distributors Association, "2020 Market Report: Personal Protective Equipment."
36. Jennifer Steinhauer, "Frustrated by Lack of Coronavirus Tests, Maryland Got 500,000 from South Korea," *New York Times*, April 20, 2020.
37. Adam Cancryn, "'We're Going to Get It Done': Governors Mobilize to Fill Trump's Testing Gap," *Politico*, August 18, 2020.
38. Rockefeller Foundation, "Governors of Six States Announce Major Bipartisan Compact for Three Million Rapid Antigen Tests," August 4, 2020, https://www.rockefellerfoundation.org/news/governors-of-six-states-announce-major-bipartisan-compact-for-three-million-rapid-antigen-tests/.
39. Erin Cox, "There's No National Testing Strategy for Coronavirus. These States Banded Together to Make One.," *Washington Post*, August 4, 2020.
40. Scott Gottlieb, "The States Are Laboratories for Covid Control," *Wall Street Journal*, November 8, 2020.
41. Asher Klein, "COVID Appears to Be Spiking in Boston-Area Sewage, and It Has Experts Worried," NBC, October 25, 2020.
42. US Centers for Disease Control and Prevention, "The Laboratory Response Network Partners in Preparedness," April 10, 2019, https://emergency.cdc.gov/lrn/.
43. US Food and Drug Administration, "FDA, CDC, and CMS Launch Task Force to Help Facilitate Rapid Availability of Diagnostic Tests during Public Health Emergencies," February 26, 2019, https://www.fda.gov/media/120328/download.
44. Atul Gawande, "We Can Solve the Coronavirus-Test Mess Now—If We Want To," *New Yorker*, September 2, 2020.
45. Stephanie Armour, Alexandra Berzon, and James Grimaldi, "Nation's Top Emergency-Preparedness Agency Focused on Warfare Threats over Pandemic," *Wall Street Journal*, July 9, 2020.
46. Sarah Karlin-Smith (@SarahKarlin), "This slide from @BARDA Acting Director Gary Disbrow is a good visual of US prioritizing funding for COVID-19 vaccines versus therapeutics," Twitter, October 27, 2020, 10:18 a.m.; "NIH to Launch Public-Private Partnership to Speed COVID-19 Vaccine and Treatment Options," US National Institutes of Health press release, April 17, 2020, https://www.nih.gov/news-events/news-releases/nih-launch-public-private-partnership-speed-covid-19-vaccine-treatment-options; and "Trump Ad-

ministration Announces Framework and Leadership for 'Operation Warp Speed,'" US Department of Health and Human Services press release, May 15, 2020, https://www.hhs.gov/about/news/2020/05/15/trump-administration-announces-framework-and-leadership-for-operation-warp-speed.html.

47. Katherine Eban, "How Jared Kushner's Secret Testing Plan 'Went Poof into Thin Air,'" *Vanity Fair*, July 30, 2020.
48. Sheryl Gay Stolberg, "Trump Vows More Coronavirus Testing, but Less Than What May Be Needed," *New York Times*, April 27, 2020.
49. Echoes of these concepts would finally emerge as part of a national strategy released by the Biden administration almost a year later, which would create a series of regional testing hubs to coordinate and oversee a sharp expansion of testing in elementary and middle schools and high-risk congregate settings like homeless shelters.

    See David Lim, "Biden Team Plots the Country's First National Covid Testing Strategy," *Politico*, March 7, 2021.
50. Eban, "How Jared Kushner's Secret Testing Plan 'Went Poof into Thin Air.'"

Chapter 10: Preparing for the Wrong Pathogen

1. Cheyenne Roundtree, "How Disease and Medical Experts across the US were Sounding Alarm about Coronavirus since January and Ridiculing Trump's Slow Response in 'Red Dawn' Email Chain," *Daily Mail*, April 12, 2020.
2. The case fatality rate or "CFR" measures the percentage of people who will die from an infection among all of those who become infected by it and develop symptoms of the disease. That's different from the infection fatality rate or "IFR," which measures the percentage of people who will die as a fraction of all of those who will become infected. For calculating the IFR, a patient didn't have to get sick from the virus, they just needed to be infected by it. The IFR accounts for people who develop symptoms, as well as those who don't. Because some people will catch a virus but not get sick from it, the CFR is often higher than the IFR, as it was in the case of COVID, which caused a fair degree of asymptomatic infection. More than 20 percent of people who were infected with COVID never developed symptoms. Yet the two metrics—the CFR and the IFR—were often confused during the COVID crisis, where people would sometimes compare the case fatality rate from seasonal flu to the infection fatality rate from COVID.
3. The higher R0 is somewhat offset by the longer generation time for COVID, so the doubling time for flu might be shorter, making it seem more explosive, but the overall attack rate in an unvaccinated population should be higher for COVID.
4. Eric Lipton, "The 'Red Dawn' Emails: 8 Key Exchanges on the Faltering Response to the Coronavirus," *New York Times*, April 11, 2020.
5. Julia Aledort et al., "Non-Pharmaceutical Public Health Interventions for Pandemic Influenza: An Evaluation of the Evidence Base," *BMC Public Health* 7, no. 1 (2007): 208; and Noreen Qualls et al., "Community Mit-

igation Guidelines to Prevent Pandemic Influenza—United States, 2017," *Morbidity and Mortality Weekly Report—Recommendations and Reports* 66, no. 1 (2017): 1–34.

6. Robert Webster et al., "Characterization of H5N1 Influenza Viruses That Continue to Circulate in Geese in Southeastern China," *Journal of Virology* 76, no. 1 (2002): 118–26.
7. D. Alexander, "A Review of Avian Influenza in Different Bird Species," *Veterinary Microbiology* 74, no. 3 (2000): 3–13.
8. US Centers for Disease Control and Prevention, "Influenza Virus Genome Sequencing and Genetic Characterization," October 15, 2019, https://www.cdc.gov/flu/about/professionals/genetic-characterization.
9. This is why the flu virus can so easily thwart our vaccines, or why we keep getting reinfected with the same basic strains of flu from year to year. The H1N1 virus that caused the 2009 pandemic probably had different HA proteins on its surface than the H1N1 that will threaten people this year. So, prior antibodies, programmed by memory B cells and other parts of our immune system in response to past infections or vaccination, may not be very effective against new variants on the same strains. This is why people don't develop enduring immunity to the flu, and why it's hard to engineer vaccines that offer protection for more than one season. By contrast, coronavirus only consists of one single subunit of RNA, and therefore undergoes less rapid mutation because it doesn't undergo reassortment and swap genetic material between different subunits.

   Samantha Lycett, Florian Duchatel, and Paul Digard, "A Brief History of Bird Flu," *Philosophical Transactions of the Royal Society B* 374, no. 1775 (2019): 20180257.
10. Anthony Fauci, "Race against Time," *Nature* 435, no. 7041 (2005): 423–4.
11. US Centers for Disease Control and Prevention, "Isolation of Avian Influenza A(H5N1) Viruses from Humans—Hong Kong, May–December 1997," *Morbidity and Mortality Weekly Report* 46, no. 50 (1997): 1204–7.
12. Paul Chan, "Outbreak of Avian Influenza A(H5N1) Virus Infection in Hong Kong in 1997," *Clinical Infectious Diseases* 34, no. S2 (2002): S58–64.
13. Ibid.
14. US Centers for Disease Control and Prevention, "Outbreaks of Avian Influenza A (H5N1) in Asia and Interim Recommendations for Evaluation and Reporting of Suspected Cases—United States, 2004," *Morbidity and Mortality Weekly Report* 53, no. 05 (2004): 97–100.
15. Alan Sipress, "Bird Flu Drug Rendered Useless," *Washington Post*, June 18, 2005.
16. World Health Organization, "Avian Influenza A (H5N1)—Update 19: Investigation of Possible Human-to-Human Transmission in Viet Nam: New Data Are Reassuring," February 6, 2004, https://www.who.int/csr/don/2004_02_06/en/.
17. Robert Roos, "Avian Flu Spreads to Central China," CIDRAP News, January 30, 2004.
18. Qing-Yu Zhu et al., "Fatal Infection with Influenza A (H5N1) Virus in China," *New England Journal of Medicine* 354, (2006): 2731-2.

19. CIDRAP News, "Report: China Had Human H5N1 Case in Late 2003," June 22, 2006.
20. Fauci, "Race against Time."
21. World Health Organization, "Influenza: FAQs: H5N1 Influenza," April 2011, https://www.who.int/influenza/human_animal_interface/avian_influenza/h5n1_research/faqs/en/.
22. Stephanie Sonnberg, Richard Webby, and Robert Webster, "Natural History of Highly Pathogenic Avian Influenza H5N1," *Virus Research* 178, no. 1 (2013): 63–77.
23. World Health Organization, "Influenza: FAQs: H5N1 Influenza"; and Fauci, "Race against Time."
24. Ahmet Faik Oner et al., "Avian Influenza A (H5N1) Infection in Eastern Turkey in 2006," *New England Journal of Medicine* 355, no. 21 (2006): 2179–85.
25. International Institute for Sustainable Development, "Summary Report, 10–11 April 2006," *Avian Influenza & Wild Birds Bulletin* 123, no. 1 (2006).
26. In January 2014, Canada reported the first human infection with HPAI Asian H5N1 virus in the Americas. This was an "imported" case occurring in a traveler who had recently returned from China.

    US Centers for Disease Control and Prevention, "Highly Pathogenic Asian Avian Influenza A(H5N1) Virus," December 12, 2018, https://www.cdc.gov/flu/avianflu/h5n1-virus.htm; and Gregory Juckett, "Avian Influenza: Preparing for a Pandemic," *American Family Physician* 74, no. 5 (2006): 783–90.
27. Andrew Pollack, "U.S. Will Miss Half Its Supply of Flu Vaccine," *New York Times*, October 6, 2004.
28. Denise Grady et al., "With Few Suppliers of Flu Shots, Shortage Was Long in Making," *New York Times*, October 17, 2004.
29. Matthew Mosk, "George W. Bush in 2005: 'If We Wait for a Pandemic to Appear, It Will Be Too Late to Prepare,'" ABC News, April 5, 2020.
30. World Health Organization, "Epidemiology of WHO-Confirmed Human Cases of Avian Influenza A (H5N1) Infection," *Weekly Epidemiological Record* 81, no. 26 (2006): 249–60; and Tran Tinh Hien et al. "Avian Influenza A (H5N1) in 10 Patients in Vietnam," *New England Journal of Medicine* 350, no. 12 (2004): 1179–88.
31. Michael Lewis, *The Premonition* (New York: W.W. Norton & Company, 2021).
32. Betsy McKay and Phred Dvorak, "A Deadly Coronavirus Was Inevitable. Why Was No One Ready?," *Wall Street Journal*, August 13, 2020.
33. The report had made some grim assumptions. If a pandemic struck, they calculated that it would sicken 90 million Americans and claim the lives of about two million people. The pandemic planning was anticipated to be a multiyear effort, with the initial plan just laying out the opening blueprint. White House, Homeland Security Council, "National Strategy for Pandemic Influenza," November 1, 2005, https://georgewbush-whitehouse.archives.gov/homeland/pandemic-influenza.html#section9; and CNN, "Transcript of Bush Speech on Pandemic Flu Strategy," November 1, 2005.
34. Anthony F. Rock, "Meeting on Avian Influenza & Human Pandemic In-

fluenza," November 8, 2005, https://web.archive.org/web/20080720165521/http://geneva.usmission.gov/Press2005/1108AvianInfluenza.htm.

35. Lucia Mullen et al., "An Analysis of International Health Regulations Emergency Committees and Public Health Emergency of International Concern Designations," *British Medical Journal Global Health* 5, no. 002502 (2020).
36. President Bush's plan also called for the creation of new domestic capacity to produce vaccines and the stockpiling of countermeasures. These were to include components that would be critical for manufacturing vaccines, such as "adjuvants" that could be used to boost the immune response that a vaccine generated and make them more potent. The stockpiles also included antiviral drugs. The aim was to make sure there would be enough medicines on hand to respond to the initial wave of a flu pandemic. And then, enough resilient manufacturing capacity and components to develop and deploy a vaccine that could thwart a second wave. The report envisioned the need for enough vaccine to protect the entire population within six months after a new strain emerged. To help detect a new virus, and monitor its spread, it called for creating the equivalent of a sentinel surveillance system, described as "advance mechanisms for real-time clinical surveillance in domestic acute care settings such as emergency departments, intensive care units and laboratories to provide local, state and federal public health officials with continuous awareness of the profile of illness in communities." To support these efforts, the plan leaned heavily on the development and use of point-of-care diagnostic tests "with greater sensitivity and reproducibility to allow onsite diagnosis of pandemic strains of influenza at home and abroad . . . to facilitate early warning, outbreak control and targeting of antiviral therapy." That initial report also envisioned the role of non-pharmaceutical interventions as a way to contain spread as part of the plan's third pillar, noting that "where appropriate, use governmental authorities to limit non-essential movement of people, goods and services into and out of areas where an outbreak occur," including "guidance to all levels of government on the range of options for infection-control and containment, including those circumstances where social distancing measures, limitations on gatherings, or quarantine authority may be an appropriate public health intervention." See White House, Homeland Security Council, "National Strategy for Pandemic Influenza," November 1, 2005, https://georgewbush-whitehouse.archives.gov/homeland/pandemic-influenza.html#section9.
37. Fauci, "Race against Time."
38. *Wall Street Journal*, "Secret Swine Flu Egg Farms in Amish Country," January 11, 2010, https://www.wsj.com/video/secret-swine-flu-egg-farms-in-amish-country/E8147107-2248-4893-A6A6-2D0500EA6CEE.html.
39. The government sought a "secure system to protect these birds," with "very strict conditions for the entry and exit of people and product," said Robin Robinson, an official at the Department of Health and Human Services who hatched the secret egg program. "If we had no vaccine now it would be a very bad thing."

    See Jeanne Whalen, "Shell Game: Government Hatches Secret Program to Lay Eggs," *Wall Street Journal*, January 13, 2010.

40. World Trade Organization, "Developing and Delivering COVID-19 Vaccines Around the World," December 22, 2020, https://www.wto.org/english/tratop_e/covid19_e/vaccine_report_e.pdf.
41. Ernest Milián and Amine Kamen, "Current and Emerging Cell Culture Manufacturing Technologies for Influenza Vaccines," *BioMed Research International* (2015): 50483.
42. US Centers for Disease Control and Prevention, "Summary of the 2017–2018 Influenza Season," September 5, 2019, https://www.cdc.gov/flu/about/season/flu-season-2017-2018.htm.
43. Jane Zhang, "Sanofi Gets $100 Million Contract from U.S. for Avian-Flu Vaccine," *Wall Street Journal*, September 16, 2005; and "Sanofi Awarded $226 million by US Government to Expand Pandemic Influenza Preparedness," Sanofi press release, December 19, 2019.
44. IW Staff, "First U.S. Cell-Based Flu Vaccine Plant Opens in North Carolina," *Industry Week*, December 12, 2011.
45. The vaccine, an inactivated product, is grown in mammalian cells using a particular cell-culture technology that had already been widely used for decades to produce other FDA approved vaccines, including vaccines against polio and rubella.

    Reuters Staff, "FDA Approves Novel Novartis Seasonal Flu Vaccine," Reuters, November 20, 2012; and Lisa Schnirring, "FDA Clears First Cell-Based Flu Vaccine," CIDRAP News, November 21, 2012.

Chapter 11: Stay-at-Home Orders

1. White House Homeland Security Council, "National Strategy for Pandemic Influenza," November 1, 2005, https://georgewbush-whitehouse.archives.gov/homeland/pandemic-influenza.html#section9.
2. Carter Mecher, interview; Rajeev Venkayya, interview; and Eric Lipton and Jennifer Steinhauer, "The Untold Story of the Birth of Social Distancing," *New York Times*, April 22, 2020.
3. US Centers for Disease Control and Prevention, "Cities Readiness Initiative," December 18, 2020, https://www.cdc.gov/cpr/readiness/mcm/cri.html.
4. M. Elizabeth Halloran et al., "Modeling Targeted Layered Containment of an Influenza Pandemic in the United States," *Proceedings of the National Academy of Sciences* 105, no. 12 (2008): 4639–44.
5. Nancy Tomes, "Quarantine! East European Jewish Immigrants and the New York City Epidemics of 1892," *New England Journal of Medicine* 338 (1998): 1235.
6. Howard Markel et al., "Nonpharmaceutical Interventions Implemented by US Cities During the 1918–1919 Influenza Pandemic," *JAMA* 298, no. 6 (2007): 644–54.
7. Attention to the Hatchett paper was significant. The Altmetrics attention score for the article was almost 7,000 by the end of 2020, which made it the 203rd ranking article of the 16.5 million articles in the database and the ninth-highest ranked article ever for the Proceedings of the National Academy of Sciences.

    Richard Hatchett, Carter Mecher, and Marc Lipsitch, "Public Health Interventions and Epidemic Intensity During the 1918 Influenza Pandemic," *Proceedings of the National Academy of Sciences* 104, no. 18 (2007): 7582–7.

8. Martin Bootsma and Neil Ferguson, "The Effect of Public Health Measures on the 1918 Influenza Pandemic in U.S. Cities," *Proceedings of the National Academy of Sciences* 104, no. 18 (2007): 7588–93.
9. *Philadelphia Evening Bulletin*, "Spanish Influenza Causes Death Here," September 9, 1918, 1.
10. *Philadelphia Inquirer,* "Spanish Influenza Epidemic Waning," September 23, 1918, 22.
11. Kenneth Davis, "Philadelphia Threw a WWI Parade That Gave Thousands of Onlookers the Flu," *Smithsonian Magazine*, September 21, 2018; and Christakis, Apollo's Arrow.
12. Ibid.
13. Within a month of the Sturgis gathering, states with the largest number of residents attending the gathering— including North and South Dakota, Wyoming, Minnesota, and Montana—were leading the nation in new COVID cases per capita. One estimate said that the Sturgis event may have been responsible for more than 250,000 new infections, at a total public health cost of as much as $12.2 billion.

    Dhaval Dave, Drew McNichols, and Joseph Sabia, "The Contagion Externality of a Superspreading Event: The Sturgis Motorcycle Rally and COVID-19," *Southern Economic Journal*, 10 (2020); and Melanie Firestone et al., "COVID-19 Outbreak Associated with a 10-Day Motorcycle Rally in a Neighboring State—Minnesota, August–September 2020," *Morbidity and Mortality Weekly Report* 69, no. 47 (2020): 1771–6.
14. *Philadelphia Inquirer,* "Epidemic Slows Up, Deaths Increase," October 1, 1918, 5.
15. Davis, "Philadelphia Threw a WWI Parade That Gave Thousands of Onlookers the Flu."

    Similar to what we'd see with COVID, the Spanish flu killed by triggering an extreme and inappropriate immune response in its victims, a process known as cytokine storm.
16. *Philadelphia Inquirer,* "Theatres, Saloons in Penna. Closed to Halt Influenza," October 4, 1918, 1, 13; and *Philadelphia Evening Bulletin*, "City to Be 'Dry' Tonight; Courts Close for 'Flu,'" October 4, 1918, 1, 2.
17. *Philadelphia Inquirer,* "Theatres, Saloons in Penna. Closed to Halt Influenza."
18. *Philadelphia Inquirer,* "Spanish Influenza and the Fear of It," October 5, 1918, 12; *Philadelphia Evening Bulletin*, "The Influenza Edicts," October 5, 1918, 6.
19. *Philadelphia Inquirer,* "Stop the Senseless Influenza Panic," October 8, 1918, 12.
20. *Philadelphia Inquirer,* "Ban Raised Here as Epidemic Ends," October 31, 1918, 8; and *Philadelphia Inquirer,* "Influenza No Longer Has Official Standing," November 1, 1918, 12.
21. *St. Louis Globe-Democrat*, "Doctors Here Must Report Influenza," September 20, 1918, 2; and *St. Louis Post-Dispatch*, "Pneumonia Cases 10 Per Cent of Winter Deaths; How to Avoid It," September 21, 1918, 11.

22. *St. Louis Post-Dispatch*, "Family of 7 Here Ill with Influenza," October 5, 1918, 3.
23. *St. Louis Post-Dispatch*, "To Close Schools and Theaters to Check Influenza," October 7, 1918, 1, 3; and *St. Louis Globe-Democrat*, "Influenza Quarantine Placed on City and Schools, Theaters, Churches Are to Be Closed," October 8, 1918, 1, 2.
24. Hatchett, Mecher, and Lipsitch, "Public Health Interventions and Epidemic Intensity During the 1918 Influenza Pandemic."
25. Carter Mecher, senior medical adviser at the Department of Veterans Affairs, in conversation with the author, August 16, 2020.
26. Hatchett, Mecher, and Lipsitch, "Public Health Interventions and Epidemic Intensity During the 1918 Influenza Pandemic."
27. Lewis, *The Premonition*.
28. Markel et al., "Nonpharmaceutical Interventions Implemented by US Cities during the 1918–1919 Influenza Pandemic."
29. China implemented these measures with no frills; among other things, imposing an extraordinary cordon sanitaire in the city of Wuhan on January 23, when the government severely restricted outbound traffic and confined 10 million people to their homes.
30. Markel et al., "Nonpharmaceutical Interventions Implemented by US Cities during the 1918–1919 Influenza Pandemic."
31. Carter Mecher, senior medical adviser at the Department of Veterans Affairs, in conversation with the author, August 16, 2020.
32. Lori Uscher-Pines et al., "School Practices to Promote Social Distancing in K–12 Schools: Review of Influenza Pandemic Policies and Practices," *BMC Public Health* 18, no. 1 (2018): 406.
33. Charlotte Jackson et al., "School Closures and Influenza: Systematic Review of Epidemiological Studies," *BMJ Open* 3 (2013); and Simon Cauchemez et al., "Closure of Schools During an Influenza Pandemic," *Lancet Infectious Diseases* 9, no. 8 (2009): 473–81.
34. Christophe Fraser et al., "Pandemic Potential of a Strain of Influenza A (H1N1): Early Findings," *Science* 324, no. 5934 (2009): 1557–61; Fatimah Dawood et al., "Emergence of a Novel Swine-Origin Influenza a (H1N1) Virus in Humans," *New England Journal of Medicine* 360, no. 25 (2009): 2605–15; and Gerardo Chowell et al., "Characterizing the Epidemiology of the 2009 Influenza A/H1N1 Pandemic in Mexico," *PLOS Medicine* 8, no. 5 (2011).
35. Jackson et al., "School Closures and Influenza."
36. Benjamin Cowling et al., "Effects of School Closures, 2008 Winter Influenza Season, Hong Kong," *Emerging Infectious Diseases* 14, no. 10 (2008): 1660–2.
37. Anthony Heymann et al., "Influence of School Closure on the Incidence of Viral Respiratory Diseases Among Children and on Health Care Utilization," *Pediatric Infectious Disease Journal* 23, no. 7 (2004): 675–7.
38. Maria Litvinova et al., "Reactive School Closure Weakens the Network of Social Interactions and Reduces the Spread of Influenza," *Proceedings of the National Academy of Sciences* 116, no. 27 (2019): 13174–81.
39. M. Elizabeth Halloran et al., "Modeling Targeted Layered Containment of

an Influenza Pandemic in the United States," *Proceedings of the National Academy of Sciences* 105, no. 12 (2008): 4639–44.

40. Simon Cauchemez et al., "Estimating the Impact of School Closure on Influenza Transmission from Sentinel Data," *Nature* 452, (2008): 750–4.
41. Ibid.
42. Maria Litvinova et al., "Reactive School Closure Weakens the Network of Social Interactions and Reduces the Spread of Influenza."
43. US Centers for Disease Control and Prevention, "Flu-Related Hospitalizations and Deaths in the United States from April 2009–January 30, 2010," April 16, 2010.
44. Héctor S. Izurieta et al., "Influenza and the Rates of Hospitalization for Respiratory Disease Among Infants and Young Children," *New England Journal of Medicine* 342, no. 4 (2000): 232–9; Katherine A. Poehling et al., "The Underrecognized Burden of Influenza in Young Children," *New England Journal of Medicine* 355, no. 1 (2006): 31–40, and A. S. Monto and K. M. Sullivan, "Acute Respiratory Illness in the Community. Frequency of Illness and the Agents Involved," *Epidemiology and Infection* 110, no. 1 (1993): 145–60.
45. Bernhard R. Ruf and Markus Knuf, "The Burden of Seasonal and Pandemic Influenza in Infants and Children," *European Journal of Pediatrics* 173, no. 3 (2014): 265–76.
46. Ibid.
47. Sheryl Gay Stolberg, "He Routed Smallpox, Now Tackles Bioterror," *New York Times*, November 18, 2001.
48. Thomas Inglesby et al., "Disease Mitigation Measures in the Control of Pandemic Influenza," *Biosecurity and Bioterrorism* 4, no. 4 (2006).
49. US Department of Health and Human Services, "HHS Pandemic Influenza Plan," November 2005, https://www.cdc.gov/flu/pdf/professionals/hhspandemicinfluenzaplan.pdf; and White House Homeland Security Council, "National Strategy for Pandemic Influenza: Implementation Plan," May 2006, http://www.whitehouse.gov/homeland/ nspi_implementation.pdf; and US Centers for Disease Control and Prevention, "Interim Pre-Pandemic Planning Guidance: Community Strategy for Pandemic Influenza Mitigation in the United States—Early, Targeted, Layered Use of Nonpharmaceutical Interventions," February 2007, https://www.cdc.gov/flu/pandemic-resources/pdf/community_mitigation-sm.pdf?fbclid=IwAR1sMmehOSZ8hHRKEPJEP2hUbkXbNMlS4sGRqQ7s5iLWOpyfQDErd4Wg_SE.
50. White House Homeland Security Council, "National Strategy for Pandemic Influenza."
51. "Press Briefing by Scott McClellan and Homeland Security Advisor Fran Townsend," White House press release, May 3, 2006, https://georgewbush-whitehouse.archives.gov/news/releases/2006/05/20060503-6.html; CIDRAP News, "New US Pandemic Plan Stresses Local Self-Reliance," May 3, 2006; and White House Homeland Security Council, "National Strategy for Pandemic Influenza."
52. US Centers for Disease Control and Prevention, "Interim Pre-Pandemic Planning Guidance: Community Strategy for Pandemic Influenza Mitigation in the United States."

53. One of the pandemics occurred in 1957, the other in 1968. Significant but less serious pandemics occurred in 1947 and 1977.

See Joshua Lederberg, Robert E. Shope, and Stanley C. Oaks Jr., eds., *Emerging Infections: Microbial Threats to Health in the United States* (Washington, DC: National Academies Press, 1992).

54. *New York Times*, "Crimson Contagion 2019 Functional Exercise Key Findings," October 2019, https://int.nyt.com/data/documenthelper/6824-2019-10-key-findings-and-after/05bd797500ea55be0724/optimized/full.pdf#page=1.

55. David Sanger et al., "Before Virus Outbreak, a Cascade of Warnings Went Unheeded," *New York Times*, March 22, 2020.

Chapter 12: A Plan Gone Awry

1. "SARS-CoV-2 spread rapidly throughout the United States since its 15 January start date and was likely accompanied by a large undiagnosed population of potential COVID-19 outpatients with presumably milder distribution of clinical symptoms than estimated from prior studies of SARS-CoV-2+ inpatients."

See Justin D. Silverman, Nathaniel Hupert, and Alex D. Washburne, "Using Influenza Surveillance Networks to Estimate State-Specific Prevalence of SARS-CoV-2 in the United States," *Science Translational Medicine* 12, no. 554 (2020).

2. Yasmeen Abutaleb et al., "The Inside Story of How Trump's Denial, Mismanagement and Magical Thinking Led to the Pandemic's Dark Winter," *Washington Post*, December 19, 2020.
3. Lev Facher, "President Trump Just Declared the Coronavirus Pandemic a National Emergency. Here's What That Means," STAT, March 13, 2020.
4. Abutaleb et al., "The Inside Story of How Trump's Denial, Mismanagement and Magical Thinking Led to the Pandemic's Dark Winter."
5. Dareh Gregorian, "Brazilian Official Who Met Trump at Mar-a-Lago Tests Positive for Coronavirus," NBC, March 12, 2020; and Yasmeen Abutaleb and Damian Paletta, *Nightmare Scenario: Inside the Trump Administration's Response to the Pandemic That Changed History*.
6. Lisandra Paraguassu and Anthony Boadle, "As Bolsonaro Flouts Warnings, Coronavirus Spreads in Brazil," Reuters, March 24, 2020.
7. Robert Costa and Philip Rucker, "Woodward Book: Trump Says He Knew Coronavirus Was 'Deadly' and Worse than the Flu While Intentionally Misleading Americans," *Washington Post*, September 9, 2020.
8. CBS News, "Transcript: Scott Gottlieb on 'Face the Nation,' March 8, 2020," March 8, 2020.
9. Facher, "President Trump Just Declared the Coronavirus Pandemic a National Emergency."
10. Meridith McGraw and Caitlin Oprysko, "Inside the White House During '15 Days to Slow the Spread,'" *Politico*, March 29, 2020.
11. Lawrence Gostin and Lindsay Wiley, "Governmental Public Health Powers During the COVID-19 Pandemic: Stay-at-Home Orders, Business Closures, and Travel Restrictions," *JAMA* 323, no. 21 (2020): 2137–8.

12. President's Coronavirus Guidelines for America, "15 Days to Slow the Spread," US Department of Justice, March 15, 2020, https://www.justice.gov/doj/page/file/1258511/download.
13. Ibid.
14. Jeff Mason and Steve Holland, "Trump Urges U.S. to Halt Most Social Activity in Virus Fight, Warns of Recession," Reuters, March 16, 2020.
15. Benjamin Siegel, "How Anesthesia Machines Can Help Hospitals with Ventilator Shortages Fight Coronavirus," ABC News, March 27, 2020; and Kathleen Moore, "Local Anesthesiologists Converting Machines to Ventilators," *Post Star,* March 26, 2020.
16. Jessie Yeung et al., "Pence Announces 4,000 Ventilators Will Be Shipped to New York," CNN, March 24, 2020.
17. Sean O'Kane, "How GM and Ford Switched Out Pickup Trucks for Breathing Machines," Verge, April 15, 2020.
18. Kirsten Korosec, "GM, Ford Wrap Up Ventilator Production and Shift Back to Auto Business," Tech Crunch, September 1, 2020.
19. Tamara Keith et al., "How 15 Days Became 45: Trump Extends Guidelines to Slow Coronavirus," National Public Radio, March 30, 2020.
20. Foege commented, "In retrospect it seems clear—we didn't know how to eradicate smallpox when we started. But this was not a negative. It was a characteristic of all unsolved problems. We are always faced with making sufficient decisions based on insufficient information. If we had waited until all the answers were available, the work on smallpox eradication would never have started—selecting the target helped develop the appropriate tools and strategy."
21. Amanda Watts, "Health Official's Advice on Coronavirus Response: 'Speed Trumps Perfection,'" CNN, March 13, 2020.
22. James Glanz and Campbell Robertson, "Lockdown Delays Cost at Least 36,000 Lives, Data Show," *New York Times,* May 22, 2020.
23. According to the *New York Times,* from February 29 to June 1, 2020, 203,792 COVID-19 cases were diagnosed and reported among residents of New York City, including 54,211 (26.6 percent) in persons known to have been hospitalized and 18,679 (9.2 percent) in persons who died. "New York City Coronavirus Map and Case Count," *New York Times,* April 6, 2021.
24. Jeffery C. Mays and Andy Newman, "Virus Is Twice as Deadly for Black and Latino People than Whites in N.Y.C.," *New York Times,* June 26, 2020.
25. Aaron Katersky and Ella Torres, "Black People in NYC Twice as Likely to Die from COVID as White People: Data," ABC News, April 17, 2020; and Gbenga Ogedegbe et al., "Assessment of Racial/Ethnic Disparities in Hospitalization and Mortality in Patients With COVID-19 in New York City," *JAMA* 3, no. 12 (2020): e2026881.
26. Derek Thompson, "Hygiene Theater Is a Huge Waste of Time," *Atlantic,* July 27, 2020.
27. US Centers for Disease Control and Prevention, "Science Brief: SARS-CoV-2 and Potential Airborne Transmission," October 5, 2020, https://www.cdc.gov/coronavirus/2019-ncov/more/scientific-brief-sars-cov-2.html.
28. Associated Press, "Should I Wipe Down Groceries During the Pandemic?," December 8, 2020.

29. Elizabeth Anderson et al., "Consideration of the Aerosol Transmission for COVID-19 and Public Health," *Risk Analysis* 40, no. 5 (2020): 902-7; and Nick Wilson, Stephen Corbett, and Euan Tovey, "Airborne Transmission of Covid-19," *British Medical Journal* 370, no. 3206 (2020).
30. Talib Dbouka and Dimitris Drikakis, "On Coughing and Airborne Droplet Transmission to Humans," *Physics of Fluids* 32, no. 5 (2020).
31. Zeynep Tufekci, "We Need to Talk about Ventilation," *Atlantic*, July 30, 2020.
32. "CDC Updates COVID-19 Transmission Webpage to Clarify Information About Types of Spread," Centers for Disease Control and Prevention press release, May 22, 2020, https://www.cdc.gov/media/releases/2020/s0522-cdc-updates-covid-transmission.html; Laurel Wamsley, "CDC Publishes—Then Withdraws—Guidance On Aerosol Spread Of Coronavirus," National Public Radio, September 21, 2020; and Lena H. Sun and Ben Guarino, "CDC Says Airborne Transmission Plays a Role in Coronavirus Spread in a Long-Awaited Update after a Website Error Last Month," *Washington Post*, October 5, 2020.
33. Maggie Fox, "CDC Website now Emphasizes Coronavirus Spreads in the Air," CNN, May 7, 2021.
34. Roni Caryn Rabin et al., "In a Boost to Reopening Schools, C.D.C. Says Students Can Be 3 Feet Apart," *New York Times*, March 27, 2021.
35. Scott Gottlieb, "Where's the Science Behind CDC's 6-Foot Social-Distance Decree?," *Wall Street Journal*, March 21, 2021.
36. David Zweig, "The Problem with the CDC's Six-Foot Rule for Schools," *New York Magazine*, March 9, 2021.
37. Emily Anthes, "Three Feet or Six? Distancing Guideline for Schools Stirs Debate," *New York Times*, March 16, 2021.
38. World Health Organization, "COVID-19: Physical Distancing," https://www.who.int/westernpacific/emergencies/covid-19/information/physical-distancing.
39. William Booth, "Two Meters? One Meter Plus? Social Distancing Rules Prompt Fierce Debate in U.K.," *Washington Post*, June 22, 2020.
40. European Centre for Disease Prevention and Control, "Questions and Answers on COVID-19: Prevention," February 4, 2021, https://www.ecdc.europa.eu/en/covid-19/questions-answers/questions-answers-prevention.
41. Alvin Powell, "Is Air Conditioning Helping Spread COVID in the South?," *Harvard Gazette,* June 29, 2020; Christopher Flavelle, "Coronavirus Makes Cooling Centers Risky, Just as Scorching Weather Hits," *New York Times*, May 6, 2020; and Jianyun Lu et al., "COVID-19 Outbreak Associated with Air Conditioning in Restaurant, Guangzhou, China, 2020," *Emerging Infectious Diseases* 26, no. 7 (2020).
42. Scott Gottlieb and Yuval Levin, "New Rules for Covid Summer: Be Flexible and Vigilant," *Wall Street Journal*, June 14, 2020.
43. President's Coronavirus Guidelines for America, "15 Days to Slow the Spread."
44. Keith et al., "How 15 Days Became 45: Trump Extends Guidelines to Slow Coronavirus."
45. Fiona P. Havers et al., "Seroprevalence of Antibodies to SARS-CoV-2 in 10

Sites in the United States, March 23–May 12, 2020," *JAMA* 180, no. 12 (2020): 1576–86.

46. US Centers for Disease Control and Prevention, "Large-Scale Geographic Seroprevalence Surveys," October 2, 2020, https://www.cdc.gov/coronavirus/2019-ncov/cases-updates/geographic-seroprevalence-surveys.html.
47. Anne Schuchat, "What I Learned in 33 Years at the C.D.C.," *New York Times*, June 10, 2021.
48. Keith et al., "How 15 Days Became 45: Trump Extends Guidelines to Slow Coronavirus."
49. Zachary Basu, "Trump Announces 30-Day Extension of Coronavirus Guidelines," Axios, March 29, 2020.
50. "New at-Home COVID Test: Results in Minutes," Abbott press release, December 16, 2020; and Wudan Yan, "At-Home Coronavirus Testing Is Here," *New York Times*, March 5, 2021.
51. Scott Gottlieb, "Where's the Science behind CDC's 6-Foot Social-Distance Decree?," *Wall Street Journal*, March 21, 2021.
52. Ibid.
53. Kelly Servick, "How Will COVID-19 Affect the Coming Flu Season? Scientists Struggle for Clues," *Science*, August 14, 2020.
54. Staff Writer, "'Near Extinction' of Influenza in NZ as Numbers Drop Due to Lockdown," RNZ, October 11, 2020.
55. Brian Resnick, "We Wiped Out the Flu This Year. Could We Do It Again?," Vox, February 11, 2021.
56. Sarah Kaplan and Chris Mooney, "Genetic Data Show How a Single Superspreading Event Sent Coronavirus Across Massachusetts—and the Nation," *Washington Post*, August 25, 2020.
57. Hanna Krueger, "Biogen Conference in Boston Likely Linked to as Many as 300,000 COVID-19 Cases Worldwide, Researchers Say," *Boston Globe*, December 10, 2020.
58. Anne Schuchat, "Public Health Response to the Initiation and Spread of Pandemic COVID-19 in the United States, February 24–April 21, 2020," *Morbidity and Mortality Weekly Report* 69, no. 18 (2020): 551–6.
59. Isaac Ghinai et al., "Community Transmission of SARS-CoV-2 at Two Family Gatherings—Chicago, Illinois, February–March 2020," *Morbidity and Mortality Weekly Report* 69, no. 15 (2020): 446–50.
60. Steve Almasy, "Almost 100 People in Ohio Were Infected with Coronavirus After Man Attended Church Service," CNN, August 6, 2020.
61. Zeynep Tufekci, "Why Did It Take So Long to Accept the Facts About Covid?," *New York Times*, May 7, 2021.
62. Ramanan Laxminarayan et al., "Epidemiology and Transmission Dynamics of COVID-19 in Two Indian States," *Science* 370, no. 6517 (2020): 691–7; and Dillon C. Adam et al., Clustering and Superspreading Potential of SARS-CoV-2 Infections in Hong Kong," *Nature Medicine* 26, (2020): 1714–1719.
63. Silvia Stringhini et al., "Seroprevalence of Anti-SARS-CoV-2 IgG Antibodies in Geneva, Switzerland (SEROCoV-POP): A Population-based Study," *Lancet* 396, no. 10247 (2020): 313–9; and Oriol Guell, "Spain's Macro Study

Shows Just 5.2% of Population has Contracted the Coronavirus," El Pais, June 5, 2020.

64. Russell M. Viner et al., "Susceptibility to SARS-CoV-2 Infection Among Children and Adolescents Compared with Adults," *JAMA Pediatrics* 175, no. 2 (2021): 143–56.
65. Rebecca T. Leeb et al., "COVID-19 Trends Among School-Aged Children—United States, March 1–September 19, 2020," *Morbidity and Mortality Weekly Report* 69, no. 39 (2020): 1410–5.
66. Public Health England, "Investigation of Novel SARS-CoV-2 Variant: Variant of Concern 202012/01," January 22, 2021, https://www.gov.uk/government/publications/phe-investigation-of-novel-sars-cov-2-variant-of-concern-20201201-technical-briefing-3-6-january-2021.
67. Stuart P. Weisberg et al., "Distinct Antibody Responses to SARS-CoV-2 in Children and Adults Across the COVID-19 Clinical Spectrum," *Nature Immunology* 22 (2021): 25–31.
68. Apoorva Mandavilli, "Children Produce Weaker Coronavirus Antibodies, Study Finds," *New York Times*, February 24, 2021.
69. Xiaoxia Lu et al., "SARS-CoV-2 Infection in Children," *New England Journal of Medicine* 382 (2020): 1663–5.
70. Klara M. Posfay-Barbe et al., "COVID-19 in Children and the Dynamics of Infection in Families," *Pediatrics* 146, no. 2 (2020).
71. Juanjuan Zhang et al., "Changes in Contact Patterns Shape the Dynamics of the COVID-19 Outbreak in China," *Science* 368, no. 6498 (2020): 1481–6.
72. Eric Umansky, "My Kids' School Closed Again. So I Started Calling Experts," ProPublica, April 2, 2021.
73. Kanecia O. Zimmerman et al., "Incidence and Secondary Transmission of SARS-CoV-2 Infections in Schools," *Pediatrics* 147, no. 4 (2021).
74. Editorial Board, "Coronavirus Is Past Containment, but America Can Limit Epidemic: Q&A with Former FDA Chief," *USA Today*, March 10, 2020.
75. Scott Gottlieb, "Schools Can Open Safely This Fall," *Wall Street Journal*, July 12, 2020.
76. Douglas N. Harris, Engy Ziedan, and Susan Hassig, "The Effects of School Reopenings on COVID-19 Hospitalizations," National Center for Research on Education Access and Choice, January 4, 2021.
77. Zimmerman et al., "Incidence and Secondary Transmission of SARS-CoV-2 Infections in Schools."
78. Amy Falk et al., "COVID-19 Cases and Transmission in 17 K–12 Schools—Wood County, Wisconsin, August 31–November 29, 2020," *Morbidity and Mortality Weekly Report* 70, no. 4 (2021): 136–40.

Chapter 13: The Information Desert

1. US Centers for Disease Control and Prevention, "About the Morbidity and Mortality Weekly Report (MMWR) Series," August 25, 2020, https://www.cdc.gov/mmwr/about.html.
2. Susan Scutti, "Flu Season Deaths Top 80,000 Last Year, CDC Says," CNN, September 27, 2018.
3. US Food and Drug Administration, "Statement from FDA Commissioner

Scott Gottlieb, M.D. on the Efficacy of the 2017–2018 Influenza Vaccine," February 15, 2018, https://www.fda.gov/news-events/press-announcements/statement-fda-commissioner-scott-gottlieb-md-efficacy-2017-2018-influenza-vaccine.

4. US Centers for Disease Control and Prevention, "Summary of the 2017–2018 Influenza Season," September 5, 2019, https://www.cdc.gov/flu/about/season/flu-season-2017-2018.html.
5. US Centers for Disease Control and Prevention, "Novel Coronavirus Reports," March 8, 2021, https://www.cdc.gov/mmwr/Novel_Coronavirus_Reports.html.
6. Slavitt, *Preventable: The Inside Story of How Leadership Failures, Politics, and Selfishness Doomed the U.S. Coronavirus Response*, page 28.
7. US Centers for Disease Control and Prevention, "Discontinuation of Isolation for Persons with COVID-19 Not in Healthcare Settings," February 18, 2021, https://www.cdc.gov/coronavirus/2019-ncov/hcp/disposition-in-home-patients.html.
8. Julia C. Pringle et al., "COVID-19 in a Correctional Facility Employee Following Multiple Brief Exposures to Persons with COVID-19—Vermont, July–August 2020," *Morbidity and Mortality Weekly Report* 69, no. 43 (2020): 1569–70; and KHN Morning Briefing, "What Does 'Close Contact' Mean? CDC Redefines COVID Exposure Time to Qualify," Kaiser Health News, October 22, 2020.
9. Matt Hoffman, "14-Minute Shuffle: Billings Schools Retreat from Controversial Policy after Criticism from Health Experts," *Billings Gazette*, October 21, 2020.
10. Zeynep Tufekci, "5 Pandemic Mistakes We Keep Repeating," *Atlantic*, February 26, 2021.
11. Erin Banco and Adam Cancryn, "Mask Controversy Spurs CDC to Rethink its Pandemic Response," *Politico*, May 17, 2021.
12. Lauren Camera, "DeVos: Not My Job to Track School Reopening Plans," *US News & World Report*, October 20, 2020.
13. Christakis, *Apollo's Arrow*, page 119.
14. Christopher J. L. Murray, "Why Can't We See All of the Government's Virus Data?," *New York Times*, October 23, 2020.
15. Richard A. Oppel Jr. et al., "The Fullest Look Yet at the Racial Inequity of Coronavirus," *New York Times*, July 5, 2020.
16. Nicholas Florko and Eric Boodman, "How HHS's New Hospital Data Reporting System Will Actually Affect the U.S. Covid-19 Response," STAT, July 16, 2020.
17. US Centers for Disease Control and Prevention, "About NHSN," January 25, 2021, https://www.cdc.gov/nhsn/about-nhsn/index.html.
18. US Centers for Disease Control and Prevention, "Estimated Influenza Illnesses, Medical Visits, Hospitalizations, and Deaths in the United States—2018–2019 Influenza Season," January 8, 2020, https://www.cdc.gov/flu/about/burden/2018-2019.html.
19. Sheila Kaplan, "Ex-C.D.C. Chief on Challenge of Serving Trump during Pandemic," *New York Times*, January 20, 2021.

20. Charles Piller, "Federal Hospital Data System Falters at Tracking Pandemic," *Science* 370, no. 6521 (2020): 1148–9.
21. "Final Report Confirms Remdesivir Benefits for COVID-19," US National Institutes of Health press release, October 20, 2020, https://www.nih.gov/news-events/nih-research-matters/final-report-confirms-remdesivir-benefits-covid-19.
22. Piller, "Federal Hospital Data System Falters at Tracking Pandemic."
    The HHS effort to create the new federal portal was widely and wrongly reported as a measure of Trump administration interference in the reporting of bottom-line Covid data. It came before the election, and many public health experts worried that the administration would use its new exercise of control over the collection and reporting of this information to tip the scales on what was ultimately released, "cooking" the numbers.
23. Pien Huang, "Hospitals That Don't Report COVID-19 Data Daily Will Get Warning Letter," National Public Radio, October 7, 2020.
24. Robinson Meyer and Alexis C. Madrigal, "Why the Pandemic Experts Failed," *Atlantic*, March 15, 2021.
25. Sheryl Gay Stolberg, "'Sole Source' Contract for Covid-19 Database Draws Scrutiny From Democrats," *New York Times*, August 14, 2020.
26. Patty Murray, "Murray Demands Answers Regarding Non-Competitive, Multimillion Dollar Contract for Duplicative Health Data System," June 3, 2020
27. Lewis, *The Premonition*, page 126.
28. Kiva A. Fisher et al., "Community and Close Contact Exposures Associated with COVID-19 Among Symptomatic Adults ≥18 Years in 11 Outpatient Health Care Facilities—United States, July 2020," *Morbidity and Mortality Weekly Report* 69, no. 36 (2020): 1258–64.
29. Alicia Kelso, "Dining Out Increases COVID-19 Risk More than Other Activities, CDC Report Finds," Restaurant Dive, September 11, 2020; and Molly Walker, "CDC: Dining Out Tied to Coronavirus Infection," MedPage Today, September 10, 2020.
30. Pandemic and All-Hazards Preparedness Act, Pub. L. No. 109–417, 42 USC § 319D (2006).
31. US Government Accountability Office, "Public Health Information Technology: HHS Has Made Little Progress toward Implementing Enhanced Situational Awareness Network Capabilities," September 2017; and Yuval Levin, "Biden's Pandemic-Policy Challenge," *National Review*, January 18, 2021.
32. Melissa Healy, "California's Coronavirus Strain Looks Increasingly Dangerous: 'The Devil Is Already Here,'" *Los Angeles Times*, February 23, 2021.
33. Public Health England, "SARS-CoV-2 Variants of Concern and Variants under Investigation in England: Technical Briefing 18," July 9, 2021, https://assets.publishing.service.gov.uk/government/uploads/system/uploads/attachment_data/file/1001358/Variants_of_Concern_VOC_Technical_Briefing_18.pdf.
34. Harry Stevens and Miriam Berger, "U.S. Ranks 43rd Worldwide in Sequencing to Check for Coronavirus Variants Like the One Found in the U.K.," *Washington Post*, December 23, 2020.
35. William Wan and Ben Guarino, "Why America Is 'Flying Blind' to the Coronavirus Mutations Racing Across the Globe," *Washington Post*, January 29, 2021.

36. Caitlin Rivers and Dylan George, "How to Forecast Outbreaks and Pandemics," *Foreign Affairs*, June 29, 2020.
37. Adam Rogers, "It's Time for a National Pandemic Prediction Agency," *WIRED*, February 3, 2021.
38. Dan Diamond and Nahal Toosi, "Trump Team Failed to Follow NSC's Pandemic Playbook," *Politico*, March 25, 2020.
39. Scott Gottlieb et al., "National Coronavirus Response: A Roadmap to Reopening," American Enterprise Institute, March 29, 2020, https://www.aei.org/research-products/report/national-coronavirus-response-a-road-map-to-reopening/.
40. Rev.com, "Donald Trump Coronavirus Task Force Press Conference Transcript March 30," March 30, 2020.
41. Kathryn Watson, "Trump Announces CDC Recommends Cloth Masks in Public but Says He Won't Wear One," CBS News, April 3, 2020.
42. Lynne Peeples, "Face Masks: What the Data Say," *Nature*, October 6, 2020.
43. Christopher Leffler et al., "Association of Country-Wide Coronavirus Mortality with Demographics, Testing, Lockdowns, and Public Wearing of Masks," *American Journal of Tropical Medicine and Hygiene* 103, no. 6 (2020): 2400–11.
44. Wei Lyu and George L. Wehby, "Community Use of Face Masks and COVID-19: Evidence from a Natural Experiment of State Mandates in the US," *Health Affairs* 39, no. 8 (2020).
45. Scott Gottlieb, "Some Masks will Protect you Better than Others," The Wall Street Journal, November 22, 2020.
46. *Sacramento Bee*, "Campaign Against Gauze Masks Is Without Facts," November 5, 1918.
47. Megan Garber, "Atlas Coughed," *The Atlantic*, October 9, 2020.
48. Kelly Mena and Veronica Stracqualursi, "Trump Wears a Mask during Visit to Wounded Service Members at Walter Reed," CNN, July 11, 2020.
49. Michael D. Shear et al., "Trump's Focus as the Pandemic Raged: What Would It Mean for Him?," *New York Times*, January 13, 2021.
50. Paul M. Sharp and Beatrice H. Hahn, "Origins of HIV and the AIDS Pandemic," *Cold Spring Harbor Perspectives in Medicine* 1, no. 1 (2011).

Chapter 14: Hardened Sites

1. Hyonhee Shin, "South Korea's Emergency Exercise in December Facilitated Coronavirus Testing, Containment," Reuters, March 30, 2020.
2. Ibid.
3. Eun Kyung Choi and Jong-Koo Lee, "Changes of Global Infectious Disease Governance in 2000s: Rise of Global Health Security and Transformation of Infectious Disease Control System in South Korea," *Uisahak* 25, no. 3 (2016): 489–518.
4. World Health Organization, "Middle East Respiratory Syndrome Coronavirus (MERS-CoV) in the Republic of Korea," June 2, 2015, https://www.who.int/mediacentre/news/situation-assessments/2-june-2015-south-korea/en/.

5. Sun Young Cho et al., "MERS-CoV Outbreak Following a Single Patient Exposure in an Emergency Room in South Korea: An Epidemiological Outbreak Study," *Lancet* 388, no. 10048 (2016): 994–1001.
6. BBC, "South Korea Hospital 'Is Source of Many Mers Cases,'" June 14, 2015.
7. Jeyup S. Kwaak, "South Korea MERS Outbreak Brings Focus on Samsung Hospital," *Wall Street Journal*, June 18, 2015.
8. BBC, "South Korea Hospital 'Is Source of Many Mers Cases'"; and Jack Kim, "Hospital at Center of South Korea's MERS Suspends Services; Seven New Cases," Reuters, June 13, 2015.
9. Ministry of Health and Welfare, "Measures to Reform National Infection Prevention and Control System for the Purpose of Immediate Response to Emerging Infectious Diseases," September 1, 2015, http://www.mohw.go.kr/eng/nw/nw0101vw.jsp?PAR_MENU_ID=1007&MENU_ID=100701&page=1&CONT_SEQ=326060.
10. Bruce Klinger, "South Korea Provides Lessons, Good and Bad, on Coronavirus Response," Heritage Foundation, March 28, 2020, https://www.heritage.org/asia/commentary/south-korea-provides-lessons-good-and-bad-coronavirus-response; and Joon-Young Song et al., "Covid-19 in South Korea—Challenges of Subclinical Manifestations," *New England Journal of Medicine* 382 (2020): 1858–9.
11. The Korean regulatory process provides regulators flexibility in allowing the use of unapproved medical equipment as part of a rapid response to epidemics.
12. Chad Terhune et al., "Special Report: How Korea Trounced U.S. in Race to Test People for Coronavirus," Reuters, March 18, 2020.
13. Timothy W. Martin and Dasl Yoon, "How South Korea Successfully Managed Coronavirus," *Wall Street Journal*, September 25, 2020.
14. Debra Ladner, Katsumasa Hamaguchi, and Kyuri Kim, "The Republic of Korea's First 70 Days of Responding to the COVID-19 Outbreak," Global Delivery Initiative, April 13, 2020.
15. Timothy W. Martin and Dasl Yoon, "How South Korea Successfully Managed Coronavirus," *Wall Street Journal*, September 25, 2020.
16. Terhune et al., "Special Report: How Korea Trounced U .S. in Race to Test People for Coronavirus."
17. Ibid.
18. Ladner, Hamaguchi, and Kim, "The Republic of Korea's First 70 Days of Responding to the COVID–19 Outbreak."
19. Ibid.
20. Sharon Begley, "A Plea from Doctors in Italy: To Avoid Covid-19 Disaster, Treat More Patients at Home," STAT, March 21, 2020.
21. George Baca, "Eastern Surveillance, Western Malaise, and South Korea's COVID-19 Response: Oligarchic Power in Hell Joseon," *Dialectical Anthropology* (2020): 1–7.
22. Keren Landman, "What We Can Learn from South Korea's Coronavirus Response," Elemental, June 1, 2020.
23. "Across different initial numbers of cases, the majority of scenarios with an

R0 of 1·5 were controllable with less than 50 percent of contacts successfully traced. To control the majority of outbreaks, for R0 of 2·5 more than 70 percent of contacts had to be traced, and for an R0 of 3·5 more than 90 percent of contacts had to be traced. In most scenarios, highly effective contact tracing and case isolation is enough to control a new outbreak of COVID-19 within 3 months."

See Joel Hellewell et al., "Feasibility of Controlling COVID-19 Outbreaks by Isolation of Cases and Contacts," *Lancet Global Health* 8, no. 4 (2020): 488–96.

"Taking recent estimates for COVID-19 transmission we predict that under effective contact tracing less than 1 in 6 cases will generate any subsequent untraced infections, although this comes at a high logistical burden with an average of 36 individuals traced per case. . . . The current contact tracing strategy within the UK is likely to identify a sufficient proportion of infected individuals such that subsequent spread could be prevented, although the ultimate success will depend on the rapid detection of cases and isolation of contacts."

See Matt J. Keeling, T. Deirdre Hollingsworth, and Jonathan M. Read, "Efficacy of Contact Tracing for the Containment of the 2019 Novel Coronavirus (COVID-19)," *Journal of Epidemiology and Community Health* 74, no. 10 (2020): 861–6.

24. Justin Fendos, "How Surveillance Technology Powered South Korea's COVID-19 Response," TechStream, April 29, 2020.
25. Heesu Lee, "These Elite Contact Tracers Show the World How to Beat Covid-19," Bloomberg, July 25, 2020.
26. Gregg A. Brazinsky, "South Korea Is Winning the Fight Against Covid-19. The U.S. Is Failing.," *Washington Post*, April 10, 2020.
27. Slavitt, *Preventable: The Inside Story of How Leadership Failures, Politics, and Selfishness Doomed the U.S. Coronavirus Response*, page 28.
28. Choe Sang-Hun, "South Korean Leader Said Coronavirus Would 'Disappear.' It Was a Costly Error.," *New York Times*, March 2, 2020.
29. Andrea Crisanti and Antonio Cassone, "In One Italian Town, We Showed Mass Testing Could Eradicate the Coronavirus," *Guardian*, March 20, 2020.
30. Franco Ordonez, "Ex-Officials Call For $46 Billion for Tracing, Isolating in Next Coronavirus Package," National Public Radio, April 27, 2020.
31. CNBC, "Gottlieb, AEI's Michael Strain: Congress Should Support Cities Amid Coronavirus Outbreak," March 12, 2020.
32. Scott Gottlieb (@ScottGottliebMD), "THREAD: In U.S. we face two alternative but hard outlooks with #COVID19: that we follow a path similar to South Korea or one closer to Italy. We probably lost chance to have an outcome like South Korea. We must do everything to avert the tragic suffering being borne by Italy 1/10," Twitter, March 12, 2020, 7:03 a.m.
33. Ibid.
34. Laurie McGinley, "FDA, Industry Step Up Efforts to Avert Drug Shortages after Puerto Rico Hurricane," *Washington Post*, September 29, 2017; and Katie Thomas and Sheila Kaplan, "Hurricane Damage in Puerto Rico Leads to Fears of Drug Shortages Nationwide," *New York Times*, October 4, 2017.

35. US Department of Homeland Security, "Implementing 9/11 Commission Recommendations: Progress Report 2011," 2011, https://www.dhs.gov/xlibrary/assets/implementing-9-11-commission-report-progress-2011.pdf.
36. Vijay Singh et al., "Radiation Countermeasure Agents: An Update (2011–2014)," *Expert Opinion on Therapeutic Patents* 24, no. 11 (2014): 1229–55; and Vijay Singh et al., "Medical Countermeasures for Unwanted CBRN Exposures: Part II Radiological and Nuclear Threats with Review of Recent Countermeasure Patents," *Expert Opinion on Therapeutic Patents* 26, no. 12 (2016): 1399–1408.
37. US Food and Drug Administration, "FDA Approves Radiation Medical Countermeasure," April 30, 2019, https://www.fda.gov/emergency-preparedness-and-response/about-mcmi/fda-approves-radiation-medical-countermeasure.
38. US Department of Health and Human Services, "HHS Boosts Stockpile of Products to Treat Acute Radiation Syndrome," September 26, 2013, https://www.emsworld.com/news/11178384/hhs-boosts-stockpile-products-treat-acute-radiation-syndrome.
39. Katie Thomas and Knvul Sheikh, "Estimates Fall Short of F.D.A.'s Pledge for 1 Million Coronavirus Tests," *New York Times*, March 3, 2020.
40. Ali S. Khan, "Public Health Preparedness and Response in the USA Since 9/11: A National Health Security Imperative," *Lancet* 378, no. 9794 (2011): 953–6.
41. Due to the temporal nature of Zika virus RNA in blood and urine samples, a negative blood test for the infection's genetic material (using PCR) doesn't mean that a person isn't infected. The levels of virus in the blood can wax and wane and impact the ability to detect its RNA through the kinds of tests used to screen for COVID. As a consequence, to test for infections like Zika, providers often also want to use serum tests that screen for the antibodies that the body produces in response to infection. See US Centers for Disease Control and Prevention, "Testing for Zika Virus Infections," June 13, 2019, https://www.cdc.gov/zika/laboratories/types-of-tests.html.
42. US Centers for Disease Control and Prevention, "Technology Transfer: Frequently Asked Questions," October 11, 2017, https://www.cdc.gov/os/technology/techtransfer/faq.htm.
43. "Researchers from Chiron Corp. and the Centers for Disease Control and Prevention (CDC) finally unmasked the insidious agent in 1988. The next year, they jointly published papers in *Science* describing the new virus and a way to test for it in blood samples." See Jon Cohen, "The Scientific Challenge of Hepatitis C," *Science* 285, no. 5424 (1999): 26–30; and *Bradley v. Chiron Corp.*, 136 F.3d 1317 (Fed. Cir. 1998).
44. V. Slind-Flor, "Chiron Challenged on Hepatitis-C Patent," *Science* 267, no. 5194 (1995): 23.
45. Ibid.
46. Paul Elias, "Feds Review Complaints That Chiron Hinders HCV Research," Associated Press, February 28, 2004.
47. Sarah Karlin-Smith (@SarahKarlin), "This slide from @BARDA Acting Director Gary Disbrow is a good visual of US prioritizing funding for COVID-19 vaccines versus therapeutics," Twitter, October 27, 2020, 10:18 a.m.

48. US Government Accountability Office, "Zika Supplemental Funding: Status of HHS Agencies' Obligations, Disbursements, and the Activities Funded," May 2018, https://www.gao.gov/assets/gao-18-389.pdf.

Chapter 15: Evidence Is Hard to Collect in a Crisis

1. US Centers for Disease Control and Prevention, "CDC Telebriefing—Updates On Middle East Respiratory Syndrome Coronavirus (MERS-coV) Investigation In The United States," May 17, 2014.
2. Stephanie R. Bialek et al., "First Confirmed Cases of Middle East Respiratory Syndrome Coronavirus (MERS-CoV) Infection in the United States, Updated Information on the Epidemiology of MERS-CoV Infection, and Guidance for the Public, Clinicians, and Public Health Authorities," *Morbidity and Mortality Weekly Report* 63, no. 19 (2014): 431–6.
3. Florida Department of Health Communications, "Health Officials Confirm First MERS-CoV Case in Florida," May 12, 2014, http://www.floridahealth.gov/diseases-and-conditions/mers/_documents/mers-cov-press-release.pdf.
4. Aisha M. Al-Osail and Marwan J. Al-Wazzah, "The History and Epidemiology of Middle East Respiratory Syndrome Corona Virus," *Multidisciplinary Respiratory Medicine* 12, no. 20 (2017).
5. World Health Organization, "Middle East Respiratory Syndrome Coronavirus (MERS-CoV)—Saudi Arabia," December 4, 2015, https://www.who.int/csr/don/4-december-2015-mers-saudi-arabia/en/.
6. Al-Osail and Al-Wazzah, "The History and Epidemiology of Middle East Respiratory Syndrome Corona Virus."
7. "A Chronicle on the SARS Epidemic, Chinese Law & Government," *Chinese Law & Government* 36, no. 4 (2003): 12–5.
8. Al-Osail and Al-Wazzah, "The History and Epidemiology of Middle East Respiratory Syndrome Corona Virus."
9. Nicholas Wade, "Origin of Covid—Following the Clues," Medium, May 3, 2021.
10. Stanley Plotkin and Susan Plotkin, "The Development of Vaccines: How the Past Led to the Future," *Nature Reviews Microbiology* 9 (2011): 889–93.
11. Stephen S. Morse and Ann Schluederberg, "Emerging Viruses: The Evolution of Viruses and Viral Disease," *Journal of Infectious Diseases* 162, no. 1 (1990).
12. Simon J. Anthony et al., "Global Patterns in Coronavirus Diversity," *Virus Evolution* 3, no. 1 (2017).
13. Yi Fan et al., "Bat Coronaviruses in China," *Viruses* 11, no. 3 (2019): 210.
14. McKay and Dvorak, "A Deadly Coronavirus Was Inevitable. Why Was No One Ready?"
15. World Health Organization, "Prioritizing Diseases for Research and Development in Emergency Contexts," https://www.who.int/activities/prioritizing-diseases-for-research-and-development-in-emergency-contexts.
16. The UK and EU were added to its roster of international sponsors.
    Clive Cookson and Tim Bradshaw, "Davos Launch for Coalition to Prevent Epidemics of Emerging Viruses," *Financial Times*, January 18, 2017.
17. Jason Beaubien, "Ebola Never Went Away. But Now There's a Drug to Treat It," National Public Radio, October 20, 2020.

18. World Health Organization, "Prioritizing Diseases for Research and Development in Emergency Contexts," https://www.who.int/activities/prioritizing-diseases-for-research-and-development-in-emergency-contexts.
19. Steve Usdin, "DARPA's Gambles Might Have Created the Best Hopes for Stopping COVID-19," BioCentury, March 19, 2020.
20. Meg Tirrell (@megtirrell), "Dr. Janet Woodcock, 34-year veteran of FDA, says 90% of the clinical trials run during the pandemic won't yield actionable results that will change practice. Tune in for our discussion about why at 4:30pmET today—and send Qs w hashtag #HealthyReturns.," Twitter, October 14, 2020, 3:47 p.m.
21. Congressional Research Service, "The Defense Production Act of 1950: History, Authorities, and Considerations for Congress," March 20, 2020, https://fas.org/sgp/crs/natsec/R43767.pdf.
22. Andrew Jacobs, "Despite Claims, Trump Rarely Uses Wartime Law in Battle Against Covid," *New York Times* January 20, 2021; and White House Office of Trade and Manufacturing Policy, "How President Trump Uses the Defense Production Act to Protect Americans from the China Virus," August 2020, https://www.documentcloud.org/documents/7036228-OTMP-DPA-Report-FINAL-8-13-20.html#document/p4.
23. Jacqui Wise and Rebecca Coombes, "Covid-19: The Inside Story of the RECOVERY Trial," *British Medical Journal* 370, no. 2670 (2020).
24. RECOVERY Collaborative Group, "Dexamethasone in Hospitalized Patients with Covid-19," *New England Journal of Medicine* 384 (2021): 693–704.
25. Scott Gottlieb and Mark McClellan, "Rules for Clinical Trials in a Pandemic," *Wall Street Journal*, June 21, 2020.
26. Ibid.
27. Nicholas J. DeVito, Michael Liu, and Jeffrey K Aronson, "COVID-19 Clinical Trials Report Card: Chloroquine and Hydroxychloroquine," Oxford COVID-19 Centre for Evidence-Based Medicine, May 11, 2020.
28. US Food and Drug Administration, "Electronic Reading Room: COVID," May 11, 2021, https://www.fda.gov/regulatory-information/freedom-information/electronic-reading-room; and Steve Usdin, "FDA Documents Shed Light on Chaotic COVID Decision-Making during Trump Administration," Biocentury, May 14, 2021.
29. Tina Nguyen, "How a Chance Twitter Thread Launched Trump's Favorite Coronavirus Drug," *Politico*, April 7, 2020.
30. Kevin Roose and Matthew Rosenberg, "Touting Virus Cure, 'Simple Country Doctor' Becomes a Right-Wing Star," *New York Times*, April 2; and Katherine Eban, "'I'll Send You the Contact': Documents Expose FDA Commissioner's Personal Interventions on Behalf of Trump's Favorite Chloroquine Doctor," *Vanity Fair*, May 27, 2020.
31. Wright, "The Plague Year."
32. Katherine Eban, "'A Tsunami of Randoms': How Trump's COVID Chaos Drowned the FDA in Junk Science," Vanity Fair, January 19, 2021.
33. Christopher Rowland, Debbie Cenziper, and Lisa Rein, "White House Sidestepped FDA to Distribute Hydroxychloroquine to Pharmacies, Documents

Show. Trump Touted the Pills to Treat Covid-19," *Washington Post*, October 31, 2020.

34. Office of the Assistant Secretary for Preparedness and Response, "HHS Accepts Donations of Medicine to Strategic National Stockpile as Possible Treatments for COVID-19 Patients," March 29, 2020, https://www.hhs.gov/about/news/2020/03/29/hhs-accepts-donations-of-medicine-to-strategic-national-stockpile-as-possible-treatments-for-covid-19-patients.html.
35. Yasmeen Abutaleb, Laurie McGinley, and Josh Dawsey, "Oracle to Partner with Trump Administration to Collect Data on Unproven Drugs to Treat Covid-19," *Washington Post*, March 24, 2020.
36. The Public Readiness and Emergency Preparedness (PREP) Act authorizes the secretary of the Department of Health and Human Services to issue a PREP Act declaration. The declaration provides immunity from liability (except for willful misconduct) for claims: of loss caused, arising out of, relating to, or resulting from administration or use of countermeasures to diseases; threats and conditions determined by the secretary to constitute a present, or credible risk of a future public health emergency; and to entities and individuals involved in the development, manufacture, testing, distribution, administration, and use of such countermeasures. See US Department of Health and Human Services, "Public Readiness and Emergency Preparedness Act," March 16, 2021, https://www.phe.gov/Preparedness/legal/prepact/Pages/default.aspx
37. Rowland, Cenziper, and Rein, "White House Sidestepped FDA to Distribute Hydroxychloroquine to Pharmacies, Documents Show. Trump Touted the Pills to Treat Covid-19."
38. US Food and Drug Administration, Center for Drug Evaluation and Research Office of Surveillance and Epidemiology, "Pharmacovigilance Memorandum: Hydroxychloroquine and Chloroquine," May 19, 2020, https://www.accessdata.fda.gov/drugsatfda_docs/nda/2020/OSE%20Review_Hydroxychloroquine-Cholorquine%20-%2019May2020_Redacted.pdf.
39. Rowland, Cenziper, and Rein, "White House Sidestepped FDA to Distribute Hydroxychloroquine to Pharmacies, Documents Show. Trump Touted the Pills to Treat Covid-19."
40. FDA Commissioner Steve Hahn would write in a blog post on May 29, 2020: "Since the hydroxychloroquine product is FDA-approved for other uses with a prescription, the donation agreements also allowed for it to be distributed for clinical trials and to the commercial market for use in the outpatient setting, if appropriate." See Stephen M. Hahn, "Bringing a Cancer Doctor's Perspective to FDA's Response to the COVID-19 Pandemic," May 29, 2020, https://www.fda.gov/news-events/fda-voices/bringing-cancer-doctors-perspective-fdas-response-covid-19-pandemic; and Rowland, Cenziper, and Rein, "White House Sidestepped FDA to Distribute Hydroxychloroquine to Pharmacies, Documents Show. Trump Touted the Pills to Treat Covid-19."
41. US Food and Drug Administration, "Coronavirus (COVID-19) Update: FDA Revokes Emergency Use Authorization for Chloroquine and Hydroxychloroquine," June 15, 2020, https://www.fda.gov/news-events/press

-announcements/coronavirus-covid-19-update-fda-revokes-emergency-use -authorization-chloroquine-and.

42. US Food and Drug Administration, Center for Drug Evaluation and Research Office of Surveillance and Epidemiology, "Pharmacovigilance Memorandum."
43. Laurie McGinley, "Controversy Erupts over Plan to Let Pentagon Authorize Unapproved Drugs for Battlefield Use," *Washington Post*, November 9, 2017.
44. Laurie McGinley and Mark Berman, "Justice Department Says FDA 'Lacks Jurisdiction' over Death-Penalty Drugs," *Washington Post*, May 14, 2019.
45. Leon G. Smith et al., "Observational Study on 255 Mechanically Ventilated Covid Patients at the Beginning of the USA Pandemic," medRxiv, May 31, 2021; and Harriet Alexander, "Was Trump Right About Hydroxychloroquine All Along? New Study Shows Drug Touted by Former President can Increase COVID Survival Rates by 200%," Daily Mail, June 10, 2021.
46. US Food and Drug Administration, "FDA COVID-19 Pandemic Recovery and Preparedness Plan (PREPP) Initiative," January 27, 2021, https://www.fda.gov/about-fda/reports/fda-covid-19-pandemic-recovery-and-preparedness-plan-prepp-initiative.
47. Levin, "Biden's Pandemic-Policy Challenge."
48. Richard Harris, "FDA's Hahn Apologizes for Overselling Plasma's Benefits as a COVID-19 Treatment," National Public Radio, August 25, 2020.
49. Steve Usdin, "FDA Documents Shed Light on Chaotic COVID Decision-making during Trump Administration," Biocentury, May 14, 2021.
50. Marcus Banks, "NIH Halts Outpatient COVID-19 Convalescent Plasma Trial," Scientist, March 4, 2021.
51. RECOVERY Collaborative Group, "Convalescent Plasma in Patients Admitted to Hospital with COVID-19 (RECOVERY): a Randomized Controlled, Open-Label Platform Trial," *Lancet* (2021).
52. Katie Thomas and Noah Weiland, "The Covid-19 Plasma Boom Is Over. What Did We Learn From It?," *New York Times*, April 17, 2021.
53. Matthew Perrone and Deb Reichmann, "FDA Chief Apologizes for Overstating Plasma Effect on Virus," ABC News, August 25, 2020.
54. US Food and Drug Administration, Center for Biologics Evaluation and Research, "Development and Licensure of Vaccines to Prevent COVID-19: Guidance for Industry," January 2020, https://www.fda.gov/media/139638/download.
55. COVID Tracking Project, "The Long-Term Care COVID Tracker," https://covidtracking.com/nursing-homes-long-term-care-facilities.
56. Ibid.
57. "Trump Administration Issues Call to Action Based on New Data Detailing COVID-19 Impacts on Medicare Beneficiaries," US Centers for Medicare & Medicaid Services press release, June 22, 2020, https://www.cms.gov/newsroom/press-releases/trump-administration-issues-call-action-based-new-data-detailing-covid-19-impacts-medicare.
58. Denise Chow, "Coronavirus Missteps from CDC and FDA Worry Health Experts," NBC News, August 31, 2021.

59. Matthew Herper and Nicholas Florko, "How Key Decisions Slowed FDA's Review of a Covid-19 Vaccine—but also Gave It Important Data," STAT, December 4, 2020.

Chapter 16: Getting Drugs to Patients

1. Maggie Haberman and Michael D. Shear, "Trump Says He'll Begin 'Quarantine Process' After Hope Hicks Tests Positive for Coronavirus," *New York Times*, October 1, 2020; and Luke Harding, "Hicks, Hubris and Not a Lot of Masks: The Week Trump Caught Covid," *Guardian*, October 2, 2020.
2. Michael C. Bender and Rebecca Ballhaus, "Trump Didn't Disclose First Positive Covid-19 Test While Awaiting a Second Test on Thursday," *Wall Street Journal*, October 4, 2020.
3. Rev.com, "Donald Trump Sean Hannity Interview Transcript After Debate: Condemns White Supremacists, Talks Hope Hicks Testing Positive for COVID," October 1, 2020.
4. Harriet Sinclair, "Trump Says He Is Scared of Germs and Needs to Drink from a Straw to Avoid Contamination," *Newsweek*, September 26, 2017.
5. Jill Colvin and Zeke Miller, "Trump Says He and First Lady Tested Positive for Coronavirus," Associated Press, October 1, 2020.
6. Noah Weiland et al., "Trump Was Sicker than Acknowledged with Covid-19," *New York Times*, February 28, 2021.
7. US Food and Drug Administration, "Emergency IND Timeline," February 20, 2018, https://www.fda.gov/drugs/investigational-new-drug-ind-application/emergency-ind-timeline.
8. Weiland et al., "Trump Was Sicker than Acknowledged with Covid-19."
9. D. M. Weinreich et al., "REGN-COV2, a Neutralizing Antibody Cocktail, in Outpatients with Covid-19," *New England Journal of Medicine* 384, no. 3 (2020): 238–51; and Peter Chen et al., "SARS-CoV-2 Neutralizing Antibody LY-CoV555 in Outpatients with Covid-19," *New England Journal of Medicine* 384, no. 3 (2020): 229–37.
10. Maggie Haberman (@MaggieNYT), "Potus doctor note on his treatment > [Sean Conley Statement]," Twitter, October 2, 2020, 4:15 p.m.
11. Yasmeen Abutaleb and Damian Paletta, *Nightmare Scenario: Inside the Trump Administration's Response to the Pandemic That Changed History* (New York: HarperCollins, 2021).
12. Weiland et al., "Trump Was Sicker than Acknowledged with Covid-19."
13. Kayleigh McEnany 45 Archived (@PressSec45), "A Wednesday update from President @realDonaldTrump's physician: [Sean Conley Statement]," Twitter, October 7, 2020, 12:51 p.m.
14. Ritu Pasrijaa and Mohammad Naimeb, "The Deregulated Immune Reaction and Cytokines Release Storm (CRS) in COVID-19 Disease," *International Immunopharmacology* 90 (2021).
15. Sonu Bhaskar et al., "Cytokine Storm in COVID-19-Immunopathological Mechanisms, Clinical Considerations, and Therapeutic Approaches: The REPROGRAM Consortium Position Paper," *Frontiers in Immunology* 11 (2020): 1648.

16. Siddhartha Mukherjee, "Before Virus, after Virus: A Reckoning," *Cell* 183, no. 2 (2020): 308–14.
17. Damian Paletta and Yasmeen Abutaleb, "Inside the Extraordinary Effort to Save Trump from Covid-19," *Washington Post*, June 24, 2021; and Alistair Smout, "Regeneron's Antibody Therapy Cuts Deaths Among some Hospitalised COVID–19 Patients—Study," Reuters, June 16, 2021.
18. Jim Acosta and Caroline Kelly, "Donald and Melania Trump Received Covid Vaccine at the White House in January," CNN, March 1, 2021.
19. Carl O'Donnell and Michael Erman, "Eli Lilly's Combo Therapy for COVID-19 Cuts Serious Illness and Death in Large Study," Reuters, March 10, 2021.
20. Steve Usdin, "Christie's COVID-19 Treatment Raises Fairness, Safety Questions," *BioCentury*, October 16, 2020.
21. Paul LeBlanc, "Christie Reveals He Spent 7 Days in ICU and Admits He Was 'Wrong' to Not Wear a Mask," CNN, October 16, 2020.
22. Zachary Brennan, "Regeneron's Covid mAb As a Prophylactic Injection: Reduced Risk of Symptomatic Inection," Endpoints, April 12, 2021.
23. Sabue Mulangu et al., "A Randomized, Controlled Trial of Ebola Virus Disease Therapeutics," *New England Journal of Medicine* 381 (2019): 2293–303.
24. Johanna Hansen et al., "Studies in Humanized Mice and Convalescent Humans Yield a SARS-CoV-2 Antibody Cocktail," *Science* 369, no. 6506 (2020): 1010–4.
25. Robert Langreth and Susan Berfield, "Antibody Treatments May Be the Best Hope against the Virus until a Vaccine," Bloomberg, April 20, 2020.
26. Stacy Lawrence, "Regeneron, BARDA Partner to Develop MERS Virus Treatment," Fierce Biotech, August 22, 2016; and "Antibody Treatment for MERS Coronavirus Safe in People," US National Institutes of Health press release, March 2, 2021.
27. George D. Yancopoulos, cofounder, president, and chief scientific officer of Regeneron Pharmaceuticals, in conversation with the author.
28. Steve Usdin, "What Lessons from Ebola Can Prepare Society for Other Outbreaks?," *BioCentury*, April 27, 2015.
29. Robert Langreth and Riley Griffin, "Trump Got One Covid Antibody Drug, but You Might Get This One," Bloomberg, October 29, 2020.
30. Langreth and Griffin, "Trump Got One Covid Antibody Drug, but You Might Get This One."
31. US Food and Drug Administration, "Coronavirus (COVID-19) Update: FDA Authorizes Monoclonal Antibody for Treatment of COVID-19," November 9, 2020, https://www.fda.gov/news-events/press-announcements/coronavirus-covid-19-update-fda-authorizes-monoclonal-antibody-treatment-covid-19.
32. Both Regeneron and Lilly had taken extraordinary steps to free up some of their largest, US-based manufacturing facilities to produce as much of the drug as possible. In the case of Regeneron, it had moved the production of its product for age-related macular degeneration (AMD), Eylea, out of their

New York manufacturing site and to its Irish manufacturing facility. Regeneron knew that if it intended for the drug to be used in the US, it couldn't risk manufacturing all of it oversees. Ireland might lay claim to the drug to deal with its own epidemic before it allowed it to be shipped back into the United States. Like other biotech companies, Regeneron also had about two years' worth of bulk Eylea drug product frozen for emergency use. Most drug makers would freeze eighteen to twenty-four months' worth of their key biological drugs as a "safety stock" in the event that there's a sudden disruption in their manufacturing. They need to be able to guarantee an uninterrupted supply to patients. Transferring production of a complex biologic from one facility to another was a complex matter. The manufacturing process of a biological like a monoclonal antibody is highly dependent on the characteristics of the facility and the cell line that the drug is grown in. There's no guarantee that the drug will remain identical as it's moved from one installation to another one. That frozen product would give Regeneron a cushion as it transferred the manufacturing of Eylea from its New York plant to its one in Ireland. Regeneron also struck a deal with Roche to use spare capacity that Roche maintained in a New Jersey plant to start making the Regeneron COVID antibody out of that facility as well. Some of the supply, however, would need to be used to satisfy demand outside the United States. Lilly confronted similar challenges. The company moved the production of its cancer drug Erbitux out of their large manufacturing facility in Branchburg, New Jersey, and transferred its production to their German partner Merck KgA. That freed up the company's 86,000-liter plant in New Jersey for making the new COVID-19 monoclonal antibodies. It was enough space to make up to 4 million doses of the drug a year if the plant was running at maximum capacity and efficiency. To augment the supply, Lilly struck a deal with Amgen, which agreed to free up space in its Puerto Rico facility—the one that manufactured Neupogen—to start also producing Lilly's new COVID antibody. Lilly also dedicated a 200,000-liter facility it owned in Ireland to help meet global demand for the product. Through other innovations in how these antibodies were discovered, Regeneron and Lilly were able to accelerate their development. The US Department of Health and Human Services had sought, and secured, an antitrust exemption from the Department of Justice to enable Regeneron to work with Roche, and Lilly to work with Amgen, on the manufacture of these COVID drugs.

33. Luciana Borio and Scott Gottlieb, "Antibodies Can Be the Bridge to a Vaccine," *Wall Street Journal*, July 5, 2020.
34. Scott Gottlieb and Mark McClellan, "The Trump Treatment for Covid Is Coming Soon," *Wall Street Journal*, October 8, 2020.
35. CBS News, "Transcript: Scott Gottlieb Discusses Coronavirus on 'Face the Nation,' October 11, 2020," October 11, 2020.
36. Scott Gottlieb, "Antibodies as Covid Insurance," *Wall Street Journal*, December 13, 2020.
37. US Food and Drug Administration, "Coronavirus (COVID-19) Update: FDA Revokes Emergency Use Authorization for Monoclonal Antibody

Bamlanivimab," April 16, 2021, https://www.fda.gov/news-events/press-announcements/coronavirus-covid-19-update-fda-revokes-emergency-use-authorization-monoclonal-antibody-bamlanivimab.

38. US Food and Drug Administration, "Coronavirus (COVID-19) Update: FDA Revokes Emergency Use Authorization for Monoclonal Antibody Bamlanivimab," April 16, 2021, https://www.fda.gov/news-events/press-announcements/coronavirus-covid-19-update-fda-revokes-emergency-use-authorization-monoclonal-antibody-bamlanivimab.
39. Jared S. Hopkins, "How Pfizer Delivered a Covid Vaccine in Record Time: Crazy Deadlines, a Pushy CEO," *Wall Street Journal*, December 11, 2020.
40. Gottlieb, "Antibodies as Covid Insurance."
41. CNBC, "Only 5–20% of Antibody Drug Doses Shipped to States Are Used: Operation Warp Speed," December 15, 2020.
42. Michael Kranish, "Brett Giroir, Trump's Testing Czar, Was Forced Out of a Job Developing Vaccine Projects. Now He's on the Hot Seat," *Washington Post*, April 20, 2020.
43. US Government Accountability Office, "National Preparedness: HHS Has Funded Flexible Manufacturing Activities for Medical Countermeasures, but It Is Too Soon to Assess Their Effect," March 2014.
44. Jon Swaine, "The Government Spent Tens of Millions on a Treatment for Chemical Weapons Exposure. The Company that Makes it Won't Say Whether it Works," *Washington Post*, August 18, 2020; and Chris Hamby and Sheryl Gay Stolberg, "How One Firm Put an 'Extraordinary Burden' on the U.S.'s Troubled Stockpile," *New York Times*, March 6, 2021.
45. Erin Banco and Sarah Owermohle, "Senior Trump and Biden Officials Knew for Months about Problems at Vaccine Plant," *Politico*, April 6, 2021.
46. Riley Griffin, "Emergent Plant Will Remain on Hold Following FDA Inspection," Bloomberg Law, April 21, 2021; and US Food and Drug Administration, "FDA Continues Important Steps to Ensure Quality, Safety and Effectiveness of Authorized COVID-19 Vaccines," April 21, 2021, https://www.fda.gov/news-events/press-announcements/fda-continues-important-steps-ensure-quality-safety-and-effectiveness-authorized-covid-19-vaccines.
47. Carl Zimmer, Sharon LaFraniere, and Noah Weiland, "Johnson & Johnson Expects Vaccine Results Soon but Lags in Production," *New York Times*, January 13, 2021; Sharon LaFraniere and Noah Weiland, "Johnson & Johnson's Vaccine is Delayed by a U.S. Factory Mixup," *New York Times*, March 31, 2021; and Chris Hamby, Sharon LaFraniere, and Sheryl Gay Stolberg, "U.S. Bet Big on Covid Vaccine Manufacturer Even as Problems Mounted," *New York Times*, April 6, 2021.
48. Dr. Anthony Fauci, director of the US National Institute of Allergy and Infectious Diseases, in conversation with the author, December 30, 2020.

    To help address the need for domestic capacity to produce complex biologics, I worked with a group of entrepreneurs to launch a contract manufacturing company that constructed and operated new manufacturing facilities in the US. The effort was aptly named National Resilience.

Chapter 17: The mRNA Breakthrough

1. It would be reported by multiple media outlets that Dr. Zhang Yongzhen, a researcher based in Shanghai, would share the genome against official orders.
2. Kenneth Murphy, *Janeway's Immunobiology*, 8th ed. (New York: Garland Science, 2012).
3. Ania Wajnberg et al., "Robust Neutralizing Antibodies to SARS-CoV-2 Infection Persist for Months," *Science* 370, no. 6521 (2020): 1227–30; Wilfredo F. Garcia-Beltran et al., "COVID-19-Neutralizing Antibodies Predict Disease Severity and Survival," *Cell* 184, no. 2 (2021): 476–88; "An In Vitro Study Shows Pfizer-BioNTech Covid-19 Vaccine Elicits Antibodies That Neutralize SARS-CoV-2 with a Mutation Associated with Rapid Transmission," Pfizer press release, January 8, 2021; and Alicia T. Widge et al., "Durability of Responses after SARS-CoV-2 mRNA-1273 Vaccination," *New England Journal of Medicine* 384 (2021): 80–2.
4. Dennis R. Burton, "Antibodies, Viruses and Vaccines," *Nature Review Immunology* 2 (2002): 706–13.
5. Certain immune cells that produce antibodies, known as B cells, can acquire long-term memories and have the ability to quickly produce new antibodies if your body is ever exposed to the same virus or bacteria in the future. This is how patients are able to retain some memory of an infection over many years and act quickly to counter it if they're ever exposed to the same infection a second time. Once exposed to the virus, the antibody binds to a foreign substance on the virus's surface—called an antigen. Each antibody can bind to only one specific antigen. The purpose of this binding is to help destroy the antigen. Some antibodies destroy antigens directly. Others make it easier for other immune cells to find the antigen and destroy it. An antibody is a type of immunoglobulin.
6. Wajnberg et al., "Robust Neutralizing Antibodies to SARS-CoV-2 Infection Persist for Months"; Garcia-Beltran et al., "COVID-19-Neutralizing Antibodies Predict Disease Severity and Survival"; Pfizer, "An In Vitro Study Shows Pfizer-BioNTech Covid-19 Vaccine Elicits Antibodies That Neutralize SARS-CoV-2 with a Mutation Associated with Rapid Transmission"; and Widge et al., "Durability of Responses after SARS-CoV-2 mRNA-1273 Vaccination."
7. An epitope is also known as antigenic determinant. It is the part of an antigen that is recognized by a patient's immune cells, including his or her antibodies, B cells, and T cells. In the case of antibody-mediated immunity, the epitope is the specific region on the antigen to which an antibody binds.
8. Mark Yarmarkovich et al., "Identification of SARS-CoV-2 Vaccine Epitopes Predicted to Induce Long-Term Population-Scale Immunity," *Cell Reports Medicine* 1, no. 100036 (2020); D. M. Morens, S. B. Halstead, and N. J. Marchette, "Profiles of Antibody-Dependent Enhancement of Dengue Virus Type 2 Infection," *Microbial Pathogenesis* 3 (1987): 231–7; Ann M. Arvin et al., "A Perspective on Potential Antibody-Dependent Enhancement of SARS-CoV-2," *Nature* 584, (2020): 353–63; and N. J. Sullivan, "Antibody

-Mediated Enhancement of Viral Disease," *Current Topics in Microbiology and Immunology* 260 (2001): 145–69.

9. Dennis R. Burton, "What Are the Most Powerful Immunogen Design Vaccine Strategies?," *Cold Spring Harbor Perspectives in Biology* 9, no. 11 (2017).
10. Susan L. Swain, K. Kai McKinstry, and Tara M. Strutt, "Expanding Roles for CD4+ T Cells in Immunity to Viruses," *Nature Reviews Immunology* 12 (2012): 136–48.
11. Cytokines are a broad category of small proteins important in regulating our body's inflammatory response to infection. They include interferons, interleukins, and tumor necrosis factors and work to help recruit immune cells to the site of an infection and aid in the maturation of these immune cells into disease-fighting agents.

    Penelope A. Morel and Michael S. Turner, "Designing the Optimal Vaccine: The Importance of Cytokines and Dendritic Cells," *Open Vaccine Journal* 3 (2010): 7–17.
12. Siddhartha Murkherjee provided a wonderful description of these different immune cells, the history of their discovery, and their role in providing immune protection in an essay written to accompany the announcement of the 2020 Lasker Awards, and published in the journal *Cell*. The essay was intended as a tribute in place of a keynote lecture that usually accompanies the awards, since the ceremony could not be held in person in 2020. See Siddhartha Mukherjee, "Before Virus, After Virus: A Reckoning," *Cell* 183, no. 2 (2020): 308–14.
13. The combination of higher infection rates in the West that left more of the populations with natural immunity, along with a more protective vaccine leaves the US and Western Europe perhaps better protected than China, which had comparatively very little infection to generate natural exposure (Only Wuhan saw significant spread, where about 10 percent of the population was exposed) and deployed less effective vaccines. This lower amount of population level immunity may become a more prominent consideration when China holds their Winter Olympics in 2022.
14. Joe McDonald and Huizhong Wu, "Official: Chinese Vaccines' Effectiveness Low," Associated Press, April 11, 2021.
15. US Centers for Disease Control and Prevention, "Adjuvants and Vaccines," August 14, 2020, https://www.cdc.gov/vaccinesafety/concerns/adjuvants.html.
16. Margaret A. Liu, "A Comparison of Plasmid DNA and mRNA as Vaccine Technologies," *Vaccines* 7, no. 2 (2019): 37.
17. James C. Kaczmarek, Piotr S. Kowalski, and Daniel G. Anderson, "Advances in the Delivery of RNA Therapeutics: From Concept to Clinical Reality," *Genome Medicine* 9 (2017): 60.
18. Philipp Reautschnig, Paul Vogel, and Thorsten Stafforst, "The Notorious R.N.A. in the Spotlight—Drug or Target for the Treatment of Disease," *RNA Biology* 14 (2017): 651–68; and Norbert Pardi et al., "MRNA vaccines—a New Era in Vaccinology," *Nature Reviews. Drug Discovery* 17, no. 4 (2018): 261–79.
19. Joanna Roberts, "Five Things You Need to Know About: mRNA Vaccines," *Horizon*, April 1, 2020.
20. Studies demonstrate that directly injected, nonreplicating mRNA can in-

duce protective immune responses against an infectious pathogen. See Benjamin Petsch et al., "Protective Efficacy of In Vitro Synthesized, Specific mRNA Vaccines Against Influenza A Virus Infection," *Nature Biotechnology* 30 (2012): 1210–6.

21. Sharon LaFraniere et al., "Politics, Science and the Remarkable Race for a Coronavirus Vaccine," *New York Times*, November 30, 2020.
22. Bojan Pancevski and Jared S. Hopkins, "How Pfizer Partner BioNTech Became a Leader in Coronavirus Vaccine Race," *Wall Street Journal*, October 22, 2020.
23. Nathan Vardi, "The Race Is On: Why Pfizer May Be the Best Bet to Deliver a Vaccine by Fall," *Forbes*, May 20, 2020.
24. Hopkins, "How Pfizer Delivered a Covid Vaccine in Record Time: Crazy Deadlines, a Pushy CEO."
25. Vardi, "The Race Is On."
26. Shivaani Kummar et al., "Phase 0 Clinical Trials: Conceptions and Misconceptions," *Cancer Journal* 14, no. 3 (2008): 133–7.
27. Hopkins, "How Pfizer Delivered a Covid Vaccine in Record Time: Crazy Deadlines, a Pushy CEO."
28. At the very same time that Pfizer was starting its first human trials, China was already in pivotal studies with two of its vaccines based on inactivated virus. The Chinese drug makers Sinopharm and Sinovac were further ahead than Pfizer or Moderna. However, the American drug makers would catch up, and they would ultimately bring to the market vaccines that were much more effective than the Chinese offerings.
29. Junzhi Wang, "New Strategy for COVID-19 Vaccination: Targeting the Receptor-Binding Domain of the SARS-CoV-2 Spike Protein," *Cellular & Molecular Immunology* 18 (2021): 243–4.
30. In an interview, Bancel said, "This is the 10th vaccine we've put in the clinic with Moderna's mRNA platform. We expect the safety and tolerability profile of the Moderna COVID-19 vaccine to be very similar from the other vaccines. And so I think we start to be in a place where we are cautiously optimistic." See Mark Terry, "Moderna's Zika Virus Vaccine Data Supports COVID-19 Vaccine Approach," BioSpace, April 14, 2020.

    Elie Dolgin, "How COVID Unlocked the Power of RNA Vaccines," *Nature*, January 12, 2021.
31. Caroline Hervé et al., "The How's and What's of Vaccine Reactogenicity," *NPJ Vaccines* 4, no. 39 (2019).
32. Susan Ellenberg, Thomas Fleming, and David DeMets, *Data Monitoring Committees in Clinical Trials: A Practical Perspective* (Hoboken, NJ: John Wiley & Sons Inc., 2002).
33. CNBC, "Former FDA Chief Scott Gottlieb on White House Transition of Coronavirus Task Force," November 9, 2020.
34. CNBC, "Pfizer Board Member Scott Gottlieb: '2021 Will Look Very Different from 2020,'" November 9, 2020.
35. "Pfizer and BioNTech to Submit Emergency Use Authorization Request Today to the U.S. FDA for COVID-19 Vaccine," Pfizer press release, November 20, 2020.

36. "Pfizer and BioNTech Conclude Phase 3 Study of Covid-19 Vaccine Candidate, Meeting All Primary Efficacy Endpoints," Pfizer press release, November 18, 2020.
37. US Department of Health and Human Services, "Trump Administration Purchases Additional 100 Million Doses of COVID-19 Investigational Vaccine from Pfizer," December 23, 2020.
38. "Moderna Announces it has Shipped Variant-Specific Vaccine Candidate, mRNA-1273.351, to NIH for Clinical Study," Moderna press release, February 24, 2021; and "Pfizer and BioNTech Initiate a Study as Part of Broad Development Plan to Evaluate COVID-19 Booster and New Vaccine Variants," Pfizer press release, February 25, 2021.
39. Kevin O. Saunders et al., "Neutralizing Antibody Vaccine for Pandemic and Pre-emergent Coronaviruses," *Nature* (2021).

Chapter 18: A New Doctrine for National Security

1. "There is no recent precedent for treating disease as a security threat. So unfamiliar are public health agencies with the apparatus of national defense that one early task force meeting was delayed when Co-chairwoman Sandra Thurman, whose Office of National AIDS Policy is across the street from the White House, could not find the Situation Room." See Barton Gellman, "AIDS Is Declared Threat to Security," *Washington Post*, April 30, 2000.
2. Director of the Central Intelligence Agency, National Intelligence Estimate, "The Global Infectious Disease Threat and its Implications for the United States," January 2000, https://www.dni.gov/files/documents/infectiousdiseases_2000.pdf.
3. Gellman, "AIDS Is Declared Threat to Security."
4. The 1992 report was requested and funded by the CDC and in the original agreement with Institute of Medicine, the agency requested two updates at five-year intervals.

   See Joshua Lederberg, Robert E. Shope, and Stanley C. Oaks Jr., eds., *Emerging Infections: Microbial Threats to Health in the United States* (Washington, DC: National Academies Press, 1992).
5. Ibid.
6. Institute of Medicine, *America's Vital Interest in Global Health: Protecting Our People, Enhancing Our Economy, and Advancing Our International Interests* (Washington, DC: National Academies Press, 1997).
7. Levin, "Biden's Pandemic-Policy Challenge."
8. Astrid Vabret et al., "Detection of the New Human Coronavirus HKU1: A Report of 6 Cases," *Clinical Infectious Diseases* 42, no. 5 (2006): 634–9.
9. Patrick C. Y. Woo et al., "Characterization and Complete Genome Sequence of a Novel Coronavirus, Coronavirus HKU1, from Patients with Pneumonia," *Journal of Virology* 79, no. 2 (2004): 884–95.
10. Xinhua, "21 Doctors, Nurses Infected with Pneumonia in E China," *China Daily*, June 16, 2013.
11. Monali C. Rahalkar and Rahul A. Bahulikar, "Lethal Pneumonia Cases in Mojiang Miners (2012) and the Mineshaft Could Provide Important Clues

to the Origin of SARS-CoV-2," *Frontiers in Public Health* 8 (2020): 638; and Zeynep Tufekci, "Where Did the Coronavirus Come From? What We Already Know Is Troubling.," New York Times, June 25, 2021.

12. Peng Zhou et al., "Addendum: A Pneumonia Outbreak Associated with a New Coronavirus of Probable Bat Origin," *Nature* 588, no. 6 (2020).
13. Wright, *The Plague Year*, page 222.
14. Massinissa Si Mehand et al., "World Health Organization Methodology to Prioritize Emerging Infectious Diseases in Need of Research and Development," *Emerging Infectious Diseases* 24, no. 9 (2018).
15. Nicholas Christakis, "The Long Shadow of the Pandemic: 2024 and Beyond," *Wall Street Journal*, October 16, 2020; and Christakis, *Apollo's Arrow*.
16. Arthur French, "Simulation and Modelling Applications in Global Health Security," *Global Health Security* (2020): 307–40; and Bonnie Jenkins, "Order from Chaos: Now Is the Time to Revisit the Global Health Security Agenda," Brookings Institution, March 27, 2020.
17. Julian E. Barnes and Michael Venutolo-Mantovani, "Race for Coronavirus Vaccine Pits Spy Against Spy," *New York Times*, September 5, 2020; and Julian E. Barnes, "U.S. Accuses Hackers of Trying to Steal Coronavirus Vaccine Data for China," *New York Times*, July 21, 2020.

    As *The New York Times* reports, among other acts, the two hackers tried to break into the systems of a Massachusetts biotech firm researching a COVID-19 vaccine as early as January 27, according to the indictment. On February 1, the same pair tried to find a way into the networks of a California biotech firm that had similarly announced that it was researching antiviral drugs that targeted SARS-CoV-2. Then in May, the same hackers were accused by US authorities of probing a California diagnostic firm that was developing coronavirus testing kits.
18. Julian E. Barnes, "Russia Is Trying to Steal Virus Vaccine Data, Western Nations Say," *New York Times*, December 14, 2020.
19. Michael R. Gordon and Dustin Volz, "Russian Disinformation Campaign Aims to Undermine Confidence in Pfizer, Other Covid-19 Vaccines, U.S. Officials Say," *Wall Street Journal*, March 7, 2021; and Denis Y. Logunov et al., "Safety and Efficacy of an rAd26 and rAd5 Vector-Based Heterologous Prime-Boost COVID-19 Vaccine: An Interim Analysis of a Randomised Controlled Phase 3 Trial in Russia," *Lancet* 397, no. 10275 (2021): 671–81.
20. Bret Schafer et al., "Influence-Enza: How Russia, China, and Iran Have Shaped and Manipulated Coronavirus Vaccine Narratives," German Marshall Fund, March 6, 2021; and Alexander Smith, "Russia and China Are Beating the U.S. at Vaccine Diplomacy, Experts Say," NBC News, April 2, 2021.
21. Barnes and Venutolo-Mantovani, "Race for Coronavirus Vaccine Pits Spy Against Spy."
22. Kenneth W. Bernard, "Opinion: The White House Signals That Bioterrorism and Disease Don't Matter—Again," *Washington Post*, May 22, 2018.
23. Emily Baumgaertner, "China Has Withheld Samples of a Dangerous Flu Virus," *New York Times*, August 27, 2018.
24. Ibid.

25. World Health Organization, "Factors That Contributed to Undetected Spread of the Ebola Virus and Impeded Rapid Containment," January 2015, https://www.who.int/news-room/spotlight/one-year-into-the-ebola-epidemic/factors-that-contributed-to-undetected-spread-of-the-ebola-virus-and-impeded-rapid-containment.
26. World Health Organization, "Cases of Undiagnosed Febrile Illness—United Republic of Tanzania," September 21, 2019, https://www.who.int/csr/don/21-september-2019-undiag-febrile-illness-tanzania/en/.
27. Abdi Latif Dahir et al., "Tanzania's Leader Declared the Pandemic Over. Now He's Asking the Country to Listen to Health Experts," *New York Times*, March 7, 2021; and BBC, "John Magufuli: Tanzania's President Dies Aged 61 after Covid Rumours," March 18, 2021.
28. John S. Mackenzie et al., "The Global Outbreak Alert and Response Network," *Global Public Health* 9, no. 9 (2014): 1023–39.
29. World Health Organization, "Frequently Asked Questions about the International Health Regulations (2005)," June 15, 2007, https://www.who.int/ihr/about/FAQ2009.pdf.
30. These networks included the Global Emerging Infections System of the US Department of Defense; the Global Public Health Intelligence Network, which is developed and managed by the Public Health Agency of Canada; and the WHO's Global Laboratory Network for Influenza, among other networks.

    See Mackenzie et al., "The Global Outbreak Alert and Response Network."
31. Ibid.
32. CNN, "Timeline: SARS Outbreak," April 24, 2003.
33. Selam Gebrekidan et al., "Ski, Party, Seed a Pandemic: The Travel Rules That Let Covid-19 Take Flight," *New York Times*, December 30, 2020.
34. Greg Miller, Josh Dawsey, and Aaron C. Davis, "One Final Viral Infusion: Trump's Move to Block Travel from Europe Triggered Chaos and a Surge of Passengers from the Outbreak's Center," *Washington Post*, May 23, 2020.
35. BBC, "Covid-19: 'Serious Disruption' Feared as Dover Halts Traffic to France," December 21, 2020.
36. Timothy C. Germann et al., "Mitigation Strategies for Pandemic Influenza in the United States," *Proceedings of the National Academy of Sciences* 103, no. 15 (2006): 5935-40; Paolo Bajardi et al., "Human Mobility Networks, Travel Restrictions, and the Global Spread of 2009 H1N1 Pandemic," *PLOS One* 6, no. 1 (2011): e16591; World Health Organization, "Updated WHO Recommendations for International Traffic in Relation to COVID-19 Outbreak," February 29, 2020, https://www.who.int/news-room/articles-detail/updated-who-recommendations-for-international-traffic-in-relation-to-covid-19-outbreak/; and Matteo Chinazzi et al., "The Effect of Travel Restrictions on the Spread of the 2019 Novel Coronavirus (COVID-19) Outbreak," *Science* 368, no. 6489 (2020): 395- 400.
37. Michelle Rourke et al., "Policy Opportunities to Enhance Sharing for Pandemic Research," *Science* 368, no. 6492 (2020): 716–18.

38. James W. Le Duc and Zhiming Yuan, "Network for Safe and Secure Labs," *Science* 362, no. 6412 (2018): 267; and Abutaleb et al., "The U.S. Was Beset by Denial and Dysfunction as the Coronavirus Raged."
39. Page and Hinshaw, "China Refuses to Give WHO Raw Data on Early Covid-19 Cases" and Jesse D. Bloom, "Recovery of Deleted Deep Sequencing Data Sheds More Light on the Early Wuhan SARS-CoV-2 Epidemic," *bioRxiv*, June 22, 2021, https://www.biorxiv.org/content/10.1101/2021.06.18.449051v1.
40. US Department of State, "Fact Sheet: Activity at the Wuhan Institute of Virology," January 15, 2021, https://ge.usembassy.gov/fact-sheet-activity-at-the-wuhan-institute-of-virology/.; and Michael R. Gordon, Warren P. Strobel and Drew Hinshaw, "Intelligence on Sick Staff at Wuhan Lab Fuels Debate On Covid-19 Origin," *Wall Street Journal*, May 23, 2021.
41. Holman W. Jenkins, "Wuhan Lab Theory a Dark Cloud on China," *Wall Street Journal*, March 9, 2021.
42. US Centers for Disease Control and Prevention, "Biosafety in Microbiological and Biomedical Laboratories, 6th Edition," June 2020, https://www.cdc.gov/labs/BMBL.html; and Nicoletta Lanese, "Only One Lab in China Can Safely Handle the New Coronavirus," Live Science, January 22, 2020.
43. Ping Liu, Wu Chen, and Jin-Ping Chen, "Viral Metagenomics Revealed Sendai Virus and Coronavirus Infection of Malayan Pangolins (*Manis javanica*)," *Viruses* 11, no. 11 (2019): 979.
44. David Quammen, "Did Pangolin Trafficking Cause the Coronavirus Pandemic? The Elusive Animals' Possible Involvement in the Origins of COVID-19 Gives Them a Weird Ambivalence: Threatened and, Perhaps, Dangerous," *New Yorker*, August 24, 2020.
45. Kangpeng Xiao et al., "Isolation of SARS-CoV-2-Related Coronavirus from Malayan Pangolins," *Nature* 583, no. 7815 (2020): 286–9; Tao Zhang, Qunfu Wu, and Zhigang Zhang, "Probable Pangolin Origin of SARS-CoV-2 Associated with the COVID-19 Outbreak," *Current Opinions in Biology* 30, no. 7 (2020): 1346–51; Lam, Jia, and Cao, "Identifying SARS-CoV-2-Related Coronaviruses in Malayan Pangolins"; and Quammen, "Did Pangolin Trafficking Cause the Coronavirus Pandemic?"

    Adding to the intrigue was the fact that the Wuhan Institute of Virology had a long history of poor practices. When the lab was on the cusp of being cleared for operations in 2017, an article published in the journal *Nature* raised some of the first concerns, stating that "some scientists outside China worry about pathogens escaping, and the addition of a biological dimension to geopolitical tensions between China and other nations."

    See David Cyranoski, "Inside the Chinese Lab Poised to Study World's Most Dangerous Pathogens," *Nature News*, February 23, 2017.

    In 2018, US embassy officials visiting the facility sent official warnings back to Washington about inadequate safety at the lab. One cable, from January 19, 2018, noted that the new lab "has a serious shortage of appropriately trained technicians and investigators needed to safely operate this high-containment laboratory."

See Josh Rogin, "Opinion: State Department Cables Warned of Safety Issues at Wuhan Lab Studying Bat Coronaviruses," *Washington Post*, April 14, 2020.

46. US Department of State, "Fact Sheet: Activity at the Wuhan Institute of Virology."
47. Bryan A. Johnson, Xuping Xie, and Vineet D. Menachery, "Loss of Furin Cleavage Site Attenuates SARS-CoV-2 Pathogenesis," *Nature* 591 (2021): 293–9; and Yiran Wu and Suwen Zhao, "Furin Cleavage Sites Naturally Occur in Coronaviruses," *Stem Cell Research* 50 (2021): 102115.
48. Nicholas Wade, "Origin of Covid—Following the Clues," Medium, May 3, 2021.
49. Cyranoski, "Inside the Chinese Lab Poised to Study World's Most Dangerous Pathogens."
50. Josh Rogin, "In 2018, Diplomats Warned of Risky Coronavirus Experiments in a Wuhan Lab. No One Listened.," *Politico*, March 8, 2021; U.S. Department of State, "Ensuring a Transparent, Thorough Investigation of COVID-19's Origin," press release, January 15, 2021, https://web.archive.org/web/20210116020513/https:/www.state.gov/ensuring-a-transparent-thorough-investigation-of-covid-19s-origin/; Wuhan Institute of Virology, "French Prime Minister Visits Wuhan P4 Laboratory," press release, February 27, 2017, http://english.whiov.cas.cn/ne/201802/t20180208_189991.html; and NBC News, "Meet the Press—May 30, 2021," May 30, 2021.
51. Natasha Loder, interview with Peter Daszak and Filippa Lentzos, Babbage from Economist Radio, podcast audio, April 14, 2021.
52. Filippa Lentos, "Natural Spillover or Research Lab Leak? Why a Credible Investigation is Needed to Determine the Origin of the Coronavirus Pandemic," Bulletin of the Atomic Scientists, May 1, 2020.
53. Katherine Eban, "The Lab-Leak Theory: Inside the Fight to Uncover COVID-19's Origins," *Vanity Fair,* June 3, 2021.
54. Natasha Loder, interview with Peter Daszak and Filippa Lentzos.
55. Yi Fan et al., "Bat Coronaviruses in China," *Viruses* 11, no. 3 (2019): 210.

    The Wuhan lab was a key part of that effort. Among the pathogens it was known to be working with was RaTG13, a bat coronavirus identified by the lab in January 2020. The strain was the closest known sample to SARS-CoV-2, with a 96.2 percent homology. WIV researchers had started working with this strain in 2016, after sampling it from a cave in Yunnan Province in 2013. It was found after several miners in that region died of a SARS-like illness. The WIV was later revealed to have sampled in 2015 the full sequences of eight viruses in Yunnan province that were believed to be similar to RaTG13 and "may hold evolutionary clues" to the origin of RaTG13. See Cyranoski, "Inside the Chinese Lab Poised to Study World's Most Dangerous Pathogens"; and Editorial Board, "We're Still Missing the Origin Story of This Pandemic. China Is Sitting on the Answers," *Washington Post*, February 5, 2021.
56. Ibid.
57. CBS News, "Transcript: Matt Pottinger on 'Face the Nation,' February 21, 2021," February 21, 2021.

58. Vineet D. Menachery et al., "A SARS-Like Cluster of Circulating Bat Coronaviruses Shows Potential for Human Emergence, Nature 21, (2015): 1508-13; Ren-Di Jiang et al., "Pathogenesis of SARS-CoV-2 in Transgenic Mice Expressing Human Angiotensin-Converting Enzyme 2," *Cell* 182, no. 1 (2020): 50-8; and Eban, "The Lab-Leak Theory: Inside the Fight to Uncover COVID-19's Origins."
59. US National Institutes of Allergy and Infectious Diseases of the National Institutes of Health, "Understanding the Risk of Bat Coronavirus Emergence," 2014–19, https://reporter.nih.gov/search/xQW6UJmWfUuOV01ntGvLwQ/project-details/9491676; and Peter Daszak and Vincent Racaniello, "This Week in Virology 615: Peter Daszak of EcoHealth Alliance," May 19, 2020.
60. CBS News, "Transcript: Matt Pottinger on 'Face the Nation,'" February 21, 2021.
61. CNN, "Former CDC Director Believes Virus Came from Lab in China," March 26, 2021.
62. Cambridge Working Group, "Cambridge Working Group Consensus Statement on the Creation of Potential Pandemic Pathogens (PPPs)," July 14, 2014, http://www.cambridgeworkinggroup.org/.
63. Marc Lipsitch and Thomas V. Inglesby, "Moratorium on Research Intended To Create Novel Potential Pandemic Pathogens," *mBio* 5, no. 6 (2014): e02366-14.
64. "NIH Lifts Funding Pause on Gain-of-Function Research," US National Institutes of Health press release, December 19, 2017, https://www.nih.gov/about-nih/who-we-are/nih-director/statements/nih-lifts-funding-pause-gain-function-research; and Vineet D. Menachery et al., "A SARS-Like Cluster of Circulating Bat Coronaviruses Shows Potential for Human Emergence," 1508–13.
65. Editorial Board, "We're Still Missing the Origin Story of This Pandemic. China Is Sitting on the Answers."
66. Zhang Feng, "Officials Punished for SARS Virus Leak," *China Daily*, July 2, 2004.
67. Ibid.
68. Add endnote: Martin Furmanski, "Threatened Pandemics and Laboratory Escapes: Self-Fulfilling Prophecies," Bulletin of the Atomic Scientists, March 31, 2014.
69. Peter Palese, "Influenza: Old and New Threats," *Nature Medicine* 10 (2004): S82–7; and Michelle Rozo and Gigi Kwik Gronvall, "The Reemergent 1977 H1N1 Strain and the Gain-of-Function Debate," *mBio* 6, no. 4 (2015).
70. Kristian G. Andersen et al., "The Proximal Origin of SARS-CoV-2," *Nature Medicine* 26 (2020): 450–2.
71. Bryan A. Johnson et al., "Furin Cleavage Site Is Key to SARS-CoV-2 Pathogenesis," bioRxiv (2020), https://www.biorxiv.org/content/10.1101/2020.08.26.268854v1; Thomas P. Peacock et al., "The Furin Cleavage Site in the SARS-CoV-2 Spike Protein is Required for Transmission in Ferrets," *Nature Microbiology* (2021); and Vineet D. Menachery et al., "Trypsin Treatment Unlocks Barrier for Zoonotic Bat Coronavirus Infection," *Journal of Virology* 94, no. 5 (2020): e01774-19.

72. Holman W. Jenkins, "Wuhan Lab Theory a Dark Cloud on China," *Wall Street Journal*, March 9, 2021.
73. Jenkins, "Wuhan Lab Theory a Dark Cloud on China."
74. Andrea Shalal and Michael Martina, "White House Cites 'Deep Concerns' about WHO COVID Report, Demands Early Data from China," Reuters, February 13, 2021.
75. "Statement by President Joe Biden on the Investigation into the Origins of COVID-19," White House press release, May 26, 2021, https://www.whitehouse.gov/briefing-room/statements-releases/2021/05/26/statement-by-president-joe-biden-on-the-investigation-into-the-origins-of-covid-19/
76. James Gorman and Carl Zimmer, "Another Group of Scientists Calls for Further Inquiry Into Origins of the Coronavirus," *New York Times*, May 13, 2021.
77. Hudson Institute, "Virtual Event—The Origins of COVID-19: Policy Implications and Lessons for the Future," March 12, 2021.
78. Jesse D. Bloom et al., "Investigate the Origins of COVID-19," *Science* 372, no. 6543 (2021): 694.
79. David A. Relman, "Lab leaks happen, and not just in China. We need to take them seriously," *Washington Post*, June 2, 2021.
80. Scott Gottlieb, "The CIA Can Help Spot the Next Pandemic," *Wall Street Journal*, February 28, 2021.
81. Saeed Shah, "CIA Organised Fake Vaccination Drive to Get Osama Bin Laden's Family DNA," *Guardian*, July 11, 2011.
82. Editorial Board, "Polio Eradication: The CIA and Their Unintended Victims," *Lancet* 383, no. 9932 (2014): 1862.
83. Ibid.
84. Ibid.
85. W. John Kress, Jonna A. K. Mazet, and Paul D. N. Hebert, "Opinion: Intercepting Pandemics Through Genomics," *Proceedings of the National Academy of Sciences* 117, no. 25 (2020): 13852–5.
86. One of the most significant obstacles to leveraging sequencing data alone to deduce the clinical features of a virus is limits on the ability to correlate clinical information to the different virus samples. If you want to do more sophisticated analysis, by the time a viral sample makes its way to researchers for sequencing, all the clinical information has been stripped away. Much of this is to address patient privacy considerations, but those considerations can perhaps be safeguarded while still allowing anonymized clinical information to ride along with patient samples. What happens now is that researchers and providers decide that the easiest way to sidestep privacy concerns is to just strip away the information on how patients fared, even though that information is valuable in correlating sequencing with clinical outcomes. One reform is to change patient privacy laws for samples used to inform public health surveillance.
87. Stephen K. Gire et al., "Genomic Surveillance Elucidates Ebola Virus Origin and Transmission During the 2014 Outbreak," *Science* 345, no. 6202 (2014): 1369–72; and Joshua Quick et al., "Real-Time, Portable Genome Sequencing for Ebola Surveillance," *Nature* 530 (2016): 228–32.

88. National Academies of Science, Engineering, and Medicine, *Genomic Epidemiology Data Infrastructure Needs for SARS-CoV-2: Modernizing Pandemic Response Strategies* (Washington, DC: National Academies Press, 2021).
89. White House Homeland Security Council, "National Strategy for Pandemic Influenza: Implementation Plan."
90. Matthew Herper, "At Illumina, the 'Era of the Genome' Has Arrived. But What Role Will the Company Play?," STAT, May 17, 2021; and National Human Genome Research Institute, "The Cost of Sequencing a Human Genome," December 7, 2020, https://www.genome.gov/about-genomics/fact-sheets/Sequencing-Human-Genome-cost.
91. US Centers for Disease Control and Prevention, "Advanced Molecular Detection," January 8, 2021, https://www.cdc.gov/amd.
92. Dylan George and Tara O'Toole, "Tackling the Next Epidemic: Data Technology to the Rescue," In-Q-Tel, Inc., 2017, https://www.bnext.org/wp-content/uploads/2017/07/data-tech-to-the-rescue_0710207-FINAL.pdf.
93. David Cameron, "Scientific Coalition Developing Surveillance System for Detecting Emerging Pandemics in Real-Time," Broad Institute, May 11, 2020.
94. US Centers for Disease Control and Prevention, "National Wastewater Surveillance System (NWSS)," March 19, 2021.
95. World Health Organization, "WHO-Convened Global Study of Origins of SARS-CoV-2: China Part," March 30, 2021, https://www.who.int/publications/i/item/who-convened-global-study-of-origins-of-sars-cov-2-china-part; and Mackenzie, The Pandemic That Never Should Have Happened, and How to Stop the Next One.
96. Giuseppina La Rosa et al., "SARS-CoV-2 Has Been Circulating in Northern Italy Since December 2019: Evidence from Environmental Monitoring," *Science of the Total Environment* 750 (2021): 141711; and Antonella Amendola et al., "Evidence of SARS-CoV-2 RNA in an Oropharyngeal Swab Specimen, Milan, Italy, Early December 2019," *Emerging Infectious Diseases* 27, no. 2 (2021): 648–50.
97. A. Deslandes et al., "SARS-CoV-2 Was Already Spreading in France in Late December 2019," *International Journal of Antimicrobial Agents* 55, no. 6 (2020).
98. Gislaine Fongaro et al., "SARS-CoV2 in Human Sewage in Santa Catalina, Brazil, November 2019," medRxiv, June 29, 2020, https://www.medrxiv.org/content/10.1101/2020.06.26.20140731v1; and Lorenzzo Lyrio Stringari et al., "Covert Cases of Severe Acute Respiratory Syndrome Coronavirus 2: An Obscure but Present Danger in Regions Endemic for Dengue and Chikungunya Viruses," *PLOS One* 16 (2020).
99. Melody Schreiber, "If the U.S. Already Had a Covid Variant, We Wouldn't Know," *New Republic*, December 22, 2020.
100. Mackenzie, *The Pandemic That Never Should Have Happened, and How to Stop the Next One*.
101. Trevor Bedford (@trvrb), "We also now have direct serological evidence of antigenic drift in 229E from @eguia_rachel, @jbloom_lab et al, suggesting that reinfection by seasonal coronaviruses that occurs every ~3 years is in part due to evolution of the virus. 6/18," Twitter, December 19, 2020, 4:33 p.m.

102. Kirsty Needham, "Special Report: COVID Opens New Doors for China's Gene Giant," Reuters, August 5, 2020.
103. Donald G. McNeil Jr. and Thomas Kaplan, "U.S. Will Revive Global Virus-Hunting Effort Ended Last Year," *New York Times*, September 2, 2020.
104. Elisabeth Rosenthal, "On the Front: A Pandemic Is Worrisome but 'Unlikely,'" *New York Times*, March 28, 2020.
105. McNeil and Kaplan, "U.S. Will Revive Global Virus-Hunting Effort Ended Last Year."
106. Mackenzie, *The Pandemic That Never Should Have Happened, and How to Stop the Next One*.
107. Donald G. McNeil Jr., "Scientists Were Hunting for the Next Ebola. Now the U .S. Has Cut Off Their Funding.," *New York Times*, October 25, 2019; and James Rainey and Emily Baumgaertner, "Trump, Congress Scramble to Revive Virus -hunting Agency that Was Marked for Cuts," *Los Angeles Times*, April 11, 2020.
108. Some of USAID's work in the 2014 timeframe was done in collaboration with the Wuhan Institute of Virology. See USAID, "Predict 2 Factsheet," November 2014, https://www.usaid.gov/sites/default/files/documents/1864/Predict2-factsheet.pdf.
109. Scott Gottlieb, "Don't Let Covid and the Flu Team Up to Pound America," *Wall Street Journal*, August 9, 2020.
110. Jemma L. Geoghegan et al., "Virological Factors that Increase the Transmissibility of Emerging Human Viruses," *Proceedings of the National Academy of Sciences* 113, no. 15 (2016): 4170–5; Mackenzie, *The Pandemic That Never Should Have Happened, and How to Stop the Next One*; R. Carrasco-Hernandez et al., "Are RNA Viruses Candidate Agents for the Next Global Pandemic? A Review," *ILAR Journal* 58, no. 3 (2017): 343-58; Nancy H. L. Leung, "Transmissibility and Transmission of Respiratory Viruses," *Nature Reviews Microbiology*, (2021); and Amesh Adalja et al., "The Characteristics of Pandemic Pathogens," Johns Hopkins Center for Health Security, May 10, 2018, https://www.centerforhealthsecurity.org/our-work/pubs_archive/pubs-pdfs/2018/180510-pandemic-pathogens-report.pdf.

Conclusion

1. Wright, "The Plague Year."
2. Zeke Miller, "Trump Valet has Coronavirus; President Again Tests Negative," Associated Press, May 7, 2020.
3. Gottlieb and Levin, "The Trump Coronavirus Spread."
4. Katherine J. Wu, "The White House Bet on Abbott's Rapid Tests. It Didn't Work Out.," *New York Times*, October 6, 2020.

   The machine they used, the Abbott ID Now, for all of its virtues, also could forfeit some accuracy to achieve its speed and convenience. It was authorized by the FDA for use on people who were already showing symptoms of COVID-19. There was a reason that the product's label was restricted to these parameters. When patients are symptomatic, they are more likely to have higher levels of virus in secretions, and therefore be detected using the instruments.
5. Jessica L. Prince-Guerra et al., "Evaluation of Abbott BinaxNOW Rapid

Antigen Test for SARS-CoV-2 Infection at Two Community-Based Testing Sites—Pima County, Arizona, November 3–17, 2020," *Morbidity and Mortality Weekly Report* 70, no. 3 (2021): 100–5.

6. Larry Buchanan et al., "Inside the White House Event Now Under Covid-19 Scrutiny," *New York Times*, October 5, 2020.
7. Michael C. Bender and Rebecca Ballhaus, "Inside the Week That Shook the Trump Campaign," *Wall Street Journal*, October 21, 2020.
8. William Cummings, "'The Data Speak for Themselves': Dr. Anthony Fauci Says White House Held a 'Superspreader Event' for Coronavirus," *USA Today*, October 9, 2020; and "Margaret Brennan: Did anyone ever say this is a national security risk and we need to nail down who brought this in and who infected the commander in chief? Dr. Birx: I never heard those conversations. Margaret Brennan: There was no serious contact tracing that happened after the fact? Dr. Birx: I don't know if there was contact tracing or not." *See CBS News*, "Transcript: Dr. Deborah Birx on "Face the Nation," January 24, 2021," January 24, 2021.
9. Jordan Fabian (@Jordanfabian), "@JenniferJJacobs was the unofficial contact tracer of last year's WH coronavirus outbreaks, a feat made even more impressive by the fact the previous admin was not forthcoming about officials who tested positive for Covid-19. A herculean effort that rightly received recognition," Twitter, April 19, 2021, 12:16 p.m.
10. James Glanz, "Tests Show Genetic Signature of Virus That May Have Infected President Trump," *New York Times*, December 31, 2020.
11. At the time of the White House coronavirus cluster in October, more than 40,000 SARS-CoV-2 virus genomes had been sequenced and publicly shared from the United States alone. See Trevor Bedford et al., "Viral Genome Sequencing Places White House COVID-19 Outbreak into Phylogenetic Context," medRxiv, November 13, 2020, https://www.medrxiv.org/content/10.1101/2020.10.31.20223925v2.
12. Glanz, "Tests Show Genetic Signature of Virus That May Have Infected President Trump."
13. Gertner, "Unlocking the Covid Code."
14. National Academies of Science, Engineering, and Medicine, *Genomic Epidemiology Data Infrastructure Needs for SARS-CoV-2: Modernizing Pandemic Response Strategies* (Washington, DC: National Academies Press, 2021).
15. As the authors note, viral genome sequencing represents a largely novel and very effective tool for tracing the origin of outbreaks and the direction of viral spread as part of epidemiological investigations. However, as Bedford and his authors state, deploying these tools as an adjunct to traditional approaches to epidemiology generally requires that a large fraction of infections have already been sequenced and that these sequences are made publicly available. This is necessary to provide critical comparisons for the sequence or group of sequences in question. See Bedford et al., "Viral Genome Sequencing Places White House COVID-19 Outbreak into Phylogenetic Context."
16. "Pfizer and BioNTech Initiate a Study as Part of Broad Development Plan to Evaluate COVID-19 Booster and New Vaccine Variants," Pfizer press release, February 25, 2021; and "Moderna Announces It Has Shipped Variant

-Specific Vaccine Candidate, mRNA-1273.351, to NIH for Clinical Study," Moderna press release, February 24, 2021.

17. "Can a Vaccine for Covid-19 Be Developed in Record Time?," *New York Times Magazine*, June 9, 2020; and Richard Conniff, "A Forgotten Pioneer of Vaccines," *New York Times*, May 6, 2013.
18. US Centers for Disease Control and Prevention, "National Center for Health Statistics: Nursing Home Care," March 1, 2021, https://www.cdc.gov/nchs/fastats/nursing-home-care.htm.
19. Owen Dyer, "Two Strains of the SARS Virus Sequenced," *British Medical Journal* 326, no. 7397 (2003): 999; and Marco A. Marra et al., "The Genome Sequence of the SARS-Associated Coronavirus," *Science* 300, no. 5624 (2003): 1399–404.

    In April 2003, two different strains of the SARS virus were sequenced. The Tor2 strain was isolated in Toronto, Canada, the city hit hardest in the West by the epidemic while the Urbani strain was sequenced by the US CDC. The latter strain was the more common Asian strain, named after Carlo Urbani, an infectious disease specialist at the WHO's office in the Vietnamese capital, Hanoi. He helped identify the disease as a coronavirus but later died from it.
20. Maggie Fox, "Countries Re-Think Swine Flu Vaccine Orders," Reuters, January 12, 2010.

    "First, as Canada's only domestic influenza vaccine manufacturer, GSK provided priority access to a secure supply of the H1N1 vaccine for all Canadians." See Paul N. Lucas, "GlaxoSmithKline Inc. Appearance before the Standing Committee on Social Affairs, Science and Technology," October 6, 2010.

    "In Australia, the government pressured vaccine maker CSL Limited to turn over 36 million doses of H1N1 vaccine contracted for by the U.S. and produced in an Australian-based manufacturing plant. Meanwhile, in Canada, where British drug maker GlaxoSmithKline maintains its US-focused flu vaccine facility, the company had to assure the local government that Canadians would be served from that manufacturing plant before Americans could receive any of their vaccine orders." See Scott Gottlieb, "No More Pandemic Hysterics," *Forbes*, February 22, 2010.
21. Benjamin Mueller and Matina Stevis-Gridneff, "Desperate Italy Blocks Exports of Vaccines Bound for Australia," *New York Times*, March 10, 2021.
22. World Trade Organization, "COVID-19: Measures Affecting Trade in Goods," accessed March 10, 2021, https://www.wto.org/english/tratop_e/covid19_e/trade_related_goods_measure_e.htm.
23. Matina Stevis-Gridneff, "E.U. Will Curb Covid Vaccine Exports for 6 Weeks," *New York Times*, March 28, 2021.
24. Jeremy Konyndyk, "It's Time for a 'No Regrets' Approach to Coronavirus," *Washington Post*, February 4, 2020.
25. G. Dennis Shanks and John Brundage, "Did Coronaviruses Cause 'Influenza Epidemics' Prior to 1918?" *Journal of Travel Medicine* 28, no.2 (2021); and Anthony King, "An Uncommon Cold," *New Scientist* 246, no. 3280 (2020): 32–5.
26. Bloomberg, "China Culls 18,000 Chickens After H5N1 Bird Flu Cases in Hunan," February 2, 2020.

INDEX

ABOUT THE AUTHOR

Dr. Scott Gottlieb served as the twenty-third commissioner of the US Food and Drug Administration and is a resident fellow at the American Enterprise Institute. He is a regular contributor to the business and financial news channel CNBC and a partner at the venture capital firm New Enterprise Associates. Dr. Gottlieb serves on the board of directors of the pharmaceutical company Pfizer Inc. and the genomic sequencing company Illumina Inc. *Fortune* magazine has recognized him as one of the "World's 50 Greatest Leaders," and *Time* magazine has named him one of its "50 People Transforming Healthcare." A graduate of Wesleyan University and the Mount Sinai School of Medicine, Dr. Gottlieb is an elected member of the National Academy of Medicine. He lives with his family in Westport, Connecticut.